Troubled Harvest

Recent Titles in
Contributions in Latin American Studies

Family and Favela: The Reproduction of Poverty in Rio de Janeiro
Julio César Pino

Mexico in the Age of Proposals, 1821–1853
Will Fowler

Reinventing Legitimacy: Democracy and Political Change in Venezuela
Damarys Canache and Michael R. Kulisheck, editors

Sugar and Power in the Dominican Republic: Eisenhower, Kennedy, and the Trujillos
Michael R. Hall

Tornel and Santa Anna: The Writer and the Caudillo, Mexico 1795–1853
Will Fowler

Exiles, Allies, Rebels: Brazil's Indianist Movement, Indigenist Politics, and the Imperial Nation-State
David Treece

One of the Forgotten Things: Getúlio Vargas and Brazilian Social Control, 1930–1954
R. S. Rose

The Abolition of Slavery in Brazil: The "Liberation" of Africans Through the Emancipation of Capital
David Baronov

Beyond the Ideal: Pan Americanism in Inter-American Affairs
David Sheinin, editor

Oil, War, and Anglo-American Relations: American and British Reactions to Mexico's Expropriation of Foreign Oil Properties, 1937–1941
Catherine E. Jayne

Gunboats, Corruption, and Claims: Foreign Intervention in Venezuela, 1899–1908
Brian S. McBeth

"La última gaviota": Liberalism and Nostalgia in Early-Twentieth-Century Panamá
Peter A. Szok

Troubled Harvest

Agronomy and Revolution in Mexico, 1880–2002

JOSEPH COTTER

Contributions in Latin American Studies,
Number 22

PRAEGER

Westport, Connecticut
London

Library of Congress Cataloging-in-Publication Data

Cotter, Joseph.
Troubled harvest : agronomy and revolution in Mexico, 1880–2002 / Joseph Cotter.
p. cm.—(Contributions in Latin American studies, ISSN 1054–6790 ; no. 22)
Includes bibliographical references and index.
ISBN 0–313–32515–4 (alk. paper)
1. Agriculture—Mexico—History—20th century. 2. Green Revolution—Mexico. 3. Agriculture—Technology transfer—Mexico—History—20th century. 4. Mexico—Social conditions—20th century. I. Title. II. Series.
S451.7.C68 2003
630'.972'0904—dc21 2002044952

British Library Cataloguing in Publication Data is available.

Library of Congress Catalog Card Number: 2002044952
ISBN: 0–313–32515–4
ISSN: 1054–6790

First published in 2003

Praeger Publishers, 88 Post Road West, Westport, CT 06881
An imprint of Greenwood Publishing Group, Inc.
www.praeger.com

Printed in the United States of America

The paper used in this book complies with the Permanent Paper Standard issued by the National Information Standards Organization (Z39.48–1984).

10 9 8 7 6 5 4 3 2 1

To Michael, Margaret, Cheryl, and Mary

In Memoriam
Joseph Eugene Carter

Joe Cotter died suddenly and unexpectedly on September 21, 2002 after a brief illness. He was 46 years old. His family, his friends, his students, his colleagues, and many other people were profoundly shocked and deeply saddened. He was a dynamic teacher, a scholar of considerable achievement and great promise, and delightful company. Losing him cut deep.

Troubled Harvest was near the final stage of production when Joe died. It has been my honor and privilege to work with the publisher to complete the process. I would like to express my sincere gratitude to two colleagues, Jana Sandarg and Eric Perez, for their assistance in translating Spanish terms.

Wayne Mixon
Chair, Department of History, Anthropology, and Philosophy
Augusta State University

Contents

Preface xi

Abbreviations xxiii

Introduction: Agronomy and Revolution in Mexico 1

1. Liberalism and the Origins of Mexican Agronomy 19
2. Revolution, Agronomy, and the Roots of the Cultural Campaign, 1910–1930 49
3. Revolutionary Agronomy and the Cultural Campaign, 1930–1937 81
4. Agronomy under Attack: The Demise of the Cultural Campaign and the Roots of a New Agronomy, 1938–1943 125
5. *Agrónomos Técnicos*: The Rockefeller Foundation, the United States, and the Transformation of Mexican Agronomy, 1944–1950 179
6. The Mexican Revolution and the Green Revolution, 1950–1970 233
7. The Aftermath of the First Green Revolution and the Birth of a Second? 1970–2002 281

Conclusions: Agronomy and Mexico's Revolutions 321

Bibliography 341

Index 371

Preface

Like most long research projects that end with a book, mine has a complicated history with several unexpected twists and turns. After finishing my bachelor's degree and working for four years in business, I was dissatisfied with my life and sought a new direction. I renewed a daily relationship with an old friend who moved back to San Diego to begin graduate school, and he convinced me that I also missed the environment of academe. My days growing up in southern California and my many trips to Mexico already had stimulated my interest in Mexican culture and society, and I applied to graduate programs in Latin American history. From my travels I learned that Mexico was beautiful and fascinating, but also that many Mexicans lived in abject poverty, especially in rural areas, and I decided to focus on the issue of economic development in the Masters program at the University of California at Santa Barbara (hereafter UCSB).

To prepare for my Masters exams I looked at many works on the Mexican Revolution (1910–1920), which produced the political system and elite that shaped Mexico's economic policy during the twentieth century. I also began to read about technological change in rural Mexico, discovering an abundant body of literature on the "Green Revolution" in agriculture, most of it quite critical. Most of these works said nothing about Mexican research before the Rockefeller Foundation began the Mexican Agricultural Program (hereafter MAP) in 1943, but a few, such as Francis Moore Lappe and Joseph Collins' *Food First: Beyond the Myth of Scarcity* and Cynthia Hewitt de Alcantara's *Modernizing Mexican Agriculture*, told short but fascinating stories about Mexican researchers (hereafter *agrónomos*) who worked to support the peasant farmers who benefited from the sweeping agrarian reform carried out by President Lázaro Cárdenas (1934–1940). Did they try to create a peasant-based Mexican agricultural science? Were they similar to contemporary advocates of "appropriate technology"? Did Mexico reject an alternative path of economic development that might have kept more peasants on the farm and fostered the growth of a less concentrated industrial base? I also noticed that works on the Green Revolution were either non-archival or based on English-language sources, usually from the Rockefeller Foundation Archives, such as Deborah Fitzgerald's "Exporting American Agriculture: The Rockefeller Foundation in Mexico, 1943–1953" and Bruce H. Jennings' *Foundations of International Agricultural Research, Science and Politics in Mexican Agriculture*. I decided that I wanted to write my Ph.D. dissertation on *cardenista* agronomy (referring to the governmental reforms from 1934 through 1940 made by Mexican President Lázaro Cárdenas) and how it was derailed by U.S. scientists working for the Rockefeller Foundation.

In 1986 I was admitted to UCSB's doctoral program, and I began to examine sources that I could obtain in the United States. In addition to several English-language primary sources, I located the Mexican agricultural journal *Agricultura*, published by the government during the late 1930s, which ran many articles written by *agrónomos* expressing their views on the function of scientific research, the politics of rural development, and the objectives of *cardenista* agronomy. I knew about the rich record collection in the Rockefeller Archive Center, but to avoid being overly influenced by their perspective, I wanted to begin my research in Mexico. The body of evidence that I gathered there at various times during the next six years makes *Troubled Harvest* the first study of agricultural science and development in Mexico that uses Spanish-language sources to evaluate the Mexican side of the story. In the summer of 1988 I visited Mexico City for three weeks and located a number of archival records in Cárdenas' Presidential Papers at the *Archivo General de la Nación* (hereafter AGN), but they dealt with special cases that made their way up the bureaucratic ladder to his office. I failed to find the records of the *Secretaría de Agri-*

cultura y Fomento (Ministry of Agriculture and Development, SAF). Disappointed, I worried I might not be able to find enough material to write a dissertation.

Fortunately, my friend and mentor, Dr. Manuel L. Carlos of UCSB's Anthropology Department, told me about a collection of papers and published materials that prominent *agrónomo* Ramón Fernández y Fernández gave to the *Colegio de Michoacán* in Zamora and told me that this institution would accept me as a visiting researcher. I was fortunate to secure funding and pass my Ph.D. exams, and I departed for Zamora in July 1989. In the *Colegio*'s library and archives I found personal papers, publications, and manifestos issued by several agronomists' organizations or factions; government publications; monographs published by *agrónomos* from the 1900s to the 1960s, such as Oscar Brauer H.'s *Diagnóstico de las ciencias agrícolas en México*; and journals such as *Agros* (early 1920s) or *Extensión Agrícola* (early 1930s), with many articles expressing *agrónomos*' views on scientific, professional, and political issues, as well as an extensive collection of newspaper clippings of articles on agriculture and agrarian policy with many of their public statements. As I examined the *agrónomos*' discourses over a longer period of time, I began to realize that my original research questions were not the right ones to ask. The *agrónomos* often expressed concern over professional issues like job opportunities, salaries, and their public credibility, matters that became central concerns in the late 1930s and early 1940s. The Cárdenas administration was not a significant anomaly in the history of Mexican agronomy, which is the story of how a small, unorganized group of upper- and middle-class university graduates forged an alliance with the State produced by the Mexican Revolution to transform themselves into scientific researchers who rose to international prominence as a result of the Green Revolution.

Hence, the central theme of this book is an examination of how two historical events, the Mexican Revolution and the Green Revolution, shaped the development of the profession of agronomy in Mexico (people with the title and degree *ingeniero agrónomo*, or agricultural engineer). I demonstrate how the profession attained professional objectives by allying with the State created by the Mexican Revolution. By examining these issues, light is also shed on the factors that influenced Mexico's agricultural and agrarian policies and on many occasions determined their success or failure. Some works that ask similar questions and follow the same general analytical path include Frank Safford's *The Ideal of the Practical*, which describes how engineering became a profession in Colombia, and David Lorey's *The University System and Economic Development in Mexico since 1929*, a more general work on Mexican professionals.

I spent some time in the municipal archives of Zamora, then finished my research in western Mexico with a brief visit to the State archives of Michoacán in Morelia and an unplanned four-week stay in Guadalajara to examine the extensive and well-organized collection of the Jalisco Department of Agriculture's records from the 1930s and early 1940s. These documents provided insight into how federal policy reached the local level and demonstrated some of the many problems faced by the *agrónomos* of this period, not the least of which was peasant violence directed at agents of the State. I then returned to Mexico City to see if my status as a researcher at the *Colegio de Michoacán* would help open any other doors. Using the Mexico City telephone book, I found a number for an *Archivo de la Secretaría de Agricultura y Recursos Hidraulicos* (Archive of the Ministry of Agriculture and Hydraulic Resources [SARH]) with an address in the far northeast corner of the Federal District in San Juan Aragón. I went there, showed the secretary my letter from the *Colegio*, and was instructed to return the next morning to talk to her boss. When I did he was not there, but she informed me that I could not see any records because they were "not organized," and further, that most of the agency's archives were bulk-stored "in a warehouse in Ecatepec."

Dismayed, I returned to the AGN and saw workers putting up a display about a new documentary group: the Gonzalo Robles Papers. Robles, an agronomist educated in the United States and Costa Rica, served in the agricultural and education bureaucracy during the 1920s, early 1930s, and 1940s, and his papers included personal records, correspondence with other bureaucrats, and studies of various government programs. I then examined records in the presidential papers of Abelardo Rodríguez (1932–1934), Cárdenas, Manuel Avila Camacho (1940–1946), and Miguel Alemán (1946–1952). Although they have some limitations, I found examples of the *agrónomos'* professional discourses, debates, and concerns during these years, many interesting glimpses of farmers' opinions of the *agrónomos* with whom they dealt, and some examples of *agrónomos'* attitudes about peasants and their farming systems. I also found some valuable records in the archives of the Ministry of Foreign Relations, useful agricultural journals such as *Tierra* (1940s–1960s) and *Chapingo* (1950s–1960s), unpublished materials like the Ministry of Agriculture's recommendations for the Six Year Plan (1934–1940), monographs such as *agrónomo* José Luís de la Loma's 1955 *Experimentación agrícola* in the *Biblioteca* and *Hemeroteca Nacional* at the *Universidad Nacional Autónoma de México* (National Autonomous University of Mexico, hereafter UNAM), and financial records at the *Banco de México* library.

Finishing a year of research in Mexico, I was ready to examine English-language sources in the United States. Thanks to a Rockefeller Archive Center (RAC) travel grant and other financial support, I spent four weeks in North Tarrytown, New York, looking at MAP records from 1940 to 1956 that discuss its planning, inception, and programs; officers' and trustees' correspondence about Latin America during the late 1930s, oral histories taken during the 1960s from many of the MAP's leading scientists; and the papers of Nelson Rockefeller, who developed a massive agricultural assistance program for Latin America while he was Coordinator of Inter-American Affairs during World War II. I was surprised to find that many of the documentary and manuscript primary sources that I read in Mexico corroborated the opinions of the MAP's scientists. Lastly, I spent another four weeks in Washington, D.C., looking at several record groups in the National Archives, including the Bureau of Entomology and Plant Quarantine, which ran several collaborative projects with Mexico's Ministry of Agriculture during the late 1920s, 1930s, and 1940s, the Department of State, the Coordinator of Inter-American Affairs, the Department of Agriculture, the Office of Foreign Agricultural Relations (OFAR), and the Interdepartmental Office on Scientific and Technical Cooperation with Latin America. These documents provided more glimpses into the *agrónomos'* professional lives, their feelings about collaborating with U.S. scientists, and their opinions of Mexico's agricultural and agrarian policies. They also helped put the MAP into the context of World War II and Cold War era U.S. policy toward Mexico.

I returned to Santa Barbara and began to write my dissertation, but I had several more opportunities to do research. I visited Mexico City three more times for several weeks to a month in 1992, 1993, and 1994 and located more useful materials in the National Library that helped me extend my story into the 1960s and beyond. I returned to the AGN to examine the presidential papers of Adolfo Ruis Cortinez (1952–1958), Adolfo López Mateos (1958–1964), and Gustavo Díaz Ordaz (1964–1970). Although these records were spottier than those of earlier administrations, especially for the last two presidents, they still contained some information similar to what appeared in earlier documents. I also visited the Bancroft Library, which has an excellent collection of Mexican government agricultural publications from the 1880s, 1890s, and 1900s and nearly complete sets of the journals *Boletín de la Sociedad Científica Antonio Alzate* and *Boletín de la Sociedad Mexicana de Geografía y Estadística,* which contain articles written by the leading Mexican scientists of the Porfirian era. This information helped complete my discussion of the French contributions to Mexican agronomy.

My committee signed my dissertation in 1994, and I left to take a visiting position at the University of West Florida in Pensacola. After two years I accepted another temporary job at the University of Arizona and finally secured a tenure-track professorship at Augusta State University in 1997. During my three years as an itinerant academic I found more works related to my research, such as *agrónomo* Marte R. Gómez's *Biografías de agrónomos,* a collection of anecdotes about his colleagues, or *agrónomo* Edmundo Flores' *Historias de Edmundo Flores, autobiografía 1919–1950,* which describes his experience in the profession. I continued to read in my field, discovering that many Mexicanists, informed by the works of Italian sociologist Antonio Gramsci, were studying the "cultural" dimension of the Mexican Revolution. The State's attempt to attack peasant belief systems and farming practices was another part of this process, and I incorporated this school of analysis to evaluate the *agrónomos* and their programs. I also began to work on an article for *Science, Technology, and Society,* and the editors and reviewers suggested that I examine theoretical works from the sociology of science and the professions and the sociology of knowledge. I found that the story of the *agrónomos,* the Mexican Revolution, and the Green Revolution provided a useful case study to test these theories. I also borrowed ideas from rural sociology, political science, and peasant studies, such as James

Scott's work and John Tutino's application of it to rural Mexico and Norman Long's model for evaluating rural development programs as an interface, to help explain why peasant farmers reacted in varied ways to the *agrónomos* and their programs. Lastly, I put the Green Revolution into the context of literature on Cold War diplomacy to examine how Mexico contributed to U.S. "containment policy."

After I moved to Augusta, I began to rewrite my dissertation, adding materials that I examined since it was signed and reducing my examples of issues and events before 1950. I continued to examine useful primary works such as Leobardo Jiménez Sánchez's *Las ciencias agrícolas y sus protagonistas*, a collection of interviews of prominent *agrónomos* taken during the early 1980s. I also read more secondary works in the fields of history and cultural anthropology and gathered as much information as I could on agricultural policy since 1970. This monograph benefits greatly from previous research on this subject by Harry Greenberg, Merilee Grindle, Marilyn Gates, and Alberto Arce. With archival materials unavailable, Chapter 7 relies heavily on the works of anthropologists, newspaper articles, and books and articles written by *agrónomos*. Although I have examined a considerable body of archival and published primary material, as well as many secondary works, these sources do have some limitations. I was able to include the voices of prominent *agrónomos*, government officials, and even many peasants, but the opinions of the rank and file of the agricultural research community were harder, but not impossible, to discern. For example, the letter from José del Riego and others (a group of young *agrónomos*) to Avila Camacho written on July 9, 1941, voices their opinion of the older "leaders" of their profession. Similar sources discussed throughout the book demonstrate that the *agrónomos* did not act as a monolithic entity; rather, they were divided by matters such as age, ethnicity, the school that they attended, and even their opinions of peasant agriculture and land reform. Documentary and published primary sources demonstrate that some *agrónomos* resented working under the supervision of U.S. scientists, but most of them appreciated the professional and economic rewards that this collaboration produced. Further, most *agrónomos* recognized the need to establish, improve, and maintain their credibility as professionals. Their motives were both social (to be perceived in society as university-trained professionals similar to doctors, architects, or lawyers) and economic (to reap the financial rewards of membership in Mexico's Revolutionary Family).

Personal information about the *agrónomos*, such as the profession of their father (or mother) or their birthplace and residence, comes from the anecdotal *Biografías de agrónomos*, several professional society membership rosters, school directories, and U.S. records on *agrónomos* who studied in the United States. Records from the *Escuela Nacional de Agricultura* (ENA), the *Escuela Agrícola "Antonio Narro,"* and the *Escuela Particular de Agricultura* (EPA) may exist. In the first case, the school's librarian told me that there was no archive of documents there. I did not have time to visit the other schools, but I suspect that student records would not be open for public examination. The evidence that I examined indicates that they shared many ideas typical of their social class (middle-class *mexicanos'* desire for respectability and upward economic mobility, their preference for urban life, and their negative opinion of the rural lower class). Additional documentary evidence may eventually come to light when the Ministry of Agriculture catalogs their internal records and opens their archive to the public, and I hope that some future historian will use the materials to reexamine these issues.

I incurred many debts during the years that I researched and wrote this manuscript. Thanks to the UC MEXUS, the Rockefeller Archive Center, the UCSB General Affiliates, and UCSB's History Department for graciously giving me financial assistance for my research trips. At UCSB, Dr. David Rock stimulated my interest in Latin America, and his fascinating seminars sharpened my ability to think historically, ask research questions, and defend my ideas. Dr. Manuel Carlos taught me many lessons about *lo mexicano* (Mexican culture) that became clearer and clearer as the project progressed, played a crucial role in securing the funding that let me spend 14 months doing research in Mexico and the United States, and helped me make ends meet with research jobs when money became scarce. Dr. Lawrence Badash showed me how scientists have dealt with the social implications of their work. Dr. Michael Osborne encouraged me, collaborated with me, helped me get to Paris, and told me how my research helped to bridge the gap between the history of science and the rising interest in environmental history.

Dr. Alexander Callow offered his friendship and emotional support, as well as a good time as his teaching assistant in his popular courses on the history of American urban crime. Dr. J. Sears McGee, then Chair of the History Department, generously offered me employment as a lecturer that helped keep me going while I completed my dissertation and looked for work. Many of my fellow graduate students at UCSB gave me emotional support and solid academic advice, most notably Daniel Lewis, Peter Neushul, Duane Mosser, and Stanley Shadle. Julie Garfinkel, a recipient of the University of California President's Fellowship for undergraduates, kindly shared material that she collected in Chiapas on organic agriculture, and I regret that we did not succeed in getting our article published in the Opinion-Editorial section of the *Los Angeles Times*. At the Rockefeller Archive Center the extensive knowledge of Director Darwin Stapleton and Archivist Erwin Levold helped me review a substantial body of records in a short period of time.

I also received assistance in Mexico. Thanks to Dr. Larissa Adler of UNAM for agreeing to be my collaborator in my UC MEXUS Project Development Grant. In Zamora, Brigitte Boehm de Lamerias of the *Colegio de Michoacán* gave me encouragement and opened the door to their excellent library and the documents in the *Archivo de Ramón Fernández y Fernández* (ARFF). Alvaro Ochoa gave me useful suggestions and loaned me materials from his personal library. Martín Sánchez Rodríguez graciously opened the *Archivo Municipal de Zamora* to give me a glance at a local archive and took me to Morelia, where I briefly examined the State archive of Michoacán. Edmundo Davila, an *agrónomo* and former Rockefeller Foundation fellow who lived in Zamora, befriended me and told me many interesting stories about his profession. Despite his busy schedule and advanced age, Dr. Enrique Beltrán graciously agreed to an interview while I was in Mexico City. The efficiency and gracious hospitality of the archivists at the *Archivo Histórico de Jalisco* (AHJ), the Ministry of Foreign Relations Archive, and the *Colegio de Michoacán*'s library helped me use my time in Mexico effectively.

There are several other debts that I must acknowledge. Special thanks to Enrique Ochoa, Professor of History at California State University, Los Angeles, whom I met while we were doing our graduate research in Mexico City in 1990. Our topics (his is the history of Mexico's food distribution and marketing policy) meshed well, and we shared our knowledge of archives and sources. Our many meetings became a one-on-one seminar where we discussed and defended our ideas. My connection to Dr. Ochoa led to invitations to two PROFMEX (Consortium for Research on Mexico; Professors and Professionals Who Study Mexico) conferences sponsored by Professor James Wilkie of the University of California, Los Angeles. At these meetings I presented papers that were published and received much useful input from the participants, most of them Mexican scholars. I especially want to acknowledge a comment made by the distinguished and recently deceased Mexican economist Sergio de la Peña, who shook my hand and said my paper was "*muy importante*." At the conferences and again more recently Dr. Wilkie has made many comments and suggestions that influenced the course of my research and my analysis of the issues and events discussed in this manuscript. I met Marcos Cueto at a conference sponsored by the Rockefeller Archive Center; we corresponded and collaborated afterwards, and his ideas have also influenced my work.

Thanks to James Miklovich of the University of West Florida and Helen Nader of the University of Arizona for keeping my career afloat while I searched for a permanent position. At the University of Arizona I received useful comments in discussions with Bert Barickman, Oscar Martínez, and several of the History Department's graduate students. At Augusta State University, my Department Chair, Wayne Mixon, encouraged me to rewrite my dissertation and submit it to publishers, and he graciously kept demands on my time at a minimum. I also want to thank Greenwood Press and its editorial staff, especially Heather Ruland Staines, for helping turn this raw manuscript into a publishable book. Acknowledging the importance of the many intellectual contributions listed above, I accept complete responsibility for the accuracy of this manuscript, and the onus for any errors of fact or interpretation is completely my own.

Lastly, I want to make several acknowledgments of a personal nature. Thanks to my adoptive parents, Michael and Margaret Cotter, whose love and support over many years, both emotional and financial, got me through several periods when things looked rather bleak. I am convinced that getting a Ph.D. is as much or more a test of persistence and tolerance than of intellectual brilliance. They raised me with character that enabled me to persevere. My aunt, Marie Rhodes, also gave me emotional and financial support at several crucial points in my academic career and personal life. Thanks to my biological mother, Mary Johnson of Northern Bay, Newfoundland, who gave me the chance to learn about a very unfamiliar time in my life and discover that I, raised as an only child, actually had five full biological siblings. Special thanks to my wife, Cheryl. Since we met and married in 1997, I have finally started to remember why I decided to become an academic in the first place. The joy that she has brought me, the many fascinating places that we have visited, and her unending patience and encouragement while I spent countless hours organizing old note cards and writing in my office have turned the hard times of the past into fading memories.

My husband, Joseph Cotter, passed away on September 21, 2002. This book is the culmination of many years of hard work on his part. He was very proud of the book, and I know that he would be as happy as I am to see it published.

I would like to express my deepest thanks to Dr. Wayne Mixon, Chair of the Department of History, Anthropology, and Philosophy at Augusta State University. Wayne graciously accepted the task of shepherding Joe's book to completion. I would also like to thank Joe's co-workers at Augusta State University for their friendship and support.

Cheryl Cotter

Abbreviations

AGN	*Archivo General de la Nación*
AHJ	*Archivo Histórico de Jalisco*
AHSRE	*Archivo Histórico de la Secretaría de Relaciones Exteriores*
AID	Agency for International Development
ARFF	*Archivo de Ramón Fernández y Fernández*
BAR	*Bloque de Agrónomos* (Revolutionary Agronomists Block)
BNCA	*Banco Nacional de Crédito Agrícola* (National Agricultural Credit Bank)
BNCE	*Banco Nacional de Crédito Ejidal* (National Ejidal Credit Bank)
CIAA	Coordinator of Inter-American Affairs
CIANO	Construction Industry Association of New Orleans
CIMMYT	*Centro Internacional del Mejoramiento de Maiz y Trigo* (International Center for Corn and Wheat Improvement)
CINVESTAV	*Centro de Investigación y de Estudios Avanzados* (Center of Investigation and Advanced Studies)

CNA	*Comisión Nacional Agraria* (National Agrarian Commission)
CNAG	*Consejo Nacional de Agricultura* (National Council on Agriculture)
CNC	*Confederación Nacional Campesino* (National Peasant Confederation)
CNI	*Comisión Nacional de Irrigación* (National Irrigation Commission)
CONACYT	*Consejo Nacional de Ciencia y Technología* (National Council of Science and Technology)
CONASUPO	*Compania Nacional de Subsistencias Populares* (National Company of Popular Subsistences)
CREA	Center for Reflection, Education, and Action
CTM	*Confederación de Trabajadores Mexicanos* (Confederation of Mexican Workers)
DGA	*Dirección General de Agricultura* (General Office of Agriculture)
ENA	*Escuela Nacional de Agricultura* (National School of Agriculture)
EPA	*Escuela Particular de Agricultura* (Private Agricultural School)
EZLN	*Ejército Zapatista de Liberación Nacional* (Zapatista Army for National Liberation)
FAO	Food and Agriculture Organization
FEA	Foreign Economic Administration
FRAM	*Frente Revolucionario de Agrónomos Mexicanos* (Revolutionary Front of Mexican Agronomists)
IFSF	*Imprenta y Fototipía de la Secretaría de Fomento*
IHD	International Health Division
IIA	*Instituto de Investigaciones Agrícolas* (Agricultural Research Institute)
INIA	*Instituto Nacional de Investigaciones Agrícolas* (National Agricultural Research Institute)

IRRI	International Rice Research Institute
LAS	*Liga de Agrónomos Socialistas* (Socialist Agronomists League)
MAP	Mexican Agricultural Program
MIT	Massachusetts Institute of Technology
NA	National Archives
NAFTA	North American Free Trade Agreement
NS	Natural Sciences
OFAR	Office of Foreign Agricultural Relations
PIDER	*Proyecto Internacional de Desarrollo Rural* (International Rural Development Project)
PNR	*Partido Nacional Revolucionario* (National Revolutionary Party)
PRI	*Partido Institucional Revolucionario* (Institutional Revolutionary Party)
PROFMEX	Consortium for Research on Mexico (Professors and Professionals Who Study Mexico)
RAC	Rockefeller Archive Center
RFA	Rockefeller Family Archives
RG	Record Group
SAF	*Secretaría de Agricultura y Fomento* (Ministry of Agriculture and Development)
SAM	*Sociedad Agronómica Mexicana* (Mexican Agronomic Society)
SAN	*Sociedad Agronómica Nacional* (National Agronomic Society)
SARH	*Secretaría de Agricultura y Recursos Hidraulicos* (Ministry of Agriculture and Hydraulic Resources)
SEP	*Secretaría de Educación Pública* (Ministry of Public Education)
UFCO	United Fruit Company
UN	United Nations

UNAM	*Universidad Nacional Autónoma de México* (National Autonomous University of Mexico)
UNFASA	*Union de Fabricantes de Conservas S.A.* (Union of Canned Goods Manufacturers)
USAID	U.S. Agency for International Development
USDA	U.S. Department of Agriculture
YMCA	Young Men's Christian Association

Troubled Harvest

Introduction: Agronomy and Revolution in Mexico

During the twentieth century two "revolutions" transformed rural Mexico: the Mexican Revolution (1910–1920) and the Green Revolution (1950–1970). Socioeconomic and political forces produced the first revolution, but the revolutionaries also saw science and technology as tools to attain their objectives. Scientists who created new technologies caused the second, and these revolutionaries called their revolution apolitical, but it had profound social, economic, and political consequences. The Mexican Revolution captured the attention of political radical John Reed, President Woodrow Wilson, the social activist scholar Frank Tannenbaum, and famous writers Mariano Azuela and Graham Greene. The Green Revolution attracted the interest of prominent social scientists such as economist Keith Griffin and anthropologist Marvin Harris, science writer Stanley Johnson, Presidents Richard Nixon and Gustavo Díaz Ordaz (1964–1970) of Mexico, the United Nations (UN) Food and Agriculture Organization (FAO), environmentalists such as Barry Commoner, and the Nobel Peace Prize Committee, which gave the 1970 award to Norman Borlaug for his hybrid dwarf wheat. Since the 1980s the prospects of a second Green Revolution based on genetic engineering have stimulated international discussion and debate. Its first phase already indicates that it will bring another series of profound transformations to rural Mexico as it unfolds.

Both revolutions influenced the evolution of Mexican agronomy profoundly, and the struggle to define and control the agenda for scientific research to "modernize" agriculture has been a central issue in Mexico's economic development. The contradictions of Porfirio Díaz's age of liberalism (1876–1911) spawned a violent revolution (1910–1920), the new political elite that governed Mexico until 2000 through the *Partido Institucional Revolucionario* (Institutional Revolutionary Party, hereafter PRI), and a period of socioeconomic reform (1920–1940) that included the distribution of a significant amount of farmland to peasants (*campesinos*). After 1940 the government redefined the Revolution to mean industrial and urban growth, transferred control of basic food crop research to the Rockefeller Foundation to foster these goals, and reduced its support for land reform communities. In the long run, these policies did not improve the lot of most of Mexico's peasants. Reacting to rising domestic tensions that culminated in the Tlatelolco Massacre (1968), the State undertook various projects to address the problems of peasant farmers but then canceled them when Mexico entered the economic crisis of the "lost decade" (1982–1988). To address the situation, President Carlos Salinas de Gortari (1988–1994) adopted neoliberal economic policies and "deconstructed" many pillars of the Mexican Revolution in the process, including its land reform program.

A few scholars have demonstrated the significant role that federal bureaucrats play in Mexican politics, including how agronomists (*agrónomos*) implemented agricultural policies from 1940 to 1970 and especially after 1970.[1] But *agrónomos* and their programs before 1940 have received little attention, and some scholars suggest strong discontinuities before and after this year that most consider a key turning point in Mexico's history. Enhancing the present-centered approach of recent works on the agrobureaucracy, this study demonstrates how the historical forces that shaped the evolution of Mexican agronomy as a scientific profession before 1940 profoundly affected the *agrónomos* and their contributions to agricultural policy and development during the years studied in more detail by other scholars. *Agrónomos* played a role in both Revolutions, their transformation from "revolutionaries" into "scientific professionals" reflected broader trends in Mexico's history, and the process exerted and continues to exert a significant influence on the outcomes of the government's agricultural policy. The story of Mexican agronomy also provides a case study to test sociological theories about the nature of professions, the development of expertise, and how professionals acquire a clientele that accepts their credibility as "experts." Peter Cleaves claims that the *agrónomos*, like other Mexican professionals, failed to organize as a profession, relying only on "university alumni associations" to advance their interests, but evidence proves that they acted collectively to ally with the State and validate their expertise during all of these historical periods.[2]

After attaining independence in 1821, Mexico experienced many years of political turmoil and economic stagnation. Díaz's long reign brought political stability without democracy and sustained economic growth fostered by foreign investment, including the expansion of Mexico's haciendas (large estates), often by taking land from peasant farmers. During the *Porfiriato* (the regime of Porfiro Díaz from 1876 to 1911) the *Escuela Nacional de Agricultura* (National School of Agriculture, hereafter ENA) trained *agrónomos* not to be independent researchers but rather, like their counterparts in other "underdeveloped" regions, to transmit the research and technologies of European and North American scientists to Mexico.[3] Like other Porfirian elites who loved French culture, the *agrónomos* believed that France led the world in the field of agronomy. But by the end of the *Porfiriato*, propinquity and direct contacts stimulated their interest in U.S. agricultural science, especially plant breeding. Mexican agronomy began as a "liberal" profession of private practitioners. As purveyors of Western science seeking to establish credibility as "experts," they tried to invalidate the knowledge of Mexico's *hacendados* (hacienda owners), *rancheros* (owners of small farms, *ranchos*), and *campesinos* by defining their farming practices as "backward, Colonial, and routine" and thus inferior to the European "scientific agriculture" that they studied at the ENA. The *agrónomos* understood the link between "scientific truth" and "power."[4] But *hacendados* saw *agrónomos* as urban intellectuals who knew nothing about agriculture, refuted their accusations of incompetence, and rejected their "expertise" by refusing to hire them. Sociologists of science claim that professionals often promote their "economic self-interest" in the name of advancing the "common good."[5] To obtain government support, professionals promise to help politicians foster material prosperity. Like scientists in other countries, the *agrónomos* responded to the *hacendados* by rejecting the "liberal" model of professions, claiming that their knowledge could solve Mexico's agricultural problems and promote capitalist development, and exhorting the State to hire them.[6]

In recent years many Mexicanists turned their attention to the process of State formation after the Revolution created the system and elite that governed Mexico since 1920. Informed by the ideas of Italian sociologist Antonio Gramsci, they analyzed the revolutionaries' "cultural campaign" to transform the *campesinos*. Some stress the continuity of ideas and motivations that influenced its architects from the agenda of the Díaz regime to the project of the "State-building" revolutionaries of the 1920s and 1930s. These scholars studied issues such as rural education and schoolteachers, the construction and use of "revolutionary symbols," and the manipulation of Mexico's indigenous culture and history. But they have ignored the State's use of "agricultural science" to construct a new society, one of the "lesser projects" to which historian Alan Knight alluded in 1990. World events, historians, and sociologists have demonstrated how science is useful (or even essential) to twentieth-century nation-states, and they have begun to explore its role in the process of State formation and the construction of ideological hegemony. Like rural teachers, after the Mexican Revolution *agrónomos* became another of Benedict Anderson's "pilgrims of the imagined national community," or, in historian Mary Kay Vaughan's words, "middle class products of Mexico's public schools and of the civic celebratory culture of the prerevolutionary modernizing society."[7] Historian Adrian Bantjes describes the revolutionaries' attempt to form an "invisible religion" to supplant the masses' Catholicism.[8] During the 1920s and 1930s the State incorporated science, particularly agronomy, into the new creed's canon and thus tried to use "agricultural science," broadly defined, as a tool to construct its hegemony.

The Mexican Revolution began as a movement for democratic reform to open Díaz's closed political system, but his fall unleashed an insurrection against the inequities caused by Porfirian economic growth, including the alienation of the *campesinos'* lands, as well as foreign investors' prominent and privileged position in Mexico. Peasant revolutionary Emiliano Zapata voiced the *campesinos'* grievances by demanding "*¡Tierra y Libertad!*" (Land and Liberty!), but other revolutionaries assassinated him before the conflict ended. Mexicans who worried about foreign economic domination put Article 27 in the Constitution of 1917, asserting government control over the nation's resources. This measure also co-opted Zapata's agenda by granting the State a right to expropriate and distribute rural properties. As Hannes Siegrist observes, professionals can benefit by "participating in revolutionary activities."[9] Rejected by the *hacendados* and seeking political influence to help establish their role as "experts," *agrónomos* joined the ranks of the revolutionaries and participated in the first land reform programs, thus earning the right to claim membership in what political scientist Frank Brandenberg called the "Revolutionary Family": people who gained political, social, and economic power from the State created by the Mexican Revolution.[10]

In 1920 middle-class revolutionaries from the northern states led by future presidents Alvaro Obregón (1920–1924) and Plutarco Elias Calles (1924–1928) "won" the Revolution, defeating the *campesinos* who followed Zapata, Pancho Villa, and a host of lesser "*caudillos*" (populist rural leaders), but the winners courted the support of *agraristas* (advocates of land reform) who survived the conflict by carrying out a limited agrarian reform, creating the *ejido* (communities of land reform beneficiaries). The *Comisión Nacional Agraria* (National Agrarian Commission, hereafter CNA) distributed land to create a clientele of loyal *campesinos* and establish control over the many independent peasant movements that the Revolution spawned. After the Revolution *agrónomos* served in the agrarian reform and claimed that they became *agraristas* to bring "justice" to the *campesinos*. To "reach exclusive societal advantages" (in this case a place in the Revolutionary Family that dominated Mexican politics) and attain "preference of interpretation within their special field of knowledge and practice" (the right to define the problems of Mexican agriculture), in 1921 *agrónomos* who served in the Revolution organized the *Sociedad Agronómica Nacional* (National Agronomic Society, hereafter SAN).[11] The organization supported *agrarismo*, helped members get CNA jobs that helped ameliorate their employment problem by transforming the profession from liberal practitioners into bureaucrats, and assisted the State's project to establish control over independent agrarian movements. The *agrónomos'* "science" fits William T. Lynch's definition of an ideology: "any aspect of knowledge . . . that plays a *causal* role in maintaining or creating power disparities in society" (author's emphasis).[12]

Reflecting Obregón's and Calles' interest in capitalist modernization, the government promoted commercial agriculture by building roads and forming the *Comisión Nacional de Irrigación* (National Irrigation Commission, hereafter CNI) to construct irrigation projects and channeled some credit to farmers by establishing the *Banco Nacional de Crédito Agrícola* (National Agricultural Credit Bank, hereafter BNCA). The State also undertook a "cultural campaign" to mold *campesinos*, Native Americans, and other "subaltern groups," whom the revolutionaries defined as backward, "superstitious," and "fanatically" Catholic relics of the colonial past, into "modern," secular, and loyal citizens of the new regime. These projects provided jobs for *agrónomos* as extension workers, teachers, researchers, surveyors, and agrarian organizers in the CNI, the *Secretaría de Agricultura y Fomento* (Ministry of Agriculture and Development, hereafter SAF), the CNA, the *Secretaría de Educación Pública* (Ministry of Public Education, hereafter SEP), and also as agents for the BNCA. As historian David E. Lorey observed, the rise of an activist state provided more jobs for professionals like the *agrónomos*. According to *agrónomo* Edmundo Flores, the profession "constituted the technical personnel for a new state administrative structure to support the *campesinos* and agriculture."[13]

The administration of President Lázaro Cárdenas (1934–1940) was a crucial juncture in the history of Mexican agronomy because it established the conditions that led the *agrónomos* to seek foreign training to become scientific professionals. This study supports Richard Harvey Brown's analysis of how scientists and institutions use a "knowledge discourse" to "achieve monopoly domination or hegemony in its market sector or cognitive domain" by convincing "dominant groups" that their expertise, which only they are qualified to discuss, will advance their interests.[14] During the early 1930s the right wing of the Revolutionary Family, led by ex-president and *Jefe Máximo* (literally supreme boss) Calles, agreed with critics who blamed the land reform and by extension the *agrónomos* for stagnant farm productivity, threatening their alliance with the State. They reacted by reconstituting their defunct professional association as the *Sociedad Agronómica Mexicana* (Mexican Agronomic Society, hereafter SAM), arguing that their knowledge (*agrarismo* and science for the *campesinos*) was specialized and useful to "dominant groups" (the left wing of the Revolutionary

Family), demanding more jobs in the bureaucracy, and claiming that they could attain many objectives of the *cardenistas'* cultural campaign. They conceptualized and attempted to implement projects to solve many of the *campesinos'* "problems" as defined by the *cardenistas*. Building on the agricultural programs of the 1920s, including the campaign to transform peasant culture, Cárdenas expanded the agrobureaucracy to carry out a massive land reform, began a more concerted effort to teach them scientific agriculture, and established the *Banco Nacional de Crédito Ejidal* (National Ejidal Credit Bank, hereafter BNCE) to lend money to land reform beneficiaries.

Following sociologist Norman Long's suggestions, to the extent that archival and published primary and secondary sources and a national rather than local study permit, I analyze Mexico's rural development programs as a "social interface" between the State that promoted "modernization," the agencies that developed programs, the bureaucrats who implemented them, and the farmers whom they targeted.[15] The *cardenistas'* projects created more government jobs for *agrónomos*, but their scope was too vast to tackle given the State's limited financial resources and the shortage of skilled professionals. Eyler N. Simpson, who studied Mexico's land reform during the 1930s, called this phenomenon "*proyectismo*": "the tendency to spin out of thin air tremendous programs for the accomplishing of anything and everything under the sun."[16] To sociologist of science Magali Sarfatti Larson, "the efficacy of 'discourse' ultimately depends on the power and efficacy of the organizations which support the production, the diffusion, and the practical application of 'discourses,' " but borrowing an expression from Knight, Cárdenas' SAF was more "jalopy" than "juggernaut."[17] Most projects to transform peasant agriculture, the agencies that implemented them, and the *agrónomos* and other bureaucrats who carried them out did not take the *campesinos'* agendas, cultural values, and economic situation into account, but rather tried to impose their own definition of modern agriculture, based on economic and political objectives, foreign textbooks, and imported technologies. Anthropologist Bille R. DeWalt claims that Mexico's agricultural agents of the early 1970s could not "communicate with people whose lifestyles, beliefs, attitudes, resources, and background are so different from their own," a statement that describes the *agrónomos* throughout their professional history.[18] To establish their credibility as experts, they continued to invalidate the *campesinos'* agricultural knowledge by defining it (and the *campesinos* themselves) as backward, debilitated, and inferior to their own expertise.[19]

The *campesinos* did not agree with the *agrónomos'* definitions of their "problems." Similar to their reaction to the State's anti-Catholic and education programs, as well as its approach to *agrarismo,* the *campesinos* resisted, rejected, or modified many elements of the *agrónomos'* campaign to transform their farming practices. Svante Beckman argues that "relations of trust are rightly thought to be central to professional status."[20] The *campesinos'* traditional distrust of outsiders, desire for autonomy and "subsistence security," and well-founded belief that their agricultural practices suited their ecological and economic conditions better than those recommended by "experts," combined with the *agrónomos'* class- and education-based disdain for the *campesinos* and their culture to inhibit relationships of mutual trust and cooperation. Anthropologist Daniel Nugent demonstrates how history gave the descendants of the military colonists of Namiquipa, Chihuahua, cultural values such as a profound desire for autonomy and a valuation of maintaining control over the farming process, rather than submitting to the authority of external actors like subjugated *campesinos* who worked on haciendas. To these farmers, the agrobureacracy's "technical advice," inputs, and loans meant the loss of autonomy and hence social status. Studying a region with a very different culture and history, David Ronfeldt found the *campesinos* who grew sugar for the Atencingo mill in Puebla favored individual *ejidal parcelas* (common areas of land in a town or cooperative cultivation) because they satisfied "long-standing desires among the peasantry for personal control over their means of making a living."[21] Discussing peasant potato growers in the Andes, sociologist Jan Douwe van der Ploeg demonstrates how the conversion from traditional cultivars to hybrid varieties turned the *campesinos* into "invisible men" by devaluing their knowledge and forcing them to live by a schedule dictated by the new plants rather than personal preference. Rather than surrender to the State's hegemony, Mexico's peasants resisted and negotiated with *agrónomos* to, in Sergio Zendejas' words, "appropriate the *ejido* for purposes other than those established by the state or to mold government programs to their own interests."[22]

The *agrónomos* did not exercise enough "professional authority" over the *campesinos* to "impose . . . their own conception of what is needed," and a regime that wanted to win the peasants' affection refused to give them the coercive power "to intimidate and oppress" them into cooperating.[23] In Vaughan's words, the *agrónomos* were part of "the educating state" that failed to impress "its model of modernity" on the *campesinos*; in Larson's, they refused to let *agrónomos* "assert their monopoly over discourse."[24] Conducting ethnographic research during the 1980s, Alberto Arce found that *ejidatarios* adopted the fertilizer application practices advised by Hungarian experts who worked for the *Sistema Alimentario Mexicano* (Mexican Food Supply Program) but rejected their order to plant in June because they knew that heavy March rains provided adequate soil humidity for seed germination. The *campesinos* of the 1920s and 1930s reacted to the government's agricultural programs in a similar fashion: they did not reject all new ideas out of an innate desire to cling to "tradition"; rather, they adopted practices that experts recommended if they produced better results in their social, economic, and environmental conditions but refused to change their methods just because *agrónomos* called them inferior. Hence, they often failed to "cooperate," occasionally used "weapons of the weak" such as vandalizing experimental plots at night, and in a few instances resorted to threats or even violence to thwart unpopular agricultural programs.[25] By 1940 *agrónomos* helped many peasants obtain land and solidified their position in the Revolutionary Family, but they failed to transform the *campesinos* or their farming practices, which prevented any attempt to institutionalize their cultural campaign. Rather, the experience convinced the *agrónomos* that modernizing peasant agriculture was exceedingly difficult or impossible. As a result of these "negotiations," the *campesinos* retained more control over how they farmed, but they also drove the *agrónomos* to find other farmers who would accept their authority as experts.

Foreign businesses' favoritism for compatriots in pay, benefits, housing, promotions, and hiring, especially for technical and managerial employees, led to resentment in Mexico. In 1938 these sentiments surged during the labor dispute that culminated when Cárdenas expropriated U.S. and British oil companies' property, including the Rockefeller family's Standard Oil, the pinnacle of post-revolutionary economic nationalism. Ignoring the *agrónomos'* historical role as transmitters of foreign science and technology, some students of the Green Revolution claim that nationalism also influenced Cárdenas' agricultural policy and the *agrónomos*, who tried to develop a peasant-oriented and uniquely Mexican agronomy, but they always relied heavily on foreign science and technology and gladly accepted technical assistance from other nations, including the United States, both before and during the Cárdenas administration.[26] Before World War II, to protect its own farmers, the United States provided some assistance to Mexican pest control programs.

In 1938 a serious agricultural crisis began that included shortages of basic foodstuffs and other crops and a serious plague that all but destroyed the banana export industry. Cárdenas' enemies claimed that the land reform and the *agrónomos*' scientific incompetence caused the crisis. The concerted attack on the *agrónomos*' professional credibility drove them to seek assistance and training to improve their programs and skills as experimental scientists, including their ability to breed plants. Before Cárdenas left office, conflict in Asia and Europe stimulated the U.S. government's desire to foster increased farm production in Latin America, including training programs for the region's agricultural scientists, which fitted the *agrónomos*' agenda. Facing resistance in Congress, the Franklin D. Roosevelt administration began by assisting "complementary crop" producers (tropical farm goods), rather than foodstuffs like corn and wheat that U.S. farmers grew and exported. In addition to national security motivations such as access to a source of rubber in the event of world war, policy makers thought agricultural assistance would combat Axis influence and forge closer economic ties with Latin America.

During the "Economic Miracle" (1940–1968) the Revolutionary Family redefined the "Revolution" to mean industrial and urban growth. President Manuel Avila Camacho (1940–1946) appointed *agrónomo* Marte R. Gómez to run the SAF, giving the profession a larger role in agricultural policy-making. To maintain the *agrónomos*' alliance with the State by helping to create the "*utopía industrial*" (industrial utopia), Gómez implemented reforms to stimulate research, especially in plant breeding, and emphasized "professional competence" rather than revolutionary credentials when filling technical positions. The *agrónomos* offered the State a technical solution to Mexico's pressing agricultural problems, helping the Revolutionary Family downplay the controversial issue of land reform. *Agrónomos* defined plant breeding as a "domain" where they asserted "a privileged or exclusive right to speak *in* and *about*" (author's emphasis).[27] In 1943 Gómez and Avila Camacho took advantage of the opportunity to obtain U.S. technical assistance by reaching an agreement with the Rockefeller Foundation that established the Mexican Agricultural Program (hereafter MAP). The MAP's U.S. scientists aided research projects and scientific training at the ENA to advance the Revolutionary Family's agenda. Reacting to critics who detested their *agrarismo* and called them incompetent professionals, *agrónomos* used the new plants that they produced with the training or guidance of the MAP's scientists to demonstrate "the objectivity of [their] knowledge" and its "freedom from social influences."[28] Armed with new knowledge, the *agrónomos* bred new plants, claimed that *campesinos* could not do it themselves, and continued to reject peasant farmers' agricultural knowledge by defining it as primitive and inferior.

During World War II, U.S. policy makers developed a new understanding of the benefits of agricultural assistance programs. Rejecting the narrow focus on complementary crops, they realized that food shortages caused political unrest that Axis agents could exploit, and the government began to assist Latin American food crop cultivators. Even before Pearl Harbor a few influential U.S. citizens believed that agricultural aid programs would promote better relations and create new markets in underdeveloped regions of the world, and more came to this realization as the war progressed. When the conflict ended, the United States planned to continue these projects and extend them beyond Latin America. With the onset of the Cold War and rising fears about famines and political unrest caused by the population explosion, by working for the U.S. government at home and abroad to increase the production of food crops, U.S. agricultural scientists became warriors in the battle to contain the spread of Communism in the developing nations. The MAP's researchers performed this role in Mexico.

During the 1950s and 1960s Green Revolution technologies contributed to Mexico's Economic Miracle. Commercial farmers, especially in the northwestern states, made significant gains in productivity that temporarily ended the need to spend foreign exchange to import basic foods, and Mexico industrialized with the help of low wages and inexpensive subsidized foodstuffs. *Agrónomos* asserted their "scientific authority in the field" by using the new plants to obtain "recognized technical competence," at least from the State, farmers who benefited from the Green Revolution, and even some *campesinos*, as well as "social power over the constitutive mechanisms of the field (definition of practices and priorities, recruitment, rewards, resources)."[29] The Rockefeller Foundation's fellowship program and the MAP's work with *agrónomos* in Mexico solidified their position in the Revolutionary Family and brought Mexican agronomy into the mainstream of international agricultural science. When the Green Revolution spread to other nations, partly exporting new varieties created in Mexico, *agrónomos* basked in the glow of international acclaim that peaked in 1970, when Borlaug won the Peace Prize.

Although the administrations of the 1940s, 1950s, and 1960s distributed either little or poor land to *campesinos*, *agrónomos* called the Green Revolution the culmination of *agrarismo* because it would make Mexico's farmers prosperous. But most *campesinos* could not use the new technologies because they required inputs such as controlled irrigation, chemical fertilizers and pesticides, and the annual purchase of seed that they could not afford, did not have access to, or were provided by agencies and private companies that they did not trust. Critics of the Green Revolution, which included a few *agrónomos*, claimed that science ignored or even attacked the *campesinos*, challenging the profession's claim to "unquestionable objectivity." But most *agrónomos* recognized that their role as "experts" and that financial security depended on their alliance with the State and engaged in "prudential acquiescence" to the Revolutionary Family's leaders, claiming that the *campesinos'* lack of education, resistance to change, and character flaws explained why they did not benefit from the Green Revolution.[30] Supporting sociologist Randall Collins' analysis of professionals' behavior, *agrónomos* directed "attention away from the work which is done, and on to the style, the honour, [and] the moral standards displayed."[31] After 1940 the *ejidos* languished but remained a cornerstone of the "Revolution" and hence an essential part of the government's claim to legitimacy.[32]

In 1968 Mexico's political system entered a crisis when soldiers and police opened fire on student demonstrators at the *Plaza de las Tres Culturas*. The Tlatelolco Massacre provided graphic evidence that the "Economic Miracle" did not bring prosperity to most Mexicans. Confirming the findings of economists, the growing shantytowns of *campesino* migrants on the outskirts of the Federal District proved that the Green Revolution did not "economically emancipate" the *campesinos*. President Luís Echeverría (1970–1976) began another project to bring modern science to peasant farmers, but his programs could not overcome the legacies of the *agrónomos*' history and professionalization. Evidence from other nations demonstrated the Green Revolution's many social, economic, and environmental shortcomings, and the 1973 energy crisis provided further proof that it could not solve the Third World's agricultural problems. The charges of the Green Revolution's critics led to rethinking and new ideas about the costs, benefits, and strategies of agricultural modernization in the developing nations. A few *agrónomos* deemed *campesino* farming worthy of study because its methods were labor-intensive, not energy-intensive, and organic and thus environmentally friendly, but most believed in agrochemicals, continued to breed new crop strains, and defended the Green Revolution vigorously.

During the "oil boom" (1977–1982) the government spent more money on peasant agriculture, but when it went bust, the *campesinos*, like most other Mexicans, suffered through a "lost decade" of economic stagnation. In 1988 Salinas de Gortari won a hotly contested and disputed election and followed neoliberal economic policies, including the North American Free Trade Agreement (NAFTA). He ended land reform but promised that foreign agribusinesses would bring advanced technology to rural Mexico, including new "biotechnologies" developed by genetic engineering, and thus create a Second Green Revolution. Salinas and his allies planned to "include the *campesinos* in the discourse of agricultural development" and claimed that "inserting the village into the global economy" would finally bring them the "economic emancipation" that the State promised but failed to deliver since 1920.[33] Some *agrónomos* continued to work with peasant farmers, and some foreign scientists taught them organic agriculture, but others continued to stress the agrochemical-based practices of the Green Revolution. Although there have been some significant advances, and agronomy became a professionalized scientific discipline, modern science has not solved the problems of Mexican agriculture. The contraction of the agrobureaucracy as the State reduced its role in the economy, the increasing presence of foreign agribusiness in Mexico under NAFTA, and a biotechnology-based Second Green Revolution threaten to drive many *agrónomos* into the ranks of the un- or underemployed. In 1994 many *campesinos* in Chiapas joined the rebellion of the *Ejército Zapatista de Liberación Nacional* (Zapatista Army for National Liberation, EZLN), proving that some peasants have rejected the Mexican Revolution and lost any faith that the State that it created will bring them prosperity and democracy. However, recent developments offer the possibility of a brighter future for at least some peasant farmers, as well as the *agrónomos*. This book follows the evolution of Mexican agronomy through the course of these events.

NOTES

1. Alberto Arce, *Negotiating Agricultural Development: Entanglements of Bureaucrats and Rural Producers in Western Mexico* (Wageningen, the Netherlands: Agricultural University Wageningen, 1993); Marilyn Gates, *In Default: Peasants, the Debt Crisis and the Agricultural Challenge in Mexico* (Boulder, Colo.: Westview Press, 1993); Merilee Grindle, *Bureaucrats, Politicians, and Peasants in Mexico: A Case Study in Public Policy* (Berkeley: University of California Press, 1977); Cynthia Hewitt de Alcantara, *Modernizing Mexican Agriculture: Socioeconomic Implications of Technological Change, 1940–1970* (Geneva: United Nations Research Institute for Social Development, 1976); Bruce H. Jennings, *Foundations of International Agricultural Research, Science and Politics in Mexican Agriculture* (Boulder, Colo.: Westview Press, 1988). For other professionals see Roderic A. Camp, "A Middle Level Technocrat in Mexico," *Journal of Developing Areas* 6, 4 (July 1972): 581; Friedrich E. Schuler, *Mexico between Hitler and Roosevelt: Mexican Foreign Relations in the Age of Lázaro Cárdenas* (Albuquerque: University of New Mexico Press, 1998), pp. 1–2.

2. Michael Burrage, "The Professions in Sociology and History," in Michael Burrage and Rolf Torstendahl, eds., *Professions in Theory and History: Rethinking the Study of the Professions* (London: Sage Publications, 1990), p. 15; Peter S. Cleaves, *Professions and the State: The Mexican Case* (Tucson: University of Arizona Press, 1987).

3. Francisco R. Sagasti, *Ciencia, tecnología, y desarrollo latinoamericano* (México, D.F.: Fondo de Cultura Económica, 1981), pp. 139–50; Eli de Gortari, *La ciencia en la historia de México* (México, D.F.: Fondo de Cultura Económica, 1963), pp. 311–18.

4. Magali Sarfatti Larson, "Expertise and Expert Power," in Thomas L. Haskell, ed., *The Authority of Experts: Studies in History and Theory* (Bloomington: University of Indiana Press, 1984), p. 64; Susantha Goonatilake, *Aborted Discovery: Science and Creativity in the Third World* (London: Zed Books, 1984), p. 88; Arce, *Negotiating Agricultural Development*, p. 3.

5. Svante Beckman, "Professionalization: Borderline Authority and Autonomy in Work," in Burrage and Torstendahl, eds., *Professions in Theory*, p. 128; Eliot Freidson, "Are Professions Necessary?" in Haskell, ed., *The Authority of Experts*, p. 4.

6. Larson, "Expertise," p. 61.

7. For continuities see Mary Kay Vaughan, *The State, Education, and Social Class in Mexico, 1880–1928* (De Kalb: Northern Illinois University Press, 1982), pp. 178–79; and William E. French, "Prostitutes and Guardian Angels: Women, Work, and the Family in Porfirian Mexico," *Hispanic American Historical Review* 72, 4 (1992): 529–53. For the "cultural campaign" see Marjorie Becker, *Setting the Virgin on Fire: Lázaro Cárdenas, Michoacán Peasants, and the Redemption of the Mexican Revolution* (Berkeley: University of California Press, 1995); Adrian A. Bantjes, "Burning Saints, Molding Minds: Iconoclasm, Civic Ritual, and the Failed Cultural Revolution," in William H. Beezley, Cheryl English Martin, and William E. French, eds., *Rituals of Rule, Rituals of Resistance: Public Celebrations and Popular Culture in Mexico* (Wilmington, Del.: Scholarly Resources, 1994); John W. Sherman, *The Mexican Right: The End of Revolutionary Reform, 1929–1940* (Westport, Conn.: Praeger, 1997), pp. 41–42; Alan Knight, "Racism, Revolution, and *Indigenismo*: Mexico, 1910–1940," in Richard Graham, ed., *The Idea of Race in Latin America, 1870–1940* (Austin: University of Texas Press, 1990), p. 82; Knight, "Revolutionary Project, Recalcitrant People: Mexico, 1910–1940," in Jamie Rodríguez O., ed., *The Revolutionary Process in Mexico: Essays on Political and Social Change, 1880–1940* (Los Angeles: UCLA Latin American Center Publications, 1990), p. 245. For science and twentieth-century States see Ashis Nandy, "Science as a Reason of State," in Ashis Nandy ed., *Science, Hegemony, and Violence: A Requiem for Modernity* (Delhi: Oxford University Press, 1988), pp. 1–2; Patrick Carroll, "Science, Power, Bodies: The Mobilization of Nature as State Foundation," *Journal of Historical Sociology* 9, 2 (1996): 139–67; and Mary Kay Vaughan, *Cultural Politics in Revolution: Teachers, Peasants, and Schools in Mexico, 1930–1940* (Tucson: University of Arizona Press, 1997), p. 10.

8. Adrian A. Bantjes, *As If Jesus Walked on Earth: Cardenismo, Sonora, and the Mexican Revolution* (Wilmington, Del.: Scholarly Resources, 1998), p. 29.

9. Hannes Siegrist, "Professionalization as a Process: Patterns, Progression, and Discontinuity," in Burrage and Torstendahl, eds., *Professions in Theory*, p. 187.

10. Frank Brandenberg, *The Making of Modern Mexico* (Englewood Cliffs, N.J.: Prentice-Hall, 1964).

11. Christopher R. Boyer, "Old Loves, New Loyalities: Agrarismo in Michoacán, 1920–1928," *Hispanic American Historical Review* 78, 3 (August 1998): 423–26; Staffan Selander, "Associative Strategies in the Process of Professionalization: Professional Strategies and the Scientification of Occupations," in Burrage and Torstendahl, eds., *Professions in Theory*, p. 139.

12. William T. Lynch, "Ideology and the Sociology of Scientific Knowledge," *Social Studies of Science* 24, 2 (May 1994): 206–7.

13. David E. Lorey, *The University System and Economic Development in Mexico since 1929* (Stanford, Calif.: Stanford University Press, 1993), pp. 92–93; Edmundo Flores, *Historias de Edmundo Flores, autobiografía 1919–1950* (México, D.F.: Martín Casillas Editores, 1983), pp. 98–99.

14. Richard Harvey Brown, "Modern Science: Institutionalization of Knowledge and Rationalization of Power," *The Sociological Quarterly* 34, 1 (1993): 153.

15. Norman Long, "Sociological Perspectives on Agrarian Development and State Intervention," in Anthony Hall and James Midgley, eds., *Development Policies: Sociological Perspectives* (Manchester: Manchester University Press, 1988), pp. 108–33; Long, "Introduction: The Raison D'Être for Studying Rural Development Interface," in Long, ed., *Encounters at the Interface: A Perspective on Social Discontinuities in Rural Development* (Wageningen, the Netherlands: Agricultural University Wageningen, 1989), pp. 1–2.

16. Eyler N. Simpson, *The Ejido: Mexico's Way Out* (Chapel Hill: University of North Carolina Press, 1937), p. 580.

17. Larson, "Expertise," p. 60; Alan Knight, "Cardenismo: Juggernaut or Jalopy," *Journal of Latin American Studies* 26, 1 (February 1994): 73–107.

18. Billie R. DeWalt, *Modernization in a Mexican Ejido: A Study in Economic Adaptation* (Cambridge: Cambridge University Press, 1979), p. 200.

19. Larson, "Expertise," p. 64.

20. Boyer, "Old Loves," p. 454; Vaughan, *Cultural Politics*, p. 7. For a study of this phenomenon during the 1970s see DeWalt, *Modernization*, p. 60; Beckman, "Professionalization," p. 128.

21. John Tutino, *From Insurrection to Revolution in Mexico: Social Bases of Agrarian Violence* (Princeton, N.J.: Princeton University Press, 1986, pp. 26–28; Daniel Nugent, *Spent Cartridges of Revolution: An Anthropological History of Namiquipa, Chihuahua* (Chicago: University of Chicago Press, 1993), p. 157; David Ronfeldt, *Atencingo: The Politics of Agrarian*

Struggle in a Mexican Ejido (Stanford, Calif.: Stanford University Press, 1973), p. 102.

22. Jan Douwe van der Ploeg, "Knowledge Systems, Metaphor and Interface: The Case of Potatoes in the Peruvian Highlands," in Long, *Encounters at the Interface*, pp. 156–62; Sergio Zendejas, "Appropriating Government Reforms: The Ejido as an Arena of Confrontation and Negotiation," in Sergio Zendejas and Pieter De Vries, eds., *Rural Transformations Seen from Below: Regional and Local Perspectives from Western Mexico* (La Jolla: Center for U.S.–Mexican Studies, University of California, San Diego, 1995), p. 25.

23. Freidson, "Are Professions Necessary?", p. 23.

24. Vaughan, *Cultural Politics*, p. 10; Larson, "Expertise," p. 35.

25. Arce, *Negotiating Agricultural Development*, pp. 154–56; James C. Scott, *Weapons of the Weak: Everyday Forms of Peasant Resistance* (New Haven, Conn.: Yale University Press, 1985), pp. 28–36.

26. Hewitt de Alcantara, *Modernizing*, pp. 19–20; Francis Moore Lappé and Joseph Collins, *Food First: Beyond the Myth of Scarcity* (New York: Ballantine Books, 1977), pp. 121–24; Jennings, *Foundations*, p. 99; George I. Sánchez, *Mexico: A Revolution by Education* (New York: Viking Press, 1936), pp. 145–46. These ideas also influenced Arce; see *Negotiating Agricultural Development*, p. 40, and Angus Wright; see *The Death of Ramón González: The Modern Agricultural Dilemma* (Austin: University of Texas Press, 1991), p. 186.

27. Blanca Torres, *Hacia la utopía industrial* (México, D.F.: El Colegio de México, 1984); Larson, "Expertise," pp. 35, 56, 64.

28. Michael Mulkay, "Knowledge and Utility: Implications for the Sociology of Knowledge," *Social Studies of Science* 9 (1979): 68, as cited in Larson, "Expertise," p. 55.

29. Larson, "Expertise," p. 56.

30. Cleaves, *Professions and the State*, p. 13; Larson, "Expertise," p. 64.

31. Randall Collins, "Market Closure and the Conflict Theory of the Professions," in Burrage and Torstendahl, eds., *Professions in Theory*, p. 26.

32. John Gledhill, *Casi Nada: A Study of Agrarian Reform in the Homeland of Cardenismo* (Albany: State University of New York, Institute for Mesoamerican Studies, 1991), p. 9.

33. Gustavo Gordillo de Anda, "La inserción de la comunidad rural en la sociedad global," *Comercio Exterior* 40, 9 (September 1990): 813; Jaime de la Mora Gómez, "La banca de desarrollo en la modernización del campo," *Comercio Exterior* 40, 10 (October 1990): 943–50.

Chapter 1

Liberalism and the Origins of Mexican Agronomy

Mexico's geography and climate create a diverse array of environments that allow many crops to flourish. The *campesinos'* ancestors took advantage of these conditions by developing farming practices and technologies such as the famous *chinampas* (fertile lands constructed by the Aztecs, and made of mud dredged from canals) of the Valley of Mexico, maize, and the digging stick. When the Spanish conquered and settled Mexico after 1519, they brought European notions of land tenure like the hacienda and technology such as the wooden plow. The Catholic Church played important roles in education and the introduction of new farming practices. After 1750 the scientific developments and political changes that swept Europe fascinated "enlightened" citizens of New Spain, but in 1810 intellectual trends, European wars, and the stresses produced by its late-eighteenth-century economic boom sparked a violent struggle for (and against) independence. In 1821 Mexicans broke their bond with Spain, but the fight devastated the economy. For years the nation suffered from political turmoil as liberals tried to demolish the economic and intellectual power of the Catholic Church and conservatives fought to maintain the structure of colonial society, culminating with a U.S. invasion that took half of Mexico's territory (1846–1848), the bloody War of La Reforma (1858–1861), and the French occupation and imposition of a European king (1862–1867). With violence, political strife, and other problems, economic development languished.

A few Mexican farmers tried to improve their practices, but conditions were not favorable for scientific research or the development of new professions.

In 1867 the Liberals triumphed, starting a period of political stability that culminated in the long rule of Porfirio Díaz, which attracted foreign investors and fostered rapid economic growth. New mines and industries needed sober, punctual, and productive workers who accepted the industrial discipline of a capitalist economy, and Díaz and his allies constructed a campaign to transform their traditional culture. Railroads built by foreign investors fostered profound rural change by opening access to internal and foreign markets. Haciendas prospered and grew, often by acquiring village lands. Díaz's cultural project ignored the *campesinos*, who were controlled by the discipline of the hacienda system. Some haciendas tried new farming technologies developed abroad, but many used traditional methods, especially food producers who provisioned the internal market. Porfirian stability and prosperity created a favorable climate for science and the professions, and a career in agronomy attracted more young Mexicans. Reflecting elites' interest in French culture and Auguste Comte's positivist philosophy, Mexico's scientists worshiped France. Some historians claim that experimental science languished during the *Porfiriato*, while others argue that agricultural research progressed, but *hacendados* did not seek the assistance of *agrónomos*, and the profession suffered from a lack of clients who accepted their expertise.[1]

By 1905 food prices soared when farmers could not feed the growing population that Porfirian stability fostered. Workers went on strike, and some elites realized that Díaz's model of economic development had shortcomings. To seize control of agricultural innovation and thus establish their status as professionals with useful specialized knowledge, *agrónomos* allied with other Mexicans who wanted an activist State to modernize agriculture. They examined other scientific models, especially U.S. agronomy, obtained some foreign assistance to improve their skills as scientists, and convinced Díaz to expand the agricultural bureaucracy. The *agrónomos* made no significant advances that made Mexico's farmers more productive, but they began to forge a bond with the State that served the profession well after the Mexican Revolution. Díaz's cultural project did not produce a docile workforce and had no impact on peasant farmers, but post-revolutionary regimes borrowed from it when they tried to transform *campesino* culture. Precedents and trends established in the nineteenth century influenced Mexican agronomy and agricultural policy after the Revolution overthrew the Díaz regime.

MEXICAN AGRICULTURE IN 1880: *HACENDADOS, RANCHEROS,* AND *CAMPESINOS*

Rugged geography gives Mexico a diverse array of climates with varied agricultural potentials. In 1941 Elvin C. Stakman, Paul C. Mangelsdorf, and Richard Bradfield, the scientists who surveyed Mexican agriculture for the Rockefeller Foundation, wrote:

> Mexico is blessed with many types of land. Wide variations in altitude, temperature, and rainfall, superimposed upon a complicated geological background, have produced a diversity of soils and plant life that is unique. Practically every crop now used by man for food, clothing, shelter, medicinal, and industrial purposes can be grown successfully in some part of Mexico.[2]

The southern states have coastal jungles, tropical scrub and savannah, highland cloud forests, the shallow soils of the Yucatán Peninsula, and the eroded hills and canyons of Oaxaca. In the north there are vast deserts that irrigation can transform into productive farmland, and mountain ranges with pine forests and high-altitude chaparral. Central Mexico has high plateaus, mountain peaks above the tree line, and steep slopes where frosts and hail destroy crops. Heavy rains erode the soil or fail to fall, causing devastating droughts. Deep subtropical and tropical valleys where crops like sugarcane thrive divide the plateaus and mountain ranges. Wheat, citrus fruits, vegetables, and other crops flourish in the lower elevation of the Bajío (west-central Mexico), but rainfall is uncertain there as well. Unlike many Latin American countries where a few crops predominate, Mexicans grow a wide variety of plants in many different environments. Examples include the corn and bean growers of the Central Plateau; the wheat farmers of the northern deserts, who also grow tomatoes for export; and cultivators of tropical crops such as bananas in Tabasco, coffee in Veracruz, or sugarcane in Morelos, as well as many others. Mexican agriculture presents a vast array of problems, and many of them are different from those faced by the farmers of Europe, the United States, and Canada.

Mesoamerican corn-bean-squash agriculture originated several thousand years ago. Historian Jeffrey M. Pilcher calls the *campesinos* "people of corn," and anthropologist Alan R. Sandstrom's informants in Amatlán, Veracruz, told him that "corn is our blood."[3] The *campesinos'* distant ancestors were Mexico's first agricultural scientists. In tropical forests farmers created slash-and-burn and fallowing systems to restore the fertility of their fragile soils and protect them from downpours. In the arid north some Native Americans planted crops near rivers and built irrigation systems. In the Central Highlands many *campesinos* cultivated hillsides instead of valleys because they were less vulnerable to frosts and easier to till with a digging stick. Franciscan Friar Bernardino de Sahagún recorded the Aztecs' knowledge of agriculture and foods. Their *nahuatl* language has 19 names for cultivated trees, 12 for beans, and 11 for corn. The Aztecs also possessed sophisticated knowledge of Mexico's soils, describing 14 fertile types, 11 unfertile, 1 specifically for maguey, and 8 with unspecified characteristics. They knew that *cacateteztli* grass grew in poor soil, and they fertilized with humus. Similarly, the Maya of Campeche had 11 categories for local soils that describe their agricultural potential so effectively that *agrónomos* use them today. Some indigenous farming practices damaged the environment, but in central and southern Mexico they produced enough to feed large populations except during extreme climatic events like droughts.[4]

The *campesinos'* ancestors believed that supernatural forces determined their success as farmers. To ensure a bountiful harvest, the Aztecs asked Tlaloc, god of rain and thunder, to send ample rainfall rather than damaging hail. The sixteenth-century missionaries and the priests who came to New Spain after them failed to eradicate many ancient beliefs and rituals regarding agriculture. Anthropologist Paul Friedrich found that late-nineteenth-century Tarascan peasants in Naranja, Michoacán, asked saints to protect their crops, giving Catholic forms to their ancient deities. In the early twentieth century *campesinos* resisted another campaign against these attitudes and practices conducted by rural teachers, *agrónomos*, and other representatives of the post-revolutionary regime. Vaughan claims that in the early 1930s the peasants of

Puebla "resorted to prayer, pilgrimages, and witchcraft" when droughts and freezes threatened their crops, and in 1941 the foundation's survey team saw Native American farmers spilling the blood of sacrificed animals on their cornfields to ensure a good harvest.[5] Today anthropologists and journalists still see these practices in rural communities, especially where Native American identity and culture are strong. For example, in Oaxaca Mixtecs bury candy, *pulque,* and cigarettes in their fields to give thanks to the land. After centuries of economic and political exploitation, violence, and forced cultural change, *campesinos* often see outsiders, including representatives of the State, with profound suspicion and mistrust. In Native American communities cultural and linguistic isolation reinforce these attitudes.[6]

Mexican farmers use many traditional practices developed by centuries of practical experience. Some are very productive and use available energy or other resources efficiently. According to historian William Beezley, nineteenth-century farmers knew that sowing corn with a digging stick did not maximize yields, but the plants survived storms that uprooted taller ones grown in deep furrows, and the peasants of Tepotzlán, Morelos, used hoes to obtain twice the maize produced by tilling with iron plows. In Tlaxcala corn-bean-squash intercropping prevents insect damage as effectively as pesticides used on separate plots as well as maize yields comparable to hybrids grown with chemical fertilizers on the State's arid hillsides. Maguey, cactus, and olive and fruit tree field borders retain soil moisture and add nutritious foods to the diet. Peasant farmers create and maintain their own plant varieties, as the MAP's scientists knew when they began their research. In 1943 corn geneticist Edwin J. Wellhausen recognized that the *campesinos* of the Valley of Mexico and Puebla bred maize to survive early frosts. In 1948 the MAP's Mexican and U.S. soil scientists found that native corn tolerated events and conditions that killed hybrids. During the 1950s Wellhausen learned that *campesinos* crossbred MAP maize with traditional strains to make it more "suitable to their conditions."[7] In Campeche, Mayan farmers planted local yellow and white corn in separate fields to keep the varieties pure. During the 1970s in Nealticán, Puebla peasants grew four local strains of maize identified by the color of their kernels, but they also differed in weight, texture, taste, yields, and growing periods. They used the highest-yielding yellow corn for animal feed because they disliked its taste and preferred to plant the less productive white strain, but they could also sow fast-maturing blue or red varieties if white failed early in the growing season. The MAP's scientists understood peasant farmers' primary objective: to avoid starvation by guaranteeing at least a minimal harvest in Mexico's capricious climate and often difficult environmental conditions, a goal that historian John Tutino calls "subsistence security."[8]

Like other European colonists, Spaniards saw Native Americans as heathen savages. Conquistadores like Hernando Cortés and missionaries like Sahagún admired complex urban societies like that of the Aztecs to an extent but thought that Satan corrupted them. They had greater contempt for the Native Americans of northern Mexico, like the migratory Apache, calling them uncivilized barbarians. Cultural imperialism influenced the evolution of Mexican agriculture and fostered Pilcher's "tortilla discourse." Some Spaniards admired indigenous farming systems like *chinampas* but considered them inferior to European practices. Spanish immigrants and their descendants tried to transfer their homeland's cuisine to the New World. After 1493, when Christopher Columbus brought European livestock and seeds to Hispañola, the Spaniards who explored, conquered, and settled in the Americas introduced animals, wheat, grapes, sugarcane, coffee, and olive trees with varying degrees of success. If the environment made production difficult or impossible, like in sixteenth-century Florida, Spaniards complained about diets of "herbs, fish, scum, and vermin."[9] Native Americans adopted some Spanish foods but also clung to their traditional corn, beans, squash, and chiles. During the colonial period the racially mixed population (*castas*) grew. A few *castas* prospered, but most joined the large lower class, which New Spain's elite viewed with fear and contempt. Eating wheat bread indicated upper-class status and European origins, but corn represented poverty and Native American ancestry, attitudes that persisted long after Mexico's independence.

During the sixteenth century, replacing the *encomenderos* (Spanish colonists who held authority of Native territories) who received farm goods grown by Native American peasants as tribute, the Conquistadores and the immigrants who followed them acquired land (haciendas) to satisfy city dwellers' demands for a Spanish diet of wheat bread, olive oil, wine, and meat. *Hacendados* used European technologies such as "Egyptian plows" and began to grow indigenous crops like corn to sell to the poor. Some acquired vast estates, such as the marquis de Aguayo, whose holdings covered 11 million hectares in the northern province of Coahuila. Owning a hacienda brought social status, but *hacendados* also tried to turn a profit and exploited crises like droughts by withholding their crops from the market until scarcity drove prices very high. In New Spain, Catholic clerics played a leading role in agricultural innovation, introducing European wheat and alfalfa varieties and supervising the construction of irrigation works. The professional Spanish and *criollo* hacienda managers whom the Jesuits hired used European methods and practical knowledge to run very productive farms. The friars knew that the *campesinos'* traditional farming practices were based on years of experience and felt that it would be unwise to change them.[10]

The chaos that plagued Mexico after independence discouraged agricultural innovation, and regimes had no money for programs to modernize farming. Some scholars, Díaz's critics, and many of the revolutionaries who overthrew him, including influential *agrónomos* like Marte Gómez, claimed that the haciendas' size fostered extensive farming practices and the exploitation of labor, making innovation unnecessary. Some were self-sufficient institutions with vast properties whose owners and *mayordomos* (foremen) practiced extensive agriculture based on experiential knowledge and saw no need to modernize their practices. For example, in 1937 the SAF claimed Mexican farmers grew 35 wheat strains that Spanish missionaries imported in colonial times. In 1940 some haciendas still used Egyptian plows pulled by oxen with ropes attached to their horns rather than the more efficient neck-yoke. Hacienda peasants heeded their priests, who sometimes said that the Devil possessed new farm technologies like threshing machines. Historian Margaret Channing found that although many entrepreneurial *hacendados* were reluctant to try new farming methods in the tumultuous years after independence, a few did innovate. For example, the owners of the hacienda *Plan de la Barranca* in Colima bought U.S. sugar-processing equipment, but they lost money because they had no foundry or machinists to repair it. Tutino discovered that after the Mexican War *hacendados* from Chalco, near Mexico City, began to select seed scientifically and tested new ways to maximize the effectiveness of organic fertilizers. In 1862, ignoring the turmoil caused by the French invasion, the Ortega brothers, *hacendados* from Zacapoaxtla, Puebla, tried China grass, a new fiber crop. In 1867 *hacendados* in Chiapas' Soconusco district planted rubber trees.[11]

Notwithstanding their critics' claims, with the more stable conditions of the *Porfiriato*, more *hacendados* modernized their practices or diversified their crops. Some took advantage of Díaz's effort to promote exports. In 1880 Manuel Janet planted Roatan bananas in Tabasco, but his plantation failed because he could not find a market for his fruit. In 1895 Julian Dueñas brought more bulbs from Honduras, and by 1906 he located U.S. buyers, founding a lucrative industry. *Hacendados* planted potatoes near León, Guanajuato, establishing the crop in the State. Ranchers bought purebred livestock in Europe and the United States to improve their herds. Diego Moreno Leñero, proprietor of the hacienda La Guracha, imported new sugarcane varieties and Cuban technicians. The aristocratic García Teruel family of Mexico City hired a foreign engineer to manage their haciendas. Some *hacendados* tested U.S. technologies. During the 1880s farmers in the newly irrigated La Laguna began to import cotton seed, and in 1908 a potato grower from Silao, Guanajuato, tried the variety "Early White." In tropical regions some *hacendados*, such as the coffee growers of Salina Cruz, Oaxaca, used chemical fertilizers. Like U.S. farmers, they did not trust the "book farming" advocated by university-trained "experts," but some *hacendados* in the northern states admired U.S. agriculture, abandoned their Egyptian plows, and copied their counterparts across the Rio Grande.[12]

The *rancho* also emerged in New Spain when poorer colonists acquired small farms. *Rancheros* experienced good and bad times as the colonial economy ebbed and flowed. Economic booms sometimes caused conflicts with *hacendados*, as occurred during the late eighteenth-century in the Bajío. As property owners, *rancheros*, like the eighteenth-century colonists in the Rio Grande Valley, distinguished themselves from peasants who lived and worked on haciendas or traveled from estate to estate as wage laborers and from Native Americans who farmed village *ejidos*, but their humble dress, demeanor, and financial resources gave them less status than the urbane and wealthy *hacendados*, and urban people often saw them as simple *campesinos*. Similar labels exist today in Jalisco, where *rancheros'* private property, vehicles, clothing, and behavior give them more status than *ejidatarios*. To the military colonists of Namiquipa, land-ownership meant pride, autonomy, good citizenship, and *machismo* earned by years of fighting hostile Apache and hard work farming the desert. The *rancheros* of San José de Gracia, who used the same agricultural practices as local *campesinos*, viewed their property through similar lenses. Like their counterparts in Namiquipa and the Rio Grande Valley, they were deeply suspicious of government officials who entered their lives only to tax or arrest them. Some *rancheros* innovated, like the farmers of Pisaflores, Hidalgo, who planted more sugarcane and coffee to take advantage of the expanding local market during the *Porfiriato*. In some districts ranchos were more productive than haciendas. Like the *campesinos*, many *rancheros* had strong religious beliefs and thought supernatural forces influenced the outcome of their agricultural and pastoral endeavors.[13]

AGRONOMÍA AFRANCEADA: THE FRENCH CONTRIBUTION TO MEXICAN AGRONOMY

During the eighteenth century Bourbon kings, blood relatives of the French dynasty, ruled Spain. Many Latin American intellectuals studied Enlightenment works, including José Antonio Alzate y Ramírez, the distinguished scientist who brought European agronomy to Mexico. To reinforce the bonds between New Spain and France, in 1771 the Paris Academy of Sciences made him a corresponding member. The School of Mines (*Instituto Minería*), where European professors taught chemistry and students read Spanish translations of Antoine Lavoisier's *Traite elementaire de chemie*, also transmitted French science to Mexico. After independence Mexican scientists rejected their old metropolis and looked to France for scientific inspiration. Reflecting Mexico's faith in French scientists, President Guadalupe Victoria (1824–1828) hired French natural historian Jean Louis Berlandier to conduct scientific surveys, but political turmoil and economic stagnation retarded the development of agronomy. In 1854 President-for-Life Antonio López de Santa Anna founded the *Escuela Nacional de Agricultura* (ENA), but the first *agrónomo* did not graduate until 1864. During the French Intervention scientists with the *Commission Scientifique du Mexique* (Mexican Scientific Commission) accompanied the occupation army and made several important contributions to the development of Mexican science, especially medicine, geography, and natural history. *Juarista* rebels almost killed Mexican medical doctor Manuel M. Villada when they captured him while he gathered plants for the *Commission*. After President Benito Juárez (1858–1872) pardoned his collaboration, Villada reinforced Mexico's scientific ties with France as one of the founding members of the *Sociedad Mexicana de Historia Natural* (Mexican Society for Natural History). He also taught botany at the ENA and served as the school's director several times in the 1880s and 1890s.[14]

The French invasion caused a crisis in Franco–Mexican relations but did not disrupt the scientific connection between the two nations. The popularity of philosopher Auguste Comte's positivism helped sustain these bonds. Gabino Barreda, who heard Comte lecture in Paris, disseminated his ideology at Mexico's School of Medicine in his physics and natural history classes. Hired by Juárez to improve the school system, positivism influenced Barreda's 1867 *Ley Orgánica de Instrucción Pública* (Organic Law of Public Education), which founded the *Escuela Nacional Preparatoria* (hereafter National Preparatory School), where many children of upper- and middle-class families received their secondary education. The National Preparatory School's scientific curriculum followed Comte's hierarchy of the sciences and used many French texts, including botany by Sach, Duchartée, Cauvet, Richard, Henfrey, Daguillón, and Colomb and zoology by Henri Milne Edwards and Claus. Prominent Porfirian *agrónomos* such as Alejandro Brambila Sr., Andres Basurto Larrainzar, and Agustin Aragón León attended the National Preparatory School. Some went to elementary schools modeled on French institutions, like Brambila Sr., a graduate of the *Colegio Francés de De Fosey*, and Julio Riquelme Inda, who attended Mexico City's *Lycée Fournier*. In 1895 Jesús Díaz de León, a member of the *Sociedad Científica Antonio Alzate*, recommended using French methods of agricultural education in other primary and secondary schools.[15]

Barreda influenced the development of Mexican agronomy when he reformed the ENA's curriculum. Classes relied heavily on French texts, and all new students had to take two years of French, followed by two of English and one of German. In 1883 the ENA added a course on Rational Mechanics based on the works of Delavnay and in 1890 adopted *agrónomo* Rafael Barba's

1890 text *Mecánica Agrícola*, which discussed the use of plows in France. A rural economy dissertation cited the *agronome* Lezcano from Gembloux, Goertz's *Cours d'Economie Rural*, and Brassuer de Bourburg's *Les choses de Yucatán*. In 1893 the ENA adopted a curriculum "almost the same as the School of Grignon."[16] The ENA trained hacienda administrators and advisors, not scientists who conducted research on Mexico's agricultural problems. Like other scientific professionals in the Third World, *agrónomos* transmitted knowledge created in Europe and North America to Mexico. The government's agricultural publications relied on the results of experiments performed in Europe and the United States and translations of foreign technical articles. Mexican experts used French sources heavily or predominantly. For example, to teach farmers how to grow hops, *agrónomo* Eduardo Rodríguez consulted French and Spanish texts. Luís Fernández del Campo's 1913 monograph on sugarcane cited the "notable chemists" Payon, Damseaux, N. Bassett, and Boname, as well as several German and U.S. sources.[17]

Like Mexico City's *Museo de los Bellas Artes* and *Paseo de la Reforma*, Porfirian Mexico reflected the elite's love of French culture. Along with other members of the upper class and the middle class who tried to copy their betters, many scientists became positivists and devout Francophiles who looked to France for scientific inspiration, maintained close contacts with French colleagues, and tried to emulate their achievements. Distinguished biologist Alfonso Herrera and other Mexican contributors published articles in the *Sociedad Científica Antonio Alzate*'s journal in French. Reflecting their education, *agrónomos* deeply admired French agriculture and agronomy. The prominent Lauro Viadas wrote that "professors of the stature of Muntz, Girard, Ringelman, Schribaux, and all those experts that for us, are familiar with French agricultural literature more than any other, are idols who represent all knowledge ["*suma sabidura*"] of farming," and Leopoldo Palacios said that scientists made France "the preeminent agricultural nation" that Jesús Díaz de León called "the brain of the civilized world."[18] In 1897 Aragón León went to France to read a paper at a celebration of the 100th anniversary of Comte's birth, and from 1900 to 1914 he edited *La Revista Positivista*, "one of the most steadfast defenders of positivism in Mexico."[19]

To counterbalance the influence of U.S. investors, Díaz sought French capital and trade. Scientific cooperation was another way to forge closer relations. During the *Porfiriato* some of Mexico's most prestigious scientists obtained advanced training in France. Basurto Larrainzar excelled at the ENA and won a government scholarship to the Paris Institute of Agronomy, where he took courses from Pasteur in 1883 and 1884. In 1885 he graduated, returned to Mexico, and conducted some research for the government, but he preferred teaching and became a respected ENA professor who passed his French training on to the next generation of *agrónomos* in his agricultural chemistry, agronomy, and meteorology courses. In 1891, to reward his distinguished academic record, *agrónomo* Juan E. Contreras won a scholarship to study the wine industries in France, Spain, and Italy. Other prominent Mexican scientists in fields related to agriculture, such as noted conservationist Miguel Angel Quevedo, also studied in France. French scientists and scientific institutions located in the metropolis also found ways to reinforce their ties with their Mexican colleagues. For example, biologists M. Georges Renaudet and Albert et Alexandre Mary published in the journal of the *Sociedad Científica Antonio Alzate*, one of Mexico's most prestigious scientific organizations. In 1906 more than two-thirds of the *Sociedad*'s 173 foreign corresponding members were French, while Germany was second with 12. In 1909 the Paris Academy of Sciences donated 154 volumes of its Acts and 95 of its *Memorias* to the organization's library. These contacts extended to the *agrónomos*. For example, in 1885 Basurto Larrainzar brought French carbon fever vaccine to Mexico, and later José del Carmen Segura went on a similar mission. At the 1900 Paris Exposition France gave Gabriel Gómez the *Insignia de Caballero de Mérito Agrícola* (Knight's Badge of Agricultural Excellence).[20] Steeped in the high science of French *agronomie*, *agrónomos* looked abroad for examples of excellence in agriculture.

As Lorey notes, there are no comprehensive statistical data on the social class origins of Mexican university students, but anecdotal evidence compiled by Marte Gómez indicates that most Porfirian ENA enrollees came from upper- and middle-class families. Some were elites, such as Alberto García Teruel of Mexico City, whose father drove a fine automobile, the Jaimes brothers of Temascaltepec, Mexico State, whose wealthy parents owned a number of rural properties, or Carlos Fernández Jr., whose family had a large coconut plantation in Colima and a palatial home. *Hacendados* fathered at least seven other Porfirian *agrónomos*. Others had connections to elites' blood relatives or in-laws, like Angel Posada, whose father worked for the son-in-law of Luis Terrazas, political boss of Chihuahua. Several *agrónomos'* fathers were part of Díaz's political machine, like Juan Francisco Noyola's, secretary-general of San Luis Potosí, or Manuel Mesa A.'s, a judge in Guerrero. Others had middle-class origins. Some had fathers, and in one case a mother, in the professions, including three teachers, three engineers, one lawyer, two Mexico City *licenciados* (lawyers), and a physical therapist. Some *agrónomos* came from military officers' families, like Marte Gómez and Manuel Vera. Six *agrónomos'* fathers were postal inspectors. Several were children of *rancheros*, like Carlos Noriega, whose family owned a small dairy farm near Mexico City, or Nardo Mansilla Cortés, whose parents' *rancho* in Güemes, Tamualipas, had a few head of cattle and irrigated land that made them well-off by small-town standards. Some *agrónomos'* families ran small businesses, like Enrique Morán, whose parents had a clothing shop on a corner of the town square in Ciudad Lerdo, Durango, or Enrique Martínez de Alba, whose father's hardware and machinery store was in Zacatecas. Some upwardly mobile poor families, like Ambrosio López's, who grew up in a working-class neighborhood near Mexico City's San Lázaro train station, saved money to send their children to elite schools like the National Preparatory, where they won scholarships to the ENA. Nine other *agrónomos* came from backgrounds that Gómez called modest or humble. For example, the ancestors of Gilberto Fábila, who became prominent in the profession and politics after the Revolution, sharecropped on the La Gavia hacienda in Mexico State. One Afro-mestizo earned admission to the ENA, and two students were Native Americans. For them, an ENA degree provided a chance to enter the middle class.[21]

Most Porfirian *agrónomos* were urban people. Of 148 ENA students from 1900 to 1910 for whom information is available, 43 came from the Federal District (33), Guadalajara, or Monterrey, 51 from state capitals or important provincial towns like Torreón or León, and 54 from small towns, villages, or rural areas. Reflecting the influence of U.S. agriculture, 29 were from the northern border states of Coahuila, Chihuahua, Nuevo León, Sonora, and Tamualipas. The Bajío, a more traditional rural area that includes parts of Aguascalientes, Guanajuato, Jalisco, Michoacán, and Queretaro, sent 20, and 21 attended from the southern states of Campeche,

Chiapas, Guerrero, Morelos, Oaxaca, Tabasco, Veracruz, and Yucatán. Eleven sons of La Laguna farmers studied at the ENA. Despite the suggestion of the director of Mexico City's preparatory school for women, the profession of agronomy was a male domain. In feudal Spain and the colonial society that it produced, working the soil was stigmatized by association with peasantry and servitude. Immigrant scientist Othon E. Brackel-Welda criticized Mexican elites' disdain for manual labor, as well as *agrónomos* who "feared grasping a farm implement."[22] He urged the ENA to admit rural children rather than sons of the urban upper class. Díaz de León felt that elites had no interest in agriculture, but even *agrónomos* from rural families disliked farming and country life after they went to the city for an education. Like the Colombian engineers whom Frank Safford studied, as middle- and upper-class people with university degrees, *agrónomos* felt that working alongside *campesinos* was demeaning and saw fieldwork like manually crossbreeding plants as beneath their social position. For example, Rafael M. Tello, who worked for a German fertilizer firm, told farmers to analyze their soil before buying his products, rather than doing so himself. *Agrónomos* wore suits and ties, the garb of intellectuals, and associated khakis, boots, and dirty hands with peonage. In Beezley's words they "sat in the shade."[23]

After independence Mexican society remained racially and socially stratified, and elites believed that Native Americans and *castas* (ethnic combinations consisting of mestizos [Native Americans and Europeans], mulattos [Africans and Europeans], and lobos [Native Americans and Africans]) were inferior. Like other members of the Porfirian upper and middle classes, *agrónomos* professed many negative stereotypes about the "Mexican race," especially Native Americans. Some claimed that Mexico's tropical climate made the masses lazy. Gustavo Durán lamented that "one of the characteristics of our race is the tendency to indifference, apathy, lack of interest in enterprise, fear of applying new things that science or industry produce, and this tendency hinders advancement and harms our country grievously," and Bernardo Mallen claimed that the people's "glacial passivity" inhibited Guanajuato's campaign to develop the silk industry.[24] As Pilcher demonstrates, rejection of Native American culture extended to food, especially the consumption of corn, a sign of lower-class status to elites who preferred wheat bread over the maize *tortilla*. After 1900 interest in nutritional science reinforced elites' disdain for the corn-based diet of the *campesinos* and the urban poor. The intellectual Francisco Bulnes called maize "an inferior food to those of civilized peoples, like wheat."[25] Informed by a neo-Lamarckian view of heredity and evolution, in which better nurture uplifted the inferior, the tortilla discourse influenced *agrónomos* interested in social reform. For example, Romulo Escobar, who founded the *Escuela Particular de Agricultura* (Private Agricultural School, hereafter EPA) in Ciudad Juárez in 1906, urged Mexicans to eat wheat and vegetables rather than corn. In 1908, to Europeanize the masses, Díaz wanted to transform their diet. After the Revolution, notwithstanding the rise of *indigenismo*, an ideology that celebrated Native Americans' contributions to Mexican culture and history, the tortilla discourse had a profound impact on agronomy and agricultural policy.

Sometime during the 1880s Segura met Native Americans from Chalco who told him about *teocintle* and maize, inspiring him to cross-pollinate the plants and stimulating interest in the genetic origins of corn. In 1911 Eugenio Banó said that during his 16 years farming Mexico's "dark jungles" he grew coffee, tobacco, vanilla, cacao, sugar cane, and rubber by studying the knowledge of local Indians. But like Mexico City's public health experts, who responded to an outbreak of mushroom poisoning by consulting German texts rather than the *curanderos* (folk healers) who knew the local varieties from the experience of countless generations, Porfirians who wanted to modernize farming looked to foreign science and technologies. These models informed Mexican agronomy, not the knowledge of *hacendados*, *rancheros*, or *campesinos*, a pattern that persisted after the Revolution.[26]

A NEW PROFESSIONAL CONSCIOUSNESS: U.S. AGRONOMY AND THE PORFIRIAN BUREAUCRACY

French agronomy was an abiding ideal of Mexico's agricultural scientists, but profound reliance on it began to erode before the Revolution. Elites were interested in philosophical discourse, pseudoscientific social Darwinism, and modernity, but Barreda failed to establish an experimental tradition in Mexican science. Laboratories and research languished in agronomy and other fields like physics and chemistry. Before 1883 the ENA granted only 33 degrees, and most graduates pursued military careers, like Segura (1870), who rose to the rank of lieutenant colonel, or became teachers. With the growth of the elite and middle class, by 1907 another 241 *agrónomos* left the ENA with uncertain professional futures, and many young Mexicans enrolled during the last years of the *Porfiriato*. The parents of a few, like García Teruel, sent them to the ENA to learn to manage family-owned haciendas, but many *agrónomos* complained that *hacendados* refused to hire them or listen to their advice. Even Basurto Larrainzar could not win the confidence of the *hacendados* for whom he worked briefly in 1906, and Viadas lost his job at the hacienda Calpulalpán in his home town after managing it for only two years. When Luís L. León began his ENA studies, he learned that *hacendados* "did not need us or want us, and usually had no respect for us," and Viadas and Basilio Romo claimed that they would not hire "people outside their own families."[27] With urban upper- and middle-class origins, a university education that stressed book knowledge, and little to no practical farming experience, like U.S. farmers' view of agricultural college professors during the nineteenth century, Mexican cultivators did not accept the *agrónomos'* claim to expertise. Escobar said that *rancheros* "laughed" at *agrónomos* because their knowledge was purely theoretical. Convinced that "the agriculture of Montpelier and Grignon" "failed" in Mexico, he thought that *agrónomos* could become credible scientific professionals by studying German and U.S. agronomy. Other intellectuals urged the ENA to teach practical agriculture and experimental science.[28] Many *agrónomos* had to find work outside their profession as surveyors or construction and mining engineers. For example, after studying wine production in Europe, Juan E. Contreras' only offers came from a railroad and a mine. In 1899 Brambila Sr. found work at a mine in Ameca, Jalisco, and in 1907 he got a government job as a surveyor for the U.S.–Mexico Boundary Commission. Virgilio Figueroa chose a political career, which many *agrónomos* favored after the Mexican Revolution. In 1894 he became *Presidente Municipal* (Municipal President) of Tuxtla Gutiérrez, Chiapas. Similarly, from 1900 to 1910 Aragón León was a *Diputado Federal* (hereafter Federal Deputy).[29]

In 1882 President Manuel González (1880–1884) sent Agustín C. Aguilar to Tamualipas to study the cotton industry, but before

1900 the State provided little support for agricultural research, extension, or pest control. Good students like Gabriel Gómez could become professors and administrators at the ENA. With his training in France, the *Secretaría de Fomento* (Ministry of Development) hired Basurto Larrainzar. When Manuel Vera graduated in 1891, this agency offered him a job preparing exhibits for international fairs, but most *agrónomos* had no chance to enter government service. Economic self-interest provided the first stimulus to increased State support for agricultural programs. Responding to a U.S. fruit fly (*anastrepha Ludens*) quarantine, in 1900 Díaz established the *Comisión Nacional de Parasitología* (National Parasitology Commission) and hired Herrera to run it. The agency published several works on the pest but failed to implement control procedures that convinced the United States to lift the ban. The *Comisión* also hired Gándara for pest and disease control studies of oranges in Yucatán and cotton in Oaxaca, but these employed few *agrónomos*.[30]

As businessmen and technicians brought U.S. culture like baseball to Mexico, U.S. agricultural science and farming practices exerted more influence as the *Porfiriato* progressed, especially in the north. To take advantage of its experience with cotton, President González sent *agrónomo* Alberto Ruiz y Sandoval to the southern United States to study the industry. U.S. agribusiness conducted research in Mexico. For example, the Richardson Construction Company built irrigation works and hired W.W. Mackie to run an experimentation station in Sonora to develop the export vegetable industry. He also tested bananas, corn, cotton, dates, and garbanzos and promoted "Defiance" wheat, a high-yielding strain. U.S. scientists helped develop Mexico's guayule industry and cotton cultivation in the Mexicali Valley. A few *agrónomos* from the northern states attended primary or secondary schools in the United States. Segura, Vera, and Viadas represented Mexico and the *Secretaría de Fomento* in the United States at events like the 1893 Chicago World Fair. Before 1900 one Mexican, Salome Sastré of Mexico City, earned an agricultural science degree from a U.S. university, in his case Massachusetts State College, but in the next decade 13 studied agriculture in the United States, and 11 graduated, including Juan A. González, who went to Iowa State and ran Mexico's extension service during the 1920s. After 1900 many *agrónomos* praised the U.S. Department of Agriculture's (USDA) research, extension, and pest and disease control programs, which employed many professionals. The *Secretaría de Fomento* translated U.S. agricultural publications. Propinquity and mutual self-interest stimulated cooperation in pest control. Entomologist Riquelme Inda collaborated with Leland Ossian Howard, distinguished head of the USDA's Office of Entomology, in a 1905 study of an insect pest of henequen. Direct contact also stimulated Mexican interest in U.S. plant breeding. That year Chihuahua's governor Enrique C. Creel visited several experimentation stations, met Luther Burbank, and returned with his *New Creations and Fundamental Principles of Plant Breeding*. In 1908 ENA students took three years of English, then two years of French.[31]

Historian Clifton Kroeber claims that private agricultural development programs and food shortages after 1905 convinced many influential Mexicans that the State should foster increased farm productivity. To assert control over agricultural innovation and thus establish their position as professionals with useful specialized knowledge, *agrónomos* joined them. For example, Viadas proposed many government projects to modernize agriculture. Díaz's interest in agricultural development created jobs for a few *agrónomos* and opened a window of opportunity for the profession to solve its employment problem. In 1905 the *Secretaría de Fomento* hired Pablo Córdova and Eduardo Chávez when they graduated from the ENA. A trip to France, Italy, Hungary, England, Spain, the United States, and Canada to study agricultural programs convinced Viadas and Romo that agronomy should be a profession of government service rather than a "liberal" one of private practitioners like medicine or law. Combining their prejudices about rural Mexicans with their training in French agricultural science and their need to establish professional status, *agrónomos* said that Mexico's cultivators used backward practices. Gándara told them to emulate the more advanced farmers of Europe, the United States, and Canada. In 1908, using the same language as the revolutionaries who overthrew Díaz, Mallen urged him to found experimentation stations to "destroy the empiricism and routine of our rural people."[32]

Díaz agreed with the *agrónomos*. In 1907 he appointed henequen *hacendado* Olegario Molina to run the *Secretaría de Fomento*, told him to found experimentation stations, and funded scholarships to send promising ENA students abroad for postgraduate studies.

Reflecting the persistent dominance of European agronomy and the scarcity of Mexican researchers, Molina hired French chemists to conduct research for the henequen industry. To improve scientific education the ENA hired Italian agronomists Silvio Bonacee and Mario Calvino and French agronomists León Fourtón and Gabriel Ítie. Immigrant scientist Pehr Olson-Seffer hoped that the students who went abroad would learn to run research and extension programs and end the need to hire foreign scientists. Viadas complained that the ENA taught "a foreign agriculture that is not based on our traditions and necessities."[33] They wanted Mexico to attain scientific autonomy, an issue that other presidents confronted after the Revolution.

Accepting Viadas' recommendations, in 1908 Molina established the *Estación Agrícola Central* (Central Agricultural Station) at the ENA, hired *agrónomo* Virgilio Figueroa and then Romo to run it, and hired Calvino to supervise the research. Responding to federal mandate, Yucatán, Oaxaca, Tabasco, and San Luís Potosí also founded experimentation stations. At the *Estación* Calvino crossbred foreign and native corn using methods developed in the United States. He grew fruit trees using practices recommended by a USDA expert and evaluated many other imported, high-yield crops, such as Italian soybeans and 107 wheat varieties from France, Spain, Italy, Canada, and the United States, distributing some to *hacendados*. In 1910 he published a book on the development of hybrid tomatoes in the United States, and he urged Mexican farmers to use plant breeders' methods of selecting seed and germinating a sample to test for viability. Impressed by the productivity of hybrids, to create their own improved varieties, *agrónomos* told farmers to use scientific criteria to select their seed rather than folk or experiential knowledge. After the Revolution this became the *Campaña en Pro del Maiz*, an important part of the *agrónomos'* campaign to transform the *campesinos'* culture and farming practices. In 1909 Molina appointed Romo ENA director, and he reformed the curriculum by borrowing from the foreign institutions that he visited. He told his best professors, such as Basurto Larrainzar, and talented *agrónomos* and students like Brambila Sr., Contreras, Córdova, and Juan Ballesteros Acevedo to study Calvino's methods at the *Estación*. The *Secretaría de Fomento* also ran extension programs. In 1907, to promote rubber tree cultivation, Molina organized the *Primera Convención Nacional Hulera* (First National Rubber Convention). The *Secretaría de Fomento* tried to convince farmers to change their practices by publishing bulletins and pamphlets that summarized research conducted in other countries or translated foreign articles about subjects such as the USDA's "dry farming" system for the Great Plains. In 1909 Calvino gave classes to hacienda foremen in the Federal District and told *hacendados* in Guanajuato, Jalisco, and Veracruz how to improve their cultivation and pest control practices. He tested green manure crops, dung, and chemical fertilizers and urged farmers to use organic methods because of the scarcity and high cost of agrochemicals. *Agrónomos* learned these ideas from their Porfirian mentors and made the same suggestions after the Revolution.[34]

In 1910 Olson-Seffer urged Díaz to support an extensive research and extension program that included "demonstration haciendas" and campaigns to establish new crops. Pointing to the impact of political corruption on the USDA, he warned that "for scientific labor to be efficient, obviously it should be independent from politics and politicians."[35] In December, as revolution erupted in Mexico, the *Secretaría de Fomento* sent Gándara to the United States to study the training of pest control specialists and methods to fight plant diseases. He visited the University of Florida and Louisiana State University and worked in Washington, D.C., with W.E. Safford, the head of the Office of Plant Industry's Botany Department. Reflecting the ambivalence of change, when he returned, he urged his colleagues to read Erwin F. Smith and Charles J. Chamberlain but also the *Manuel de Technique Botanique* by Paul y Gautier. Similarly, when the *Secretaría de Fomento* decided to move the ENA from its San Jacinto campus in the Federal District to the *Hacienda de Chapingo* on the eastern shore of Lake Texcoco in Mexico State, the agency used the Paris Institute of Agronomy as a model, believing that such institutions should be close to urban centers of scientific development.[36] Díaz initiated the government's relationship with the *agrónomos*; the Mexican Revolution solidified the bond.

CONCLUSIONS

The early years of Mexican agronomy were shaped by nineteenth-century liberals' faith that modernity could be attained by eradicating the legacies of Spanish colonialism. This included reducing the Catholic Church's intellectual power by creating secular scientists and reducing its economic power by fostering private agricultural enterprises, haciendas, and *ranchos*. In this utopia, as market-oriented property owners, *hacendados* and *rancheros* would hire and follow the advice of experts who studied state-of-the-art European agriculture and modernize their farming practices to end what the liberals labeled a technological backwardness that Mexico inherited from three centuries of Spanish rule. However, neither group accepted the *agronomos'* credibility as experts. Other legacies of Mexico's colonial experience also influenced the development of Mexican agronomy. Reflecting racism and the subordination of Native Americans, like other elite and middle-class Porfirians, *agrónomos* rejected indigenous agricultural technologies as inferior and took a colonial perspective that defined things European as innately superior and worthy of emulation.

U.S. influence in Mexico grew more pronounced as the *Porfiriato* progressed, and U.S. agronomy began to supplant French agricultural science as the *agrónomos'* source of inspiration. Facing the stresses that would bring down his regime, by 1905 Díaz and his allies recognized the need for an activist approach to agricultural development. *Agrónomos* justified his ideas by claiming that *hacendados* resisted modernization, and the profession began to forge an alliance with the State that secured a few government jobs. Reflecting the Porfirian social agenda for the masses, the *agrónomos* also considered ways to transform the *campesinos'* culture. After 1910 conflict disrupted these projects, and the discourse of revolution modified them, but the foundation of Mexico's twentieth-century agricultural development, which was defined by urban elite and middle class, sustained by the State, and informed by foreign scientists, science, and technologies, was laid before the Revolution.

NOTES

1. William E. French, "Prostitutes and Guardian Angels: Women, Work, and the Family in Porfirian Mexico," *Hispanic American Historical Review* 72, 4 (1992): 529–53; Eli de Gortari, *La ciencia en la historia de México* (México, D.F.: Fondo de Cultura Económica, 1963), pp. 317–18; Elias Trabulse, *El circulo roto: Estudios históricos sobre la ciencia en México* (México,

D.F.: Fondo de Cultura Económica, 1982), p. 23; Enrique Beltrán, "La Dirección de Estudios Biológicos de la Secretaría de Fomento y el Instituto de Biología de la Universidad Nacional Autónoma," *Anales de la Sociedad Mexicana de Historia de la Ciencia y de la Tecnología* 1 (1969): 105–41; Nicolas Sánchez Durón, "La investigación agrícola y el desarrollo agropecuario de México," *Anales de la Sociedad Mexicana de la Historia de la Ciencia y de la Tecnología* 2 (1970): 176; Ramón Fernández y Fernández, *Chapingo hace 50 años* (Chapingo, México: ENA, Colegio de Postgraduados, 1976), p. 44.

2. Elvin C. Stakman, Paul C. Mangelsdorf, and Richard Bradfield, "Agricultural Conditions and Problems in Mexico, Report of the Survey Commission of the Rockefeller Foundation," 1941, p. 7, RFA, RG 1.1, Series 323, Box 5, Folder 37, RAC.

3. Jeffrey M. Pilcher, *¡Que vivan los tamales! Food and the Making of Mexican Identity* (Albuquerque: University of New Mexico Press, 1998), p. 6; Alan R. Sandstrom, *Corn Is Our Blood: Culture and Ethnic Identity in a Contemporary Aztec Indian Village* (Norman: University of Oklahoma Press, 1991).

4. Eric Wolf, *Sons of the Shaking Earth: The People of Mexico and Guatemala—Their Land, History, and Culture* (Chicago: University of Chicago Press, 1959), pp. 60–61; Bernardino de Sahagún, *General History of the Things of New Spain*, Charles E. Dibble and Arthur J.O. Anderson, trans. (Provo: University of Utah Press, 1963), Volume 12, Book 11, "The Earthly Things," pp. 251–52, 279–85, and Book 8, "Kings and Lords," pp. 38–40; Lane Simonian, *Defending the Land of the Jaguar: A History of Conservation in Mexico* (Austin: University of Texas Press, 1995), pp. 20–25; Marilyn Gates, *In Default: Peasants, the Debt Crisis and the Agricultural Challenge in Mexico* (Boulder, Colo.: Westview Press, 1993), pp. 110–11.

5. Pilcher, *¡Que vivan los tamales!*, p. 141; Paul Friedrich, *Agrarian Revolt in a Mexican Village* (Englewood Cliffs, N.J.: Prentice-Hall, 1970), p. 37; Mary Kay Vaughan, *Cultural Politics in Revolution: Teachers, Peasants, and Schools in Mexico, 1930–1940* (Tucson: University of Arizona Press, 1997), p. 65; Stakman et al., "Agricultural Conditions," p. 38.

6. Sandstrom, *Corn Is Our Blood*, pp. 240–41; Judith Friedlander, *Being Indian in Hueyapan: A Study of Forced Identity in Contemporary Mexico* (New York: St. Martin's Press, 1975), pp. 99–100; Gates, *In Default*, p. 95; "Drying and Dying, Deforestation, Overfarming and Overgrazing Are Turning Once Productive Mexican Farms into Wastelands," *Albuquerque Journal*, June 18, 2000; Nathan Whetten, *Rural Mexico* (Chicago: University of Chicago Press, 1948), pp. 444–45.

7. Jesús Vanegas, "Información sobre los problemas agrícolas del Municipio de Huejacar, Jalisco" (n.d.), AHJ, Box AG-27-F, Exp. Informe de la Inspección a la 8a Zona Agrícola; Wolf, *Sons*, p. 198; Gene C.

Wilken, *Good Farmers: Traditional Resource Management in Mexico and Central America* (Berkeley: University of California Press, 1987); Angus Wright, *The Death of Ramón González: The Modern Agricultural Dilemma* (Austin: University of Texas Press, 1991), p. 156; William C. Cobb, "Interview with Edwin J. Wellhausen," June 28, 1966, pp. 7, 48, RFA, RG 13; Erlif A. Miller, John B. Pittner, Ricardo Villa J., and Carlos Romo G., "Population Density of Unirrigated Maize and Its Influence upon Fertilizer Efficiency in Central Mexico," 1948, p. 7, RFA, RG 6.13, Series 1.1, Box 46, Folder 531, RAC.

8. John Tutino, *From Insurrection to Revolution in Mexico: Social Bases of Agrarian Violence* (Princeton, N.J.: Princeton University Press, 1986), pp. 26–28; David L. Clawson and Don R. Hoy, "Nealtican, Mexico: A Peasant Community That Rejected the Green 'Revolution,'" *American Journal of Economics and Sociology* 38, 4 (October 1979): 378–79; William H. Beezley, *Judas at the Jockey Club and Other Episodes of Porfirian Mexico* (Lincoln: University of Nebraska Press, 1987), pp. 84–85; Gates, *In Default*, p. 113.

9. Francis Jennings, *The Invasion of America, Indians, Colonialism, and the Cant of Conquest* (Chapel Hill: University of North Carolina Press, 1975), pp. 43–84; Margaret C. Scarry and Elizabeth J. Reitz, "Herbs, Fish, Scum, and Vermin: Subsistence Strategies in Sixteenth-Century Spanish Florida," in David Hurst Thomas, ed., *Columbian Consequences, vol. 2, Archaeological and Historical Perspectives on the Spanish Borderlands Southeast* (Washington, D.C.: Smithsonian Institution Press, 1990), pp. 343–54; Pilcher, *¡Que vivan los tamales!*, pp. 25–43.

10. C. Harvey Gardiner, ed., *Mexico 1825–1828: The Journal and Correspondence of Edward Thornton Tayloe* (Chapel Hill: University of North Carolina Press, 1959), p. 100; Antonio García de León, "Las grandes tendencias de la producción agraria," in García de Léon, Enrique Semo, and Ricardo Gamboa Ramírez, eds., *Historia de la cuestión agraria mexicana, vol. 1: El siglo de la hacienda* (México, D.F.: *Siglo Ventinuo Editores*, 1988), p. 22; Enrique Semo, "Haciendas, campesinos, y rancheros," in Antonia García de León, Enrique Semo, and Ricardo Gamboa Ramírez, eds., *Historia*, p. 96; Adolfo Orive Alba, *La irrigación en México* (México, D.F.: *Editorial Grijalva*, 1970), p. 57; Herman W. Konrad, *A Jesuit Hacienda in Colonial Mexico: Santa Lucia, 1576–1767* (Stanford, Calif.: Stanford University Press, 1980), p. 6; Charles H. Harris III, *A Mexican Family Empire: The* Latifundio *of the Sánchez Navarros, 1765–1867* (Austin: University of Texas Press, 1975), pp. 79–93.

11. Stuart McCook, *States of Nature: Science, Agriculture and Environment in the Spanish Caribbean, 1760–1930* (Austin: University of Texas Press, 2002), pp. 15–20; Jesús Gracia Fadrique, "Enseñanza de la química en Mexico," in Gracia Fadrique, ed., *Estado y Fertilizantes en México, 1760–*

1985 (México, D.F.: Fondo de Cultura Economica, 1988), pp. 185–95; José Luís Calva Téllez and Carlos Monzón, "La economía agrícola mexicana y el consumo nacional de fertilizantes," in Gracia Fadrique, *Estado y Fertilizantes*, pp. 16–27; Wolf, *Sons*, pp. 204–5; Marte R. Gómez, *Biografías de agrónomos* (Chapingo, México: ENA, Colegio de Postgraduados, 1976), p. 514; Miguel García Cruz, "Tendencias de la enseñanza agrícola superior," in Enseñanza Agrícola Superior, BCM Bound Volume #145–43; Edmundo Flores, *Tratado de economía agrícola* (México, D.F.: Fondo de Cultura Económica, 1961, pp. 276–77; Beezley, *Judas*, pp. 72–73; SAF, *Características de los trigos cultivados en la República Mexicana* (México, D.F.: Departamento Autónomo de Prensa y Publicidad, 1937); Margaret Channing, "The Contours of the Post-1810 Depression in Mexico: A Reappraisal from a Regional Perspective," *Latin American Research Review* 27, 2 (1992): 133–37; Harris, *A Mexican Family Empire*, p. 312; Roberto Durán, "El Ramio," *Agricultura* 1, 7 (July 1938): 25–30; "Precursores de México," *Excelsior*, August 20, 1964.

12. SAF, *Memoria 1937–1938* (México, D.F.: D.A.P.P., 1938), p. 188; Fred A. Motz and Lester D. Mallory, *The Fruit Industry of Mexico* (Washington, D.C.: Office of Foreign Agricultural Relations, OFAR Foreign Agriculture Report No. 9, 1944), p. 80; José Figueroa, "Posibilidades de la ganadería Mexicana," in Partido Nacional Revolucionario (hereafter PNR), ed., *Los problemas agrícolas de México* (México, D.F.: PNR, 1934), p. 27; John Gledhill, *Casi Nada: A Study of Agrarian Reform in the Homeland of Cardenismo* (Albany: State University of New York, Institute for Mesoamerican Studies, 1991), p. 72; Gómez, *Biografías*, p. 218; Bernardo Ávilez, *Cultivo y comercio de algodón en México* (México, D.F.: SAF, Dirección de Economía Rural, 1939), pp. 107–8; "Estudio presentado por la Delegación de Defensa Agrícola de León, Gto," in Consejo Nacional de Agricultura (hereafter CNAG), *Memoria de la Primera Convención Nacional de Productores de Papa* (México, D.F.: CNAG, 1939), pp. 23–24, 56–57; Claude I. Dawson, "Fertilizers in Mexico," September 25, 1923, NA, RG 166, Box 148; Roy V. Scott, *The Reluctant Farmer: The Rise of Agricultural Extension to 1914* (Urbana: University of Illinois Press, 1971), pp. x, 29; Beezley, *Judas*, pp. 72–73.

13. Armando C. Alonzo, *Tejano Legacy: Rancheros and Settlers in South Texas, 1734–1900* (Albuquerque: University of New Mexico Press, 1998), pp. 3, 22–23; Daniel Nugent, *Spent Cartridges of Revolution: An Anthropological History of Namiquipa, Chihuahua* (Chicago: University of Chicago Press, 1993), pp. 72–75; Luís González y González, *San José de Gracia: Mexican Village in Transition*, John Upton, trans. (Austin: University of Texas Press, 1974), pp. 31–60; Stanley F. Shadle, *Andrés Molina Enríquez, Mexican Land Reformer of the Revolutionary Era* (Tucson: University of Arizona Press, 1994), p. 12; Frans J. Schryer, *The Rancheros of Pisaflores: The*

History of a Peasant Bourgeoisie in Twentieth-Century Mexico (Toronto: University of Toronto Press, 1980), pp. 29–36; Alberto Arce, *Negotiating Agricultural Development: Entanglements of Bureaucrats and Rural Producers in Western Mexico* (Wageningen, the Netherlands: Agricultural University Wageningen, 1993), p. 122.

14. Enrique Beltrán, "Fuentes mexicanas en la historia de la ciencia," *Anales de la Sociedad Mexicana de Historia de la Ciencia y de la Tecnología* 2 (1970): 78, 95; Maurice Crosland, *Science under Control: The French Academy of Sciences 1795–1914* (Cambridge: Cambridge University Press, 1992), pp. 392–93; David Wade Chambers, "Period and Process in Colonial and National Science," in Nathan Reingold and Marc Rothenberg, eds., *Scientific Colonialism: A Cross Cultural Comparison* (Washington, D.C.: Smithsonian Institution, 1987), pp. 297–321; William L. Merrill, "The Beloved Tree: *Ilex vomitoria* among the Indians of the Southeast and Adjacent Regions," in Charles M. Hudson, ed., *Black Drink* (Athens: University of Georgia Press, 1979), p. 70; Gary S. Dunbar, " 'The Compass Follows the Flag': The French Scientific Mission to Mexico, 1864–1867," *Annals of the Association of American Geographers* 78, 2 (1988): 229–40; Adolfo D. Barriero, *Reseña histórica de la enseñanza agrícola y veterinaria en México*, (México, D.F.: Tipografía El Libro de Comercio, 1906), pp. 1–25; Gilberto Mendoza Vargas, "La enseñanza y extensión agrícola," February 1954, p. 5, in Enseñanza Agrícola Superior; BCM #145–43; Gómez, *Biografías*, pp. 516–17.

15. Ramón Eduardo Ruiz, *Triumphs and Tragedy: A History of the Mexican People* (New York: W.W. Norton & Company, 1992), pp. 260–61; Guillermo Gándara, "La enseñanza de las ciencias biológicas en México: Nacional y prenacional," *Memoria y Revista, Academía Nacional de Ciencias Antonio Alzate* 53, 11–12 (1932): 397; De Gortari, *La ciencia en la historia de México*, 2nd ed. (México, D.F.: Tratados y Manuales Grijalbo, 1980), pp. 298–308; Gómez, *Biografías*, pp. 31, 72, 440; Jesús Díaz de León, "Disertación sobre la importancia del estudio de la agricultura," *Boletín de la Sociedad de Geografía y Estadística de la República Mexicana* 3, 6–7 (1895): 399–406.

16. Barriero, *Reseña*, p. 60; Lamberto Cabañas, *Reflexiones sobre economía rural* (México, D.F.: Tipografía de la Secretaría de Fomento, 1885); Gómez, *Biografías*, pp. 47–48; Fernández y Fernández, *Chapingo*, p. 37.

17. Irving A. Leonard, *Hispanic America and Science* (Miami: University of Miami Press, 1949), pp. 20–23; Edmundo Flores, *Historias de Edmundo Flores, autobiografía 1919–1950* (México, D.F.: Martín Casillas Editores, 1983), p. 98; Frank Safford, *The Ideal of the Practical, Columbia's Struggle to Form a Technical Elite* (Austin: University of Texas Press, 1976), p. 226; Eduardo Rodríguez, *El cultivo del lupulo*, 2nd ed. (México, D.F.: Talleres Gráficos de la Nación, 1920, first published in 1909); Luis Fernández del

Campo, *Cultivo de la caña de azúcar* (México, D.F.: Imprenta y Fototipía de la Secretaría de Fomento [hereafter IFSF]), 1913.

18. Alfonso Herrera, "Sur les oxydase siliciques artificielles," *Memoria y Revista de la Sociedad Científica Antonio Alzate* 29, 7–12 (1910): 331–35; Juan D. Villarello, "Description des mines 'La Bella Unión,' " *Memoria y Revista de la Sociedad Científica Antonio Alzate* 23, 7–12 (1906): 395–411; Secretaría de Fomento, *Algunos documentos para la historia de la enseñanza agrícola en México* (México, D.F.: IFSF, 1912), p. 71; Leopoldo Palacios, "Importancia social y política de las obras de irrigación," *Memoria y Revista de la Sociedad Científica Antonio Alzate* 30, 10–12 (December 1911): 425; Díaz de León, "Disertación," p. 400.

19. Gómez, *Biografías*, pp. 31–32, 510.

20. Friedrich Katz, *The Secret War in Mexico: Europe, the United States and the Mexican Revolution* (Chicago: University of Chicago Press, 1981), pp. 21–27; Gómez, *Biografías*, pp. 56–57, 118; "Fallecío el Ing. Miguel A. de Quevedo, 'El Apóstol del Arbol,' " *Siembra* 3, 41 (August 1946): 6; "Sesión solemne," October 4, 1909, *Sociedad Científica Antonio Alzate, Revista Científica y Bibliográfica* 29, 1–4 (July–October 1909): 11; Gómez, *Biografías*, p. 240; Antonio J. Carbajal, "La fiebre carbonosa," *Boletín de la Sociedad Científica Antonio Alzate* 23, 7–12 (January–June 1906): 316–17.

21. David E. Lorey, *The University System and Economic Development in Mexico since 1929* (Stanford, Calif.: Stanford University Press, 1993), pp. 136–37; Gómez, *Biografías*, pp. 179, 184–87, 210, 221, 264, 313, 340, 346, 376, 406, 420, 446, 474, 509.

22. Gómez, *Biografías*; Anna Macías, *Against All Odds: The Feminist Movement in Mexico to 1940* (Westport, Conn.: Greenwood Press, 1982), p. 11; Othon E. Brackel-Welda, "Apuntos sociológicos leidos en la Sociedad de Geografía y Estadística," *Boletín de la Sociedad de Geografía y Estadística de la República Mexicana* 2, 11–12 (1894): 684–87.

23. Díaz de León, "Disertación," p. 399; Rafael Dormantes to Manuel Avila Camacho, December 16, 1940, AGN, Avila Camacho papers, Exp. 151.3/25; Stanley Johnson, *The Green Revolution* (London: Hamish Hamilton, 1972), p. 8; Safford, *The Ideal*, pp. 3–12, 241–42; Cobb, "Interview with Norman Borlaug," June 1967, pp. 156–57, RFA, RG 13; Pauline Madow, "Reminiscences of Elvin C. Stakman," June 2, 1970, p. 1062, RFA, RG 13; Paul C. Mangelsdorf, "Report on a Trip to Mexico for the Rockefeller Foundation," February–March 1946, RFA, RG 1.2, Series 323, Box 10, Folder 61, RAC; Rafael M. Tello, "¿Por qué no han dando un resultado satisfactorio los abonos químicos?" *Memoria de la Sociedad Científica Antonio Alzate* 29, 1–6 (July–December 1909): 29–42; Beezley, *Judas*, p. 132.

24. Pilcher, *¡Que vivan los tamales!*, p. 79; Pehr Olson-Seffer, *La agricultura en varios paises tropicales y subtropicales: Organización de la agricultura en diversos paises extranjeros* (México, D.F.: IFSF, 1910), p. 123; Gustavo

Durán, "Importancia de la agricultura y del fraccionamiento de tierras," *Memoria y Revista de la Sociedad Científica Antonio Alzate* 30, 10–12 (December 1911): 441; Bernardo Mallen, *¿En donde esta nuestra riqueza?* (México, D.F.: 1908), p. 46.

25. Francisco Bulnes, *Los grandes problemas de México* (México, D.F.: Ediciones de "El Universal," 1926), p. 120; Pilcher, *¡Que vivan los tamales!*, p. 77.

26. Nancy Leys Stepan, *"The Hour of Eugenics": Race, Gender, and Nation in Latin America* (Ithaca, N.Y.: Cornell University Press, 1991), pp. 67–73; Romulo Escobar, *The Agricultural Resources of the United States of Mexico* (Buffalo, N.Y.: The White-Ebans-Penfield Company, 1904), p. 14; Pilcher, *¡Que vivan los tamales!*, pp. 60–62; Guillermo Gándara, "Desarrollo del cultivo del maiz en Mexico," *Agricultura* 1, 3 (November–December 1937): 5–6; Eugenio Banó, *Estudio sobre varias plantas tropicales y su cultivo en México* (México, D.F.: IFSF, 1911); Beezley, *Judas*, p. 83.

27. Luís González y González, *Los dias del Presidente Cárdenas* (México, D.F.: El Colegio de México, 1981), p. 290; Francisco R. Sagasti, *Ciencia, tecnología, y desarrollo latinoamericano* (México, D.F.: Fondo de Cultura Económica, 1981), pp. 139–49; Gracia Fadrique, "Enseñanza," p. 191; George I. Sánchez, *The Development of Higher Education in Mexico* (Westport, Conn.: Greenwood Press, 1970), pp. 94–95; de Gortari, *La ciencia*, pp. 298–308; Brackel-Welda, "Apuntos," p. 683; Barriero, *Reseña*, pp. 37–43; Mendoza Vargas, "La enseñanza," p. 5; Manuel Mesa A., *Relatos autobiográficos* (México, D.F.: Editorial Nuestro Tiempo, 1981), p. 7.

28. Romulo Escobar, "Problemas agrícolas en México," *Boletín de la Sociedad Científica Antonio Alzate* 23, 1–4 (1905): 93, 105.

29. Gómez, *Biografías*, pp. 59, 72, 218, 474; Díaz de León, "Dissertación," p. 400; Secretaría de Fomento, *Algunos documentos*, pp. 73–74; Luís L. León, "El gremio agronómico en la reforma agraria," *Boletín de la Sociedad Mexicana de Geografía y Estadística* 78 (1954): 60; Fernández y Fernández, *Chapingo*, p. 40; Scott, *The Reluctant Farmer*, p. 31; Juan de Dios Bojórquez, *Los agrónomos de Mexico* (México, D.F.: Ediciones del Boi, 1959), p. 7; Gómez, *Biografías*, pp. 31–32, 72, 118, 197.

30. Alberto Ruiz y Sandoval, *El algodón en México* (México, D.F.: Oficina Tipografía de la Secretaría de Fomento, 1884), pp. 126–27; Alfonso Dampf, "La importancia de la mosca de la fruta," in CNAG, *Memoria de la Primera Convención Nacional de Productores de Naranja* (México, D.F.: CNAG, 1936), p. 2; SAF, *Monografía de la naranja* (México, D.F.: 1936), pp. 79–83; Gándara, "Enfermedades y plagas del naranjo," *Memoria de la Sociedad Cientifica Antonio Alzate* 28, 5–8 (January–April 1910): 155–92; Gómez, *Biografías*, pp. 59, 240–42, 509–13.

31. Ruiz y Sandoval, *El algodón*; Compañía Constructadora Richardson, *Indicador de productors y manual de información agrícola para El Valle*

del Yaqui, Sonora, Mexico (Esperanza, Sonora: Compañía Constructadora Richardson, 1911), p. 4; Francisco Valdes, "Dictamen acerca de algunas informes sobre la industria de hule procediente del guayule," *Boletín de la Sociedad Mexicana de Geografía y Estadística* 41, 8–9 (August–September 1929): 488; Secretaría de Fomento, *Algunos documentos*, p. 12; Gómez, *Biografías*, pp. 18, 285, 474–75, 509–13, 514–15; Lorey, *The University System*, p. 23; Division of Inter-American Educational Relations, "Alumni of United States Colleges of Agriculture Who Are Citizens of the Other American Republics," April 1944, NA, RG 59, Decimal File 1940–1944, Box 196; Mallen, *¿En donde?*, p. 173; Carbajal, "La fiebre," p. 315; William MacDonald, *El cultivo de secano*, Ignacio Flores Sánchez, trans. (México, D.F.: IFSF, 1911); Julio Riquelme Inda, "El 'Max' del henequen," *Memoria y Revista de la Sociedad Científica Antonio Alzate* 35, 5–12 (September 1921): 313–18; Enrique C. Creel, "Agricultura y agrarismo," *Memoria y Revista de la Academía Nacional de Ciencias Antonio Alzate* 51, 5–6 (1929–1930): 149–52.

32. Clifton B. Kroeber, *Man, Land, and Water: Mexico's Farmlands Irrigation Policies 1885–1911* (Berkeley: University of California Press, 1983), pp. 182–94; Leandro Martínez, *Cultivo y beneficio del cacaotero*, 2nd ed. (México, D.F.: IFSF, 1912), p. 11; Guillermo Gándara, *Uso y aplicación de la Preparación Bordelesa* (México, D.F.: IFSF, 1907), pp. 1; Mallen, *¿En donde?*, p. 174; Fernández y Fernández, *Chapingo*, pp. 35–44; Gómez, *Biografías*, pp. 125, 146, 514.

33. Fernández y Fernández, *Chapingo*, pp. 35–44; Manuel Mesa A., "Memorandum sobre la ENA," August 15, 1933, p. 9, AGN, Robles papers, Agriculture Section, Box 5, Exp. 70; Gómez, *Biografías*, 460; "Aprovechamiento del bagazo de henequen para fabricar alcohol," *El Nacional*, January 6, 1936; Olson-Seffer, *La agricultura*, p. 134.

34. Mario Calvino, *El jitomate* (México, D.F.: IFSF, 1910), pp. 4–38; Calvino, *Una revolución en arborcultura: La plantación de arboles con raices cortadas y en terrenos sin roturar, ni labrar, segun el Sr. Stringfellow* (México, D.F.: IFSF, 1910); Calvino, *Informe de la División de Horticultura 1912* (México, D.F.: IFSF, 1913), pp. 39–43; *Calvino, Segundo informe de 1913 de la División de Horticultura* (México, D.F.: IFSF, 1914), pp. 46–57; Calvino, *Cultivo de la morera y la higuera* (México, D.F.: IFSF, 1912); Calvino, *La nutrición de las plantas* (San Jacinto, D.F.: 1909); Rafael Vargas, *Procedimiento que debe seguirse para hacer menos inseguras las cosechas del maiz* (México, D.F.: IFSF, 1912); Manuel Pardo y Urbina, *Cultivo del garbanzo* (México, D.F.: IFSF, 1911), pp. 5–6; Martínez, *Cultivo y beneficio*, pp. 31–32; Scott, *The Reluctant Farmer*, p. 178; Mendoza Vargas, "La enseñanza," p. 38; Sánchez Durón, "La investigación," pp. 175–76; Fernández y Fernández, *Chapingo*, p. 37; SAF, *Características*; Gómez, *Biografías*, pp. 43, 59, 72, 460. MacDonald, *El cultivo*; Gabriel Gómez, *Empleo agrícola de algunos*

residuos industriales (México, D.F.: IFSF, 1911), pp. 3–10; Escobar, "Problemas agrícolas," p. 106.

35. Olson-Seffer, *La agricultura*, pp. 3, 121–28.

36. Guillermo Gándara, "Visita a algunas instituciones de botánica y parasitología agrícola de los Estados Unidos," *Memoria y Revista de la Sociedad Científica Antonio Alzate*, 30, 7–9 (July 1911): 341–65; Gándara, "Técnica para hacer preparaciones microscopicas, según el sistema de los laboratorios de los Estados Unidos," *Memoria y Revista de la Sociedad Científica Antonio Alzate* 32, 4–6 (August 1912): 201–7; Barriero, *Reseña histórica*, p. 68; Secretaría de Fomento, *Algunos documentos*, pp. 106–7.

Chapter 2

Revolution, Agronomy, and the Roots of the Cultural Campaign, 1910–1930

In 1911 Porfirio Díaz fled a nation engulfed in Revolution. A peaceful transition to democracy seemed possible with the election of Francisco Madero, but peasants led by Emiliano Zapata continued to rebel when he rejected their demand for land reform. In 1911 the *zapatistas* issued the *Plan de Ayala,* demanding the return of village lands seized by the *hacendados*. In 1913 General Victoriano Huerta executed Madero and made himself president, but rebellions led by "First Chief" of the Constitutionalists Venustiano Carranza and his allies Pancho Villa of Chihuahua and Alvaro Obregón of Sonora surged. In 1914, facing defeat, Huerta fled into exile, but factionalism split the Constitutionalists. Carranza fought former allies and other rebels in many regions. *Carrancista* troops assassinated Zapata, but Carranza co-opted his agenda by giving a few *campesinos* land. Obregón overthrew Carranza in 1920, and a semblance of order returned. With fellow Sonoran Calles, the "Northern Dynasty" dominated politics during the 1920s. Both presidents supported projects to modernize agriculture and distributed some land.

For *agrónomos* the Revolution and the new regime that it created opened a window of opportunity: a chance to create a new order in rural Mexico. Some *agrónomos* recognized that the *agraristas* were natural allies because both benefited from State action to transform rural society: the latter by obtaining land; the former by helping them acquire it and by managing new *ejidos*. Rejecting the *hacendados* who refused to hire them, *agrónomos* wanted *campesinos* to accept their scientific expertise. Most *agrónomos* allied with the *agraristas* and supported or contributed to the State's land reform program. As Sonorans, Obregón and Calles admired the prosperity and achievements of the United States, including its farmers and agricultural scientists. Porfirian scientists still exerted authority, and *agrónomos* admired French science, but Soviet and especially U.S. agronomy became more influential. Most *agrónomos* came from the urban middle or upper classes and viewed peasant culture, including farming practices and rural folkways, with disdain. With their education and negative opinion of hacienda agriculture, they rejected the knowledge of Mexico's farmers in favor of foreign technologies and practices. They also contributed to the SEP's campaign against peasant culture, alienating many *campesinos* who might have supported *agrónomos* who helped them obtain land.[1]

FORGING AN ALLIANCE WITH THE STATE: THE *AGRÓNOMOS* AND THE REVOLUTION

Revolution offered brighter prospects for *agrónomos*, who like other unemployed professionals and students with few job opportunities were disenchanted with the Díaz regime. Reflecting his closed political system and the *agrónomos'* class origins and quest for upward mobility, most ENA students shunned politics before the Revolution. Hilario López Ibarra, who joined Madero's Anti-Reelectionist Party in 1909 and gave speeches in Mexico City's Alameda Park, was an exception. But when the old regime tottered, at least 15 ENA students, including three with ties to Madero's family, joined his movement. They studied the French Revolution, read Andres Molina Enríquez's and Wistano Luís Orozco's critiques of the hacienda, and struck in 1911 to support Madero's rebellion. An ardent fan of Díaz, Basilio Romo resigned as ENA director after the strike and left his politicized profession to work at the National Observatory, but some Porfirian *agrónomos* accepted the new regime. Virgilio Figueroa served in Madero's Congress, and die-hard positivist Aragón León ran for governor of Morelos. Land reform opened a window of opportunity for *agrónomos* unhappy with their role in the hacienda system. In 1911, reacting to the *Plan de Ayala*, Gustavo Durán proposed expropriating the property of *hacendados* who refused to innovate to create small farms that used intensive practices. Reflecting Madero's *hacendado* outlook and training in agronomy at the University of California at Berkeley, he continued Díaz's research and extension programs. At the *Estación Agrícola Central* Calvino crossbred corn, taught field plot techniques, tested forages, clover, beans, vegetables, and soybeans, compared native *chayote* squash to imported strains, grafted fruit trees, and studied planting distances for wheat. He gave lectures to hacienda *mayordomos* and convinced *hacendado* José N. Macias, a federal deputy, to plant clover to restore the fertility of his soil. His students Leandro Mendoza Vargas and Gustavo Martínez Baca tested agrochemicals. In 1913 Calvino's hybrid corn yielded an impressive 4400 kg/ha. He asked for more funds, claimed that his new strains were almost ready for distribution, and said that more research would create "temporal corn varieties even more resistant to drought than our native strains."[2]

Calvino fled to Cuba after Huerta's coup, ending Mexico's first research and extension program. Reflecting his disillusion with the *ancién regime*, Herrera began to publish articles in the *Memorias de la Sociedad Científica Antonio Alzate* in Spanish. San Luís Potosí's Río Verde Station tested alfalfa using USDA guidelines obtained by director *agrónomo* Ramón García Oses, as well as cotton and garbanzos, and stayed open until 1916, when he hired his colleague Justino T. Martínez. Oaxaca's station grew castor beans, and Tabasco's studied banana cultivation, employing *agrónomo* Guillermo Brondo Whitt from 1916 to 1918. From 1911 to 1914 the ENA granted 59 degrees, only four in tumultuous 1915 and 1916, and another 93 from 1917 to 1920 as students who enrolled after 1908 completed their studies. When fighting escalated and the ENA closed, at least five students left Mexico to study agriculture at U.S. universities. Brambila Sr. went to Havana to work with Calvino at the *Estación Agrícola Experimental*, and in 1913 the Richardson Construction Company hired Juan de Dios Bojórquez as a draftsman, sending him to Los Angeles, California. Six ENA students joined Huerta's Federal Army. José Jorge de C. Ursua fled to Germany, and Javier Ramírez de Aguilar, another supporter of the old regime, returned home and bought a *rancho*, starting a career as a private farmer.[3]

In 1915 Carranza, another *hacendado*, hired Herrera to run the

SAF's new *Dirección de Estudios Biológicos* (hereafter Office of Biological Studies), but he employed few *agrónomos* and emphasized pure science research. Carranza also established the SAF's *Dirección General de Agricultura* (General Office of Agriculture, hereafter DGA), appointing Constitutionalist *agrónomo* de Dios Bojórquez as director. He hired Porfirians such as Juan Ballesteros Acevedo, political allies such as Gonzalo González H., and a few *agrónomos* with research experience like Brondo Whitt. In 1917 Carranza sent recent ENA graduate, Samuel Torres Elorduy, to study the USDA's Extension Service, and entomologist Alfonso Madriaga and *agrónomo* Hildebrando Montemayor Martínez to La Laguna to study pink bollworms, but he soon terminated the latter project to save money. Article 3 of Mexico's Constitution of 1917 gave the new regime a duty to train scientists "to orient and promote the progress of agriculture."[4] In 1919, borrowing an idea from U.S. railroads, the SAF ran *Trenes de Maquinaria Agrícola* (Agricultural Machinery Trains), selling U.S. farm equipment in regions that the Constitutionalists controlled. Miguel Angel de Quevedo's nurseries distributed some 80,000 trees and plants. Carranza also gave Porfirian scientists SAF jobs and reissued the old regime's bulletins on many subjects, including castor bean cultivation and locusts, but military expenses, inflation, and scant revenue left Carranza little money to hire *agrónomos*. Finding a job in the tiny agricultural bureaucracy required connections and recommendations that most *agrónomos* lacked, and many had to find other jobs. For example, Mesa A. became a surveyor for the Mexican Gulf Oil Company. *Agraristas* threatened *hacendados* and their loyal employees, and some fled their estates, hiring *agrónomos* such as Felipe Chacón Jiménez to manage them. Any return satisfied estate owners, and some *agrónomos* accepted the dangers of being a *mayordomo* to profit from the lack of supervision. A few *agrónomos* such as Rafael Batiz Paredes and Francisco MacGregor Romero bought land and tried to farm. The former became a prosperous rice and cotton grower and the first president of Sinaloa's *Unión Agrícola Industrial*, but the latter gave up in 1914 and earned his living trafficking in revolutionary paper money.[5]

Land reform offered more job opportunities. Many students left the ENA to join the ranks of the revolutionaries, and states hired them to survey haciendas and irrigation systems to distribute them to *campesinos*. From 1914 to 1920, 49 *agrónomos* and ENA students served on *Comisiones Locales Agrarias* (Local Agrarian Commissions) in Morelos, Sonora, Chihuahua, Coahuila, Yucatán, and Guanajuato. *Agrónomo* and future Governor of Chiapas Efrain A. Gutiérrez went to Morelos to join Zapata, as did his ENA colleagues Marte R. Gómez, Alfonso Cruz, and 17 others. Waldo Soberón, Gilberto Fábila, Manuel Cadeña, and 9 more *agrónomos* served in Pancho Villa's *División del Norte*. Calles appointed Luis L. León to run Sonora's first land reform commission and the state's Department of Agriculture. De Dios Bojórquez worked for León, befriended Obregón, and served in Carranza's Congress. ENA student Jesús M. de la Garza became a general in Obregón's army. Forty-five *agrónonos* and ENA students served the Constitutionalists from 1914 to 1920. A few, such as Heliodoro Rendón, lost their lives in battle or while implementing agrarian reform. In 1916 MacGregor Romero decided that the life of a *coyote* (an illegitimate middleman of bureaucratic transactions) was too dangerous and moved to Los Angeles.[6]

Porfirian and new *agrónomos* looked for a place in the new revolutionary society. After Obregón's victory in 1920, *agrónomos* used their ties to the revolutionaries to get jobs in the CNA and influence agrarian and agricultural policies. Reflecting a growing professional consciousness, in 1921 de la Garza and 51 of his colleagues founded the *Sociedad Agronómica Nacional* (National Agronomic Society, hereafter SAN). *Agrónomos* advocated land reform, reformed the ENA's curriculum, and said that their training would address the *campesinos'* needs. Borrowing Viadas' ideas, *agrónomos* who allied with the *agraristas* insisted that agronomy was a profession of government service and demanded that the State hire them. Many *agrónomos* recognized that land reform could solve their employment problem. Believing that the threat of expropriation could force frightened *hacendados* to modernize, de Dios Bojórquez wrote: "With our perpetual separation from the great farmers, it is logical that we support land reform to create small rural properties and agricultural colonies. We tell those who ask where we will work if we fight the large estates that each agronomist can manage cooperative colonies."[7] Announcing his commitment to "the small farmer," ENA director Marte Gómez proclaimed that the school opposed "large capitalist farms that need the suffering of vast multitudes of salaried laborers," said that *agrónomos* "would neglect our duty as citizens and our revolutionary convictions if we let [the ENA] train competent administrators for the large estates that still proliferate in our nation," and wanted to "redeem the rural proletariat and national agriculture" from "the stingy aristocracy that considers it sacrilege to abandon the Egyptian plow."[8] *Agrónomos* exhorted Obregón to use new irrigation districts to bring social justice to the *campesinos*. In 1923 the SAN declared war on the *hacendados*, promising to "abolish the large estate and the indirect exploitation of rural property, to replace the individual capitalist exploitation with cooperatives of small farmers or collective or communal ejidos, the first step toward socializing the land, our supreme purpose."[9]

During the 1920s many *agrónomos* helped the State distribute land, including at least 41 ENA students and graduates, but most CNA employees were political appointees without degrees. The SAF's technical personnel also contributed to the agrarian reform. For example, in 1925 Durango's extension agent organized the state's Fourth Agrarian Congress. *Agrónomos* carried out the most politically important distributions after the Revolution, such as the haciendas and *ranchos* that Obregón gave to Villa and his prominent allies and the *ejidos* granted to *zapatistas* in Morelos. In 1929 Calles ordered Jorge Rodríguez Moguel to expropriate the hacienda of a family who supported General Escobar's rebellion. Sometimes *agrónomos* intervened in *ejidal* politics, which undoubtedly alienated the losers, and many later said that they spent most of their time during the 1920s helping *campesinos* petition for land and defending them from their enemies. For example, Mesa A. tried to protect the *ejidatarios* of Sumidero, Veracruz, in a dispute with a labor leader allied with the governor, which cost him his job. These activities made *agrarismo* an integral part of the *agrónomos'* professional culture and gave them a heroic role in the Revolution, validating their membership in the Revolutionary Family. A few rose to positions of political influence, such as Ignacio L. Enríquez, who became a general during the Revolution and governor of Chihuahua from 1920 to 1924, but some were not as hostile to the *hacendados*. González H. believed that his colleagues could win their confidence by running model haciendas. ENA professor Manuel Carpio wanted to train farmers and hacienda advisers. Ismael Reyes Retana claimed that *campesinos* could not produce henequen in *ejidos* and urged his colleagues to protect these haciendas from workers' agrarian demands. In 1926 León defended *hacendados* who modernized, claiming that they should not lose their lands, but he told those who used traditional practices to abandon their estates.[10]

The *agrónomos'* alliance with the State attracted the attention of its critics, who claimed that their jobs wasted government funds. In a 1922 newspaper article Francisco Bulnes, an enemy of land reform, said that *hacendados* hated *agrónomos*, claimed that their *agrarismo* proved their ignorance of science, and urged the SAF to fire them. Some *agrónomos* were not vehement *agraristas*, but many *hacendados* identified the profession and by extension the SAF as threats to their property. For example, the *gringa* Rosalie Evans called SAF employees "Bolsheviks who are ruining us" and "hated" them "more than Hades."[11] Others claimed that the Agricultural Machinery Trains sold poor equipment to enrich the SAF's directors. *Agrónomo* Eustacio L. Contreras, who served on them, resented these accusations, called the SAF's petty politics a disgrace to his profession, said that his Porfirian teachers were inept, and urged them to quit to let a new generation make the SAF an effective agency. González H. claimed that most SAF employees got their jobs by "favoritism or Revolutionary merits," not professional competence.[12]

PORFIRIAN, SOVIET, AND U.S. AGRONOMY AND THE CONSTRUCTION OF REVOLUTIONARY *AGRONOMÍA*

During the 1920s there were few agricultural scientists in Mexico. Some Porfirians such as Manuel Vera entered the private sector rather than serve the new State, but many *agrónomos* active or trained before the Revolution were influential in the profession and the SAF, which preserved old ideas about French agronomy, peasant and hacienda agriculture, the impact of race, climate, and diet on the *campesinos'* character, and the neo-Lamarckian view of heredity and evolution. For example, when the ENA reopened, Justino Martínez became director and Basurto Larrainzar regained his professorship. From 1921 until his death in 1928 Brambila Sr. directed the SAF's Department of Agronomy, Microbiology, and Chemistry. For meteorology students, the ENA adopted "more or less the same plan of studies for the career in France."[13] In 1923 the SAN's journal *Agros* ran Arcadio Sánchez's translations of articles from *Revue Horticole* and *Journal d'Agriculture Practique*, including subjects such as fertilizers, plowing practices, and the cultivation of fruit trees. Mesa A. praised the exhibits at the French Horticulture Society's fair in Bordeaux and urged his colleagues to study them. In 1926 Gándara dedicated an article on the Valley of Mexico's plants to "distinguished French psychologist" Pierre Janet.[14]

Informed by Porfirian ideas and urban and class antipathy toward peasants, Enrique A. Cervantes called the *campesinos* of lowland Puebla carefree and lazy people who lived "without worries" by casually gathering the bounty of their lush tropical environment, and said that the "Indians" of Huauhinango worked "only to cover the insignificant necessities of their lives," preferring to sit "at the foot of their *jacales* (hovels), where most of the day they wait for the night to rest."[15] SAF poultry expert Rufino Monroy said that "the Indian with his native rooster" was "the principal enemy of poultry keeping in Mexico."[16] When ENA director Soberón met students at El Mexe Agricultural School, he said (according to an enemy "more or less in these terms") that "the day they send me these Indians, I resign!"[17] Antonio Rivas Tagle wanted to make corn growers more productive, but he also believed that maize had poor nutritional qualities. He claimed that Mexico's diet caused racial degeneration, as, he argued, had already occurred in rice-dependent Asia. He joined Romulo Escobar to urge farmers to grow and eat more vegetables, but he also believed that scientific seed selection produced more nutritious plants. He wrote that "improving the nutritive elements in seed would provide a better diet for our people, a desirable objective, because ability for physical labor, as well as for intellectual development, are so much better when the diet is better. The selection of the richest seed will improve our population."[18]

Joining Marxist labor leader Vicente Lombardo Toledano, to validate their professional expertise and justify their demand for government jobs, the urban upper- and middle-class *agrónomos* of the 1920s called the practices of peasant farmers and *hacendados* backward and mired in tradition or superstition. Monroy said that Mexican farmers believed that incubators made roosters grow up sterile. *Agrónomo* Carlos Terrazas Moro wanted *hacendados* to accept his colleagues' advice, claiming that the "ignorant majority" of Mexico's farmers "repudiate modernity, detest the principles of scientific farming with horror, and rather than seek an alliance with professionals, who should be their friends and counselors, cling to the obscurity of old routines to live egotistically and in isolation."[19] Augusto Prister said that they used "the same systems as their grandparents of the past two or three centuries," and CNA director León thought that the peasants of highland Chihuahua "do not apply their labor to water works or even initiate the basic improvements their land requires."[20] *Agrónomos* specifically denounced corn, grape, and tomato growers' "rudimentary cultivation practices." Reflecting Porfirian attitudes about peasant agriculture, they attacked corn-bean intercropping and believed that they must replace traditional methods with the scientific practices that they learned from foreign textbooks at the ENA. In Vaughan's words, like the SEP's leaders, *agrónomos* "assumed that peasants had no information to contribute in their own transformation."[21] Porfirian *agrónomos* told their junior colleagues that yields would increase significantly if farmers used scientific criteria to select their seed, disinfect it, and test it for viability, which became the foundation of the *Campaña en Pro del Maiz*, the primary extension program for corn growers. Knowing that many farmers did not trust their advice, *agrónomos* exhorted them to accept their credibility as experts.

During the 1920s *agrónomos* moved further from their Porfirian reliance on French agronomy. When Obregón reformed the SAF, he told *agrónomos* to study other nations' research and extension programs. Other foreign models, especially U.S. agronomy, influenced the SAF and Mexican *agronomía*. Costa Rican immigrant Gonzalo Robles, who graduated from the ENA in 1913 and took civil engineering courses in Indiana at Valparaiso University during the Revolution, and Ernesto Martínez de Alba designed the SAF's Extension Service after studying the USDA and collecting literature from agricultural colleges at 12 U.S. universities. In 1923 Obregón sent Robles and Ballesteros to the United States, the Soviet Union, Spain, India, China, the Philippines, Hawaii, and Cuba to study agricultural schools, experimentation stations, and extension programs. The *Escuelas Centrales Agrícolas* (Central Agricultural Schools, hereafter *Centrales*), secondary schools that Calles founded to teach *campesino* children scientific agriculture, copied U.S. institutions. Alfonso E. Bravo praised ex-Secretary of Agriculture Willet M. Hays' support for scientific agriculture, Manuel Alcazar lauded California's agricultural regulations, Monroy admired U.S. poultry farmers, Guardiola advocated dry farming methods, and Elpidio López applauded the USDA's weather service. The SAF obtained USDA information on Louisiana's sugar industry, the impact of electricity and ozone on plant growth, and using wastepaper as fertilizer. The Office of Biological Studies participated in the Third Pan American Scientific Congress and corresponded with the New York Botanical Garden, the California Academy of Sciences, the Smithsonian, and the USDA's Office of Industrial Plants.

Growers of tomatoes on the west coast, cotton in the north, alfalfa in Coahuila, and potatoes in the Bajío regularly used varieties developed by U.S. plant breeders. *Hacendados* and commercial growers, mostly in the northern states, hired U.S. experts to establish new crops or improve existing ones like rice in Sonora's Yaqui Valley and tomatoes in Sinaloa. The Richardson Construction Company helped the Valley's wheat growers get USDA literature on *chahuixtle* (stem rust), which often infested their crops. U.S. citizens such as the owners of the Los Mochis sugar hacienda and the Cayeme-Phreaner Company, which farmed in the Yaqui Valley, hired U.S. scientists to introduce technologies like agrochemicals and foreign varieties. At the El Potrero hacienda in Veracruz, Don Mc H. Forbes tested imported cane and chemical fertilizers; Spaniard Manuel Pérez conducted similar experiments for William O. Jenkins' vast Atencingo estate in Puebla. *Hacendados* who attended U.S. schools, such as Juan de Noriega, who owned two haciendas in the Central Highlands, tried U.S. varieties like "Kanred" wheat. Mexican farmers contacted the USDA for technical information on pest control, seed disinfection, planting distances, livestock feed, fruit tree cultivation, and green manure crops.[22] During the 1920s U.S. diplomats and businessmen understood the economic benefits of modernizing Mexican agriculture by introducing farm machinery and chemical fertilizers. To promote this trade, in 1928 the Pan American Union founded the Division of Agricultural Cooperation. U.S. vendors made connections in the Mexican government by contacting *agrónomos* who visited the United States. Border pest and disease control problems facilitated direct USDA–SAF cooperation. Beginning in 1917, Stakman often visited Mexico to study *chahuixtle* because wind blew spores into the United States. Worried that the pink bollworm, rarely seen in the United States, might cross the border, in

1918 the USDA sent August Busch to La Laguna to study them. In 1924 a team of inspectors conducted further research, including on control methods. In 1929 the USDA established a laboratory in Tlahualilo, Durango to test eradication practices and began to inspect cotton fields with SAF personnel. The USDA was pleased when the SAF accepted the sanitary regulations that its experts proposed. In 1928, to protect U.S. orange growers, the USDA founded a laboratory in Mexico City to study the fruit fly, helped the SAF treat groves near the Texas border with lead arsenate, and sent O.C. Trotman to assist *agrónomo* Ignacio Hernández Olmedo's search for infested trees in western Mexico. In 1929 and 1930, pleased by the SAF's cooperation, the USDA considered an extensive campaign to eradicate the pink bollworm in Mexico but decided that it was too dangerous because troops loyal to Escobar's rebellion and "antagonistic toward Americans" were active in the border area.[23] Nevertheless, these programs fostered communication between *agrónomos* and U.S. pest control experts.

In 1915 Juan Garza Morales left the ENA to study agriculture at Purdue University. He graduated in 1918 but could not find a technical job in the SAF until 1928. He worked for El Aguila Oil Company and the CNA, but his case was exceptional, as U.S.-trained *agrónomos* and scientists put a profound stamp on agricultural policy during the 1920s. Juan González came from a prominent family of wheat farmers in northern Mexico and befriended Henry A. Wallace while he attended Iowa State University. From 1924 to 1933 he was head of the SAF's Extension Service. He ran the *Campaña en Pro del Maiz* and used his connection with Wallace to import improved fruit tree parts for grafting. González sent extension agent Genero Arzave to Florida, California, and Texas to study the orange industry, and his knowledge earned the confidence of Nuevo León's growers. During the Revolution Fábila abandoned Villa's declining movement to take graduate courses in farm management at Columbia University. After the conflict he taught agricultural economics at the ENA. In 1929 he returned to the United States as Mexico's agricultural attaché and attended the annual meeting of the National Association of Nurserymen. From 1925 to 1928 Mesa A. held this position, visited rural areas in several states, studied agricultural economics at George Washington University, and used what he learned to enhance the curriculum in his ENA course. The ENA sent *agrón-*

omo Enrique Coppel Rivas to Utah to study the apple industry. The SAF hired a *ranchero* with a degree from Iowa State to run one of the *Centrales* and took advantage of several opportunities to send employees to the United States for advanced training, including González H., teachers from the *Centrales*, and Ignacio L. Figueroa of the CNI. When González H. returned, he became head of the Office of Statistics. He chose 1925 as the base year for all agricultural data, disregarding previous figures as inaccurate. Calles sent several *agrónomos* to study the U.S. Bureau of Reclamation's irrigation projects and hired engineers from the J.G. White Engineering Corporation to advise the CNI's projects. In 1928 the CNI employed a University of California professor to teach agrology at the ENA. In 1941 Marte Gómez claimed that the U.S. experience, the class, and the students who took it brought Mexican agrology up to international scientific standards.[24]

The Soviet Union used science as propaganda for its Revolution and to advance foreign policy interests. To the U.S.S.R., Mexicans were fellow revolutionaries. According to historian Ramón Eduardo Ruiz, Soviet rural education influenced José Vasconcelos, Obregón's minister of education. Contact with Russians also affected some of Mexico's agricultural scientists. Gándara met botanists S. Bukasov and S. Juzepezuck, microbiologist G. Bossé, and agronomist G.N. Voronoff, who came to Mexico to study the origin of corn. Their stories about Soviet advances in plant genetics stimulated *agrónomos'* interest, and Gándara praised the U.S.S.R.'s agricultural programs. In 1927 he accompanied Voronoff, the director of a botanical garden, on a specimen-gathering trip in Mexico. In 1929 he became a corresponding member of the Leningrad Institute of Applied Botany. Contacts continued and exerted more influence after 1930, when *cardenista agrónomos* developed their campaign to transform *campesino* culture.[25]

AGRICULTURAL PROGRAMS DURING THE 1920s: AN ABORTED CULTURAL CAMPAIGN

Obregón and Calles agreed with *agrónomos* who called the *hacendados'* and *campesinos'* farming practices backward and unscientific. An innovative farmer from Sonora, Obregón admired U.S. agricultural science, and he took a personal interest in modernizing Mexican agriculture. To woo trade unions and *agraristas* during his struggle with Carranza, he promised to support programs to modernize agriculture. His SAF operated an extension service, a research division, and an experimentation station at the ENA. Some revolutionaries urged Obregón to "refashion rural society," and the SEP built rural schools to transform *campesinos* from "superstitious, locally oriented pariahs into patriotic, scientifically informed commercial producers."[26] Although objectives and methods changed with turnover in the SEP's leadership, the cultural campaign's architects believed that the persistence of folk beliefs contributed to the *campesinos'* poverty and oppression and wanted to inculcate a worldview based on scientific explanations of nature and society. Instruction in agriculture, including "the value of cultivation, irrigation, and rotation of crops . . . a cultural advance of hundreds of years over having seed blessed, setting up rain gods, and holding festivals for good crops while planting seed in poorly plowed ground and never cultivating, weeding, or irrigating," was integral to the project.[27] Classes incorporated principles from botany, biology, and basic agriculture, and the SEP's Cultural Missions program, begun in 1923, included *agrónomos* who gave lectures on improved practices. Robles urged rural teachers to plant school gardens to teach children basic farming, including scientific seed selection and vegetable cultivation to improve their diet. The SAN wanted not only to solve the technical problems of farming but also to address political, economic, and social issues like land reform, the organization of agricultural production, and public health and nutrition. In the SAF, *agrónomos* aided the cultural project, combining *Porfirian* ideas about diet, race, and peasants with revolutionary anti-clericism and a neo-Lamarckian view of good nurture uplifting the inferior.

In 1922 Obregón founded the *Agrónomos Regionales*, the SAF's extension service. In 1923 U.S.-educated González became director and ordered the *Agrónomos Regionales* to find new crops for "humble *campesinos*" to cultivate to attain "improvement" and "economic emancipation."[28] With little land to demonstrate new methods or varieties, they often used "*convencimiento*" (lectures) to persuade farmers to change their practices. In 1924 González told his agents to enlist rural teachers as *Agentes de Información y Propaganda Agrícola*, but many had no agricultural training or experience, and the *Agrónomos Regionales* in Guerrero, Jalisco, and Aguascalientes had to give them short courses on basic farming.

Like the teachers, some of the SAF's *agrónomos* and scientists supported the Revolution's anti-clerical agenda. For example, Herrera disliked the Catholic Church, Mesa A. admired a female teacher "free of religious prejudices," and de Dios Bojórquez proclaimed that "the redemption of the world is among us, the new Jesus Christ is work."[29] In 1924 the ENA began an extension project with peasants from Chapingo, posting a sign forbidding alcohol and shrines on the grounds. Robles urged the SAF to establish a demonstration farm and agricultural school in Cordoba, Veracruz, to improve the many crops grown in the state and introduce new ones and to "impart practical agricultural knowledge at the secondary level to prepare good farmers to live as honest and progressive citizens."[30] Tabasco's extension agent studied living conditions in two *ejidos*, and Jalisco's helped *ejidatarios* in Tecalitlán build a reservoir. To improve the *campesinos'* diet, *Agrónomos Regionales* urged them to plant vegetables. They organized producers, for example, forming a cooperative of beekeepers in Jalisco. The SAF considered undertaking an extensive study of the farming practices and crop strains used by Mexican corn growers but decided against it, rejecting the *campesinos'* knowledge. Rather, *Agrónomos Regionales* tested foreign wheat, corn, oats, barley, and potatoes, working with "progressive farmers": *hacendados, rancheros,* or commercial growers, but demonstration plots often did poorly because the strains were bred for Iowa or North Dakota rather than Mexico. In 1923 the SAF began to offer correspondence courses, which targeted affluent and literate farmers.[31]

Calles also tried to modernize agriculture. He built irrigation works and roads to help farmers bring their crops to market and founded the *Centrales,* two experimentation stations in the new irrigation districts, and an agency that studied and fought pests and diseases. With corn imports draining over 1 million pesos from the federal budget, by 1925 influential Mexicans such as Pedro de Alba urged the government to bring science to the hacienda. The SAF was under pressure to show examples of improved farming practices and considered ways to attain rapid successes. Notwithstanding *agrónomos'* rhetoric about serving the *campesinos,* extension agents often worked with commercial farmers and *hacendados* who cooperated. Rafeal Batiz Paredes advised Sonora's tomato farmers, *Agrónomos Regionales* in the northeast worked with cotton growers, Justo Lozoya L. studied the mechanized cultivation of alfalfa at the hacienda El Pilar in Coahuila, Sinaloa's *Agrónomo Regional* demonstrated a garbanzo-planting machine, and Roberto Gómez Camacho helped Jalisco's orange growers try imported varieties with export potential. In 1928, to reach the same audience, the agency held a National Tractor Contest. In 1930 the *Agrónomos Regionales* in the Bajío tested imported wheat varieties with the help "of the most progressive and enthusiastic farmers of the region."[32]

To make corn growers more productive, *Agrónomos Regionales* gave lectures on seed selection, disinfection, and germination testing. In 1926, when corn imports cost 3.7 million pesos, González ordered them to devote more attention to maize farmers. That year, to "liberate Mexico from the slavery of importing corn year after year," Calles convoked a National Corn Contest, where *agrónomos* gave speeches on seed selection, disinfection, and germination tests, but, reflecting the politicization of agricultural programs, Obregón, "Sonora's best farmer," "who works from sunup to sundown to increase production," won first prize.[33] In 1929 and 1930, with the help of the State of Guanajuato, *Agrónomos Regionales* David S. Ibarra and Rivas Tagle promoted the *Campaña en Pro del Maiz*. Rivas Tagle thought that Mexican farmers could develop hybrid corn using the USDA's methods, which he described in his 1929 monograph *El cultivo racional del maiz*, and he believed that contests would persuade farmers to produce improved strains adapted to Mexico's varied local conditions. Interested in plant genetics, *agrónomos* wanted to breed corn, potatoes, sugarcane, and tomatoes, but, reflecting the SAF's faith in foreign scientists and most *agrónomos'* limited knowledge of field plot techniques, during the 1920s the SAF helped U.S. vendors to market seed and farmers to import it for planting duty-free. In 1925 Mexico's railroads shipped U.S. seeds, bulbs, and fruit trees for planting without freight charges. The SAF imported fruit trees for its nurseries and by 1928 distributed branches for grafting.[34]

Knowing that agrochemicals were expensive, scarce, untested, and sometimes misrepresented in Mexico, many *agrónomos* urged farmers to use dung or green manure crops to enhance soil fertility and non-chemical methods to control pests, but González told the *Agrónomos Regionales* to promote nationally manufactured chemical fertilizers and test them on orange groves and sugarcane in Nuevo León and Jalisco. Alejandro Brambila Jr., Fourtón, and the sugar industry ran campaigns to popularize them. Sugarcane and coffee growers in Sinaloa and Nayarit purchased Chilean nitrates and "Nitrofoska," sold by the German-owned *Beick, Felix, y Cia*. Some *agrónomos* favored the use of biological pest controls and urged the government to study useful insects. An SAF campaign to control a 1924 locust plague provided jobs for many unemployed *agrónomos*, and one was so happy that he did not want to kill the pests. Responding to extensive losses caused by insects, in 1927 Calles established the *Oficina Federal de Defensa Agrícola* (Federal Office of Agricultural Defense) and appointed German immigrant entomologist Alfonso Dampf, who came to Mexico in 1924, to run it. Reflecting Calles' desire to support commercial farmers and his allies in the Revolutionary Family, this agency used most of its resources to solve the pest control problems of orange growers in the northeast, among them politically powerful people like General Juan Andreu Almazán.[35]

Calles used rural education in his anti-clerical campaign, and after 1926 teachers were targets of *cristeros*, peasant rebels who rejected the State's project and took up arms to defend their faith. To support Calles, SAF employees agreed not to attend religious ceremonies. Priests told the faithful not to cooperate with the government, and Catholic groups planned their own agricultural research program. The anti-clerical campaign fostered poor relations between *agrónomos* and peasant farmers. In 1928 the SEP established the Permanent Cultural Missions. Reflecting the cultural campaign's broad agenda, in Actopán, Hidalgo, a mission to the Otomí Indians included an *agrónomo* who managed experimental plots and supervised the construction of wells, school buildings, and an open-air theater, as well as a doctor and nurse, professors of sports and industries, a social worker, and a musician. The SAF decided not to collaborate with the Permanent Cultural Missions, and by 1930 the agency had little to offer the *ejidatarios*. The ENA worked with local peasants, organizing a club to give lectures on scientific agriculture, but technical support for the *ejido* usually consisted of denunciations of traditional farming implements, crop strains, and cultivation practices. The *agrónomos'* brief experience working with *campesinos* convinced them that they resisted changing their practices and preferred their "ancient systems" of cultivation. Marte Gómez called Native American *agrónomo* Ezequiel M. Gracia a man who "lived in common with the land, becoming brothers with the people," but noted that this made him a rare bird among his colleagues of the 1920s.[36] *Agrónomos* may have recognized that class and ethnic differences inhibited their communication with the *campesinos*. Demonstrations at haciendas or ranchos probably alienated *agraristas* who wanted their land. Gilberto Mendoza Vargas later claimed that there were too few *agrónomos* to reach all of Mexico's peasant farmers. As Dana Markiewicz observes, the duties given to *agrónomos* who worked with *campesinos* were overwhelming.[37]

To address the situation, in 1925 Calles established the *Centrales* to train adolescent *campesinos* as intermediate agricultural specialists who would return to their villages to teach scientific agriculture. Their education included how to select seed of many crops, use dung and chemical fertilizers, and grow corn, wheat, tomatoes, beans, alfalfa, peas, chiles, melons, sugar beets, peanuts, and watermelons. The schools were "centers of biological, social, and economic research and experimentation related to the life and development of peasants" that planned to eradicate "old rural superstitions."[38] The SAF wanted the *Centrales* to teach *campesinos* modern agricultural technique, but, reflecting an orientation that favored commercial farms, haciendas, and irrigation districts, each school was supposed to have at least 500 hectares of prime irrigated land for experimentation and teaching. Many of the *Centrales'* students did not come from peasant families. According to *agrónomo* Eduardo Limón and the SEP, most graduates, including *campesinos*, "were more interested in their personal advancement" than "life on farms in isolated rural communities" and sought bureaucratic jobs in cities.[39] A post in Mexico City provided the best chance to build alliances for upward career mobility. To train at least a few *campesinos* as *agrónomos*, Calles told the ENA to admit more graduates of the *Centrales*, but the school's directors resisted this policy and also enrolled politically connected students who did not meet admissions standards. Ramón Fernández y Fernández, an ENA student during the 1920s, later said that the attempt to "orient the school's curriculum to promote and support the agrarian reform" failed.[40] As Lorey asserts, the tumult of Revolution and its aftermath discouraged the development of the professions. From 1921 to 1927 the ENA granted just 18 degrees but awarded 89 from 1928 to 1930, which formed the "Generation of 1928," which exerted considerable influence on agricultural and agrarian policy during the 1930s. In 1927 the first three *agrónomos* graduated from the private Antonio Narro Agricultural School in Saltillo, Coahuila, and another 22 joined them by 1930. During the 1920s, 11 Mexicans studied agriculture at U.S. universities, and 6 obtained degrees. Most of the new *agrónomos* came from urban middle- or upper-class backgrounds.

While demanding more State support for research, *agrónomos* again raised the issue of scientific autonomy. Robles wanted the Extension Service to copy the USDA, which used the results of experiments as a base for its programs, but he knew that Mexico had few experimental scientists and urged the ENA to train them immediately. In the meantime, to address the "delicate question" of staffing experimentation stations, he admitted that "without intending to demean our native scientists, to organize experimentation rationally, we must resort to foreign countries in some cases."[41] Fourtón warned:

> To a certain point the research conducted in other countries is useful, but in agriculture local conditions are so important and those in Mexico are so different from the regions where most research has been conducted and published, that a local adaptation of foreign works is not sufficient; it is essential to develop in the same country the agronomic rules that foster the future progress of agriculture.[42]

ENA professor Alfonso Madriaga said that "the study of entomology must be done locally: Mexico cannot learn from foreign books because they advise methods inappropriate in our country," and Monroy denounced the SAF's "concentration on the translation of articles from North American authors, which are wrong for our situation" and lead to "disaster" when farmers follow their recommendations.[43] Similarly, Contreras claimed that *agrónomos* and the SAF appeared inept when they suggested pest control procedures developed in distant places like Egypt. Rivas Tagle even acknowledged that in Mexico's unique rural environment foreign machines and implements might be less efficient than traditional ones. *Agrónomos* knew that State-supported science provided jobs. Praising the CNI's experimentation stations, Raziel Mora Olivera urged "large and small farmers to unite and sustain these institutions run by *agrónomos*."[44]

Inspired by Soviet scientists' interest in the origins of corn and

U.S. plant breeding, Pandurang Khankhoje, an East Indian immigrant trained in Great Britain and the United States, began to breed maize at the ENA. In 1927 he crossed a local strain with varieties that he collected at the National Corn Contest. By 1930 he had enough seed of his hybrid "Maiz Granada" to distribute to peasants and schools near Chapingo, and he began to work with wheat and beans. The government sent him to Europe in 1930. When he returned in 1932, he collected ears of his corn but found that it had mixed with local strains, forcing him to start over. To find crops more profitable or nutritious than maize, the ENA also tested barley, alfalfa, wheat, beans, and vegetables. Brambila Sr. and Fourtón used a system of physiological chemistry developed by the German Newbauer to evaluate the reaction of various plants to chemical fertilizers. The Office of Biological Studies conducted research, but most was only tangentially related to agriculture, like studies of deer, iguanas, and scorpions, which led to disputes with other branches of the SAF and UNAM. Enrique Beltrán tried a biological method to kill locusts, using bacteria obtained from the Pasteur and Rockefeller Institutes, but he found that it was not pathogenic except when injected or in very humid conditions, and further, that afflicted insects did not spread the disease. *Agrónomo* Olivero Tellez of the Office of Biological Studies tested methods to manufacture products like oils, greases, soap, and fertilizer from dead locusts. In 1926 the SAF created an Office of Experimentation Stations but in 1929 operated only one in the Federal District. Reflecting Calles' interest in irrigated agriculture, the CNI founded two experimentation stations. Brambila Jr. and Rafael Garcia Galán ran one in the Pabellón, Aguascalientes Irrigation District, studying soils, testing chemical fertilizers, and planting chiles, wheat, garbanzos, oats, peas, linseed, sweet potatoes, potatoes, strawberries, alfalfa, vegetables, and peanuts to find viable crops more profitable than corn. The Rodríguez, Nuevo León Irrigation District station performed similar tests. Rivas Tagle believed that maize should be grown on irrigated lands only as a last resort, after all other crops were tested.[45] Urging *ejidatarios* to reject corn in favor of commercial crops was another part of the 1930s cultural campaign.

A few of the *Agrónomos Regionales'* projects succeeded. *Agrónomo* Nicolas Sánchez Durón, a Rockefeller Foundation fellow, cites new citrus groves in Nuevo León, apple orchards in Chihuahua, potato farms in Silao, tropical crops in Veracruz, and forage grasses in several locations as examples. By 1936 Almazán's orange trees earned national recognition, and he sold the SAF branches for grafting. The proprietors of the *Hacienda de Amalucán* developed hybrid wheats that farmers in Puebla, Oaxaca, and Mexico state planted during the 1930s. *Agrónomos* and other scientists with foreign training like Arzave and Beltrán influenced the SAF, and their research and extension programs were the most likely to have positive results. Combining lectures with demonstrations, such as Ibarra's work in the Bajío, convinced some farmers to try foreign wheats. The natural environment in the northern states facilitated the success of U.S. fruit grafts, but many programs had less successful outcomes. Telling rather than showing farmers how to improve their practices was no more effective in Mexico than in the United States. The *Agrónomos Regionales'* jurisdictions changed frequently, giving them little time to earn farmers' confidence. After 1925, budget cutbacks reduced their number from 35 to 20, and by 1933 to only 12. Job turnover was rampant and rapid. For example, from 1920 to 1930 *agrónomo* Nabor Cuervo Martínez passed through 11 different jobs in the CNA, SAF, ENA, and CNI. Transferred from Jalisco, Gómez Camacho did not realize that the imported orange trees growers bought were not pedigreed. Many trees died because recipients did not receive proper training to care for them. The shortage of trained personnel slowed the SAF's chemical fertilizer tests and plant breeding projects. The SAF's broken machinery often sat unrepaired for lack of funds. Although their reports reflect U.S. biases, U.S. consuls believed that land reform and the actions of *agraristas* made *hacendados* unwilling to modernize their practices. Armed enemies of the regime intimidated SAF employees.[46]

A few scientists worried about the SAF's programs, particularly the lack or poor quality of research. In 1927 Beltrán urged Calles to provide more support for research that had been "almost asphyxiated by other concerns in national life, among which political activities undoubtedly predominate."[47] Rivas Tagle lamented that his colleagues gave bad advice that diminished their credibility with farmers because their recommendations were not based on experimental results. Obregón and Calles considered proposals for plant-breeding programs to create improved sugarcane and tropical crop varieties and to establish a separate ministry to increase corn production, but with few Mexican-trained experimental scientists and no desire or means to hire foreigners, they rejected these proposals, and the SAF did not emphasize research during the 1920s. Mexico did not have sufficient human capital to advance a project of scientific autonomy. To address the problem, in 1923 and 1929 the ENA's directors tried to reform the curriculum by introducing specializations, but frequent turnovers of leadership and the shortage of professors with training in experimental science thwarted these efforts. Beltrán believed that *agrónomos* who graduated during the 1920s learned little science, and historians José Luis Calva Téllez and Carlos Monzón claim that at this time no school in Mexico trained "technical experts capable of bringing modern agricultural techniques to the fields."[48]

CONCLUSIONS

Agrónomos contributed to the military phase of the Revolution and cemented their place in the Revolutionary Family by helping the State use agrarian reform to assert control over independent peasant movements. From this point on, the profession was inextricably linked to "the Revolution," and this alliance helped diminish the *agrónomos'* employment problem. Even agencies with little connection to agriculture like the Ministry of Communications and Public Works hired *agrónomos* as engineers and administrators. During the 1920s some *agrónomos* earned a living outside the profession, such as Carlos J. Aguilar, who worked in Torreón for the *Loteria Nacional*, as private farmers, like Montemayor Martínez, as hacienda managers, like Justino Martínez, or for the few agroindustrial firms in Mexico like *Beick, Felix, y Cia*, which hired Alfonso Magellón Haro. After a brief career in public service with many exciting but life-threatening ups and downs, in 1930 Ramón Corral Soto decided to leave the bureaucracy to start a construction company in Mexico City, but of 296 SAN members that year, 89 found jobs with the CNA, 32 filled technical positions in the SAF, 18 worked in the agricultural banks and 11 in the CNI, 13 were ENA professors, and secondary agricultural schools hired 26 to teach. Forty-six had government jobs unrelated to agriculture, mostly gathering and interpreting statistics, 42 did not specify an occupation, 10 were unemployed, and only 2 worked in the private sector.

Demonstrating the rewards of membership in the Revolutionary Family, 2 *agrónomos* represented Mexico at the embassy in Washington, D.C., and 10 served in Congress as federal deputies (8) and senators (2). Some, like Blas E. Romo Soto, entered state or local politics. Service in the bureaucracy provided a chance to launch a political career. In 1924 Calles appointed León minister of agriculture, the first *agrónomo* to run the SAF. Using their connections to powerful *agrónomos* like Marte Gómez and service in the bureaucracy, the Jaimes brothers protected their family's wealth. The *agrónomos'* alliance with the state had costs, as Enrique Baron Obregón learned when he joined General Adolfo de la Huerta's rebellion. Similarly, in 1921 President Obregón fired Assistant CNA Director Ignacio L. Figueroa when he took personal control of the *Partido Liberal Constitucional*, which he organized during the Revolution to seize power from Carranza. In 1929 President Emilio Portes Gil chose Marte Gómez to lead the SAF, but when Pascual Ortiz Rubio took office in 1930 Gómez lost his job, as did his loyal *cliente* Martínez de Alba. The Mexican Revolution did not bring democracy but rather fostered warlords' power struggles that ended with the Revolutionary Family's domination of the political process, giving *campesinos* no control over their modernization other than the ability to resist, but contrary to Alberto Arce, *agrónomos'* contributions to the Revolution and the Northern Dynasty regimes began to construct a bureaucratic mission for the profession.[49]

The *agrónomos'* participation in the land reform won some *agraristas'* support, but their role in the cultural campaign alienated *campesinos* who resented the attack on their traditional lifestyle, religion, and other belief systems. Unlike the extension agents who worked for Seaman A. Knapp in the southern United States, *agrónomos* "presented themselves as final authorities on all agricultural matters" and "disturb[ed] prevailing ideas" like folk beliefs about farming.[50] Reflecting the *agrónomos'* class background and education, they made a halfhearted, brief, and culturally imperialistic attempt to solve the problems of peasant agriculture. Their *agrarismo* alienated the *hacendados*, who now saw the *agrónomos* as political enemies as well as incompetent professionals. *Agrónomos* began to solve their employment problem by working in the bureaucracy, but they still lacked a clientele who accepted their authority as scientific professionals. They could not develop a uniquely Mexican and effective agricultural science for peasant farmers, and many influential people realized that the SAF failed to improve *campesino* agriculture or make the *ejido* prosperous and productive, but the *agrónomos* of the 1920s charted several paths for the *cardenista* cultural campaigners of the 1930s.

NOTES

1. Douglas Bennet and Keith Sharpe, "The State as Banker and Entrepreneur: The Last Resort Character of the Mexican State's Economic Intervention 1917–1970," in Sylvia Ann Hewlitt and Richard S. Weinert, eds., *Brazil and Mexico: Patterns of Late Development* (Philadelphia: Institute for the Study of Human Issues, 1982), p. 176.

2. David E. Lorey, *The University System and Economic Development in Mexico since 1929* (Stanford, Calif.: Stanford University Press, 1993), p. 174; Stanley F. Shadle, *Andrés Molina Enríquez, Mexican Land Reformer of the Revolutionary Era* (Tucson: University of Arizona Press, 1994), p. 12; Ramón Fernández y Fernández, *Chapingo hace 50 años* (Chapingo, México: ENA, Colegio de Postgraduados, 1976), p. 39; Luís L. León, "El gremio agronómico en la reforma agraria," *Boletín de la Sociedad Mexicana de Geografía y Estadística* 78 (1954): 60; Gustavo Durán, "Importancia de la agricultura y fraccionamiento de tierras," *Memoria y Revista de la Sociedad Científica Antonio Alzate* 30, 10–12 (December 1911): 442–43; Marte R. Gómez, *Biografías de agrónomos* (Chapingo, México: ENA, Colegio de Postgraduados, 1976), pp. 32, 197–98, 248, 341, 481; Mario Calvino, *Segundo Informe de la División de Horticultura* (México, D.F.: IFSF, 1914), pp. 46–57, 151; Calvino, *Informe de trabajos, División de Horticultura, 1912* (México, D.F.: IFSF, 1913), pp. 39–43, 124; Calvino, *Informe de la División de Horticultura, 1913* (México, D.F.: IFSF, 1914), pp. 3–29.

3. Ramón García Oses, *Cultivo de la alfalfa* (México, D.F.: IFSF, 1912); Luis Marín, "La higuerilla o ricino," *Agricultura* 1, 2 (September–October 1937): 22; Alfonso Herrera, "Estudios experimentales de plasmogena," *Memorias de la Sociedad Científica Antonio Alzate* 33, 9–10 (1913): 288–94; Gómez, *Biografías*, pp. 85–86, 367–68.

4. Enrique Beltrán, "La Dirección de Estudios Biológicos de la Secretaría de Fomento y el Instituto de Biología de la Universidad Nacional Autónoma," *Anales de la Sociedad Mexicana de Historia de la Ciencia y de la Tecnología* 1 (1969): 105–41; Alfonso Madriaga, "Datos históricos acerca de la aparicion del gusano rosado (*pectinophora gossypeilla, saunder*) en la Republica Mexicana," *Agricultura* 1, 3 (December 1937): 12–14; "Article 3o Constitucional; Concepto de Enseñanza Agrícola, Requisitos de Funcionamiento y Admisión, Transitorios," 1917, in Enseñanza Agrícola Su-

perior, BCM Bound Volume #145–43; Gómez, *Biografías*, pp. 149, 255–56, 392, 489; Nancy Leys Stepan, *"The Hour of Eugenics": Race, Gender, and Nation in Latin America* (Ithaca, N.Y.: Cornell University Press, 1991), pp. 72–73.

5. Fernández y Fernández, *Chapingo*, pp. 66, 92; Roy V. Scott, *The Reluctant Farmer: The Rise of Agricultural Extension to 1914* (Urbana: University of Illinois Press, 1971), pp. 180–81; Wilfred Hardy Callcott, *Liberalism in Mexico, 1857–1929* (Hamden, Conn.: Archon Books, 1965), p. 261 (1st ed. 1931); Gabriel Gómez, *La higuerilla y el aciete de ricino*, 2nd ed. (México, D.F.: Imprenta Bucareli, 1921); Manuel Mesa A., *Relatos autobiográficos* (México, D.F.: Editorial Nuestro Tiempo, 1981), pp. 12–13; Gómez, *Biografías*, pp. 42, 60, 144, 348; Daisy Caden Pettus, ed., *The Rosalie Evans Letters from Mexico* (Indianapolis: Bobbs-Merrill Company, 1926), p. 71.

6. Comisión Agraria del Estado de Chihuahua, *Memorandum para los topógrafos y agrónomos* (Chihuahua: Imprenta del Gobierno, 1915); Linda B. Hall, *Alvaro Obregón: Power and Revolution in Mexico, 1911–1920* (College Station: Texas A&M University Press, 1981), pp. 160, 192, 256; Fernández y Fernández, *Chapingo*, pp. 57–59; "El problema agrario en la zona cafetera," *Hoy*, April 29, 1939; Gilberto Mendoza Vargas, "La enseñanza y extensión agrícola," February 1954, in Enseñanza, Division of Inter-American Educational Relations, "Alumni of United States Colleges of Agriculture Who Are Citizens of the Other American Republics," April 1944, NA, RG 59, Decimal File 1940–1944, Box 196; Gómez, *Biografías*, pp. 72, 149, 348, 422–23, 434.

7. Juan de Dios Bojórquez, "Momento decisivo para el gremio agronómico," December 1, 1922, p. 9, in Agrónomos, BCM Bound Volume #168–40 (hereafter Agrónomos); Mesa A., *Relatos*, p. 74; Gómez, *Biografías*, p. 432.

8. César Martino, "El próximo Congreso Agronómico," *El Nacional*, October 17, 1935; "La cruzada de los agrónomos," *Hoy*, April 26, 1941; Fernández y Fernández, *Chapingo*, pp. 70, 84; Bernardo L. Arrieta M., "La educación agrícola como fundamental de reorganización de nuestra economía rural," in Partido Nacional Revolucionario (hereafter PNR), ed., *Los problemas agrícolas de México* (México, D.F.: PNR, 1934), p. 253; "Acta de inauguración de la ENA en terrenos de la Hacienda de Chapingo, México," in ENA, BCM Bound Volume #168–2 (hereafter ENA).

9. Rebeca de Gortari Rabiela, "Educación y conciencia nacional: Los ingenieros despues de la revolución Mexicana," *Revista Mexicana de Sociología* 49, 3 (1989): 129; Fernández y Fernández, *Chapingo*, p. 71.

10. Fernández y Fernández, *Los agrónomos* (México, D.F.: Impresa Secretaría de Maquinas, 1933), p. 5; SAF, *Memoria 1924–1925* (Tacubaya, D.F.: Imprenta de la Dirección de Estudios Geográficos y Climatológicos, 1927), p. 69; Julián Rodríguez Adame, "El problema agrario Mexicano y

la maquinización agrícola," in Centro de Estudios Sociales, ed., *Cuestiones Agrarias de México* (México, D.F.: El Colegio de México, 1945), p. 53; Leandro Mendoza Vargas, "Como proceder para que la campaña de abolición del arado de palo sea eficiente," April 1941, in S.A.M. Segundo Consejo Nacional Directivo, BCM Bound Volume #186–5; Leonel Robles, *Agricultural Education, Investigation, and Extension in Mexico: A Preliminary Study* (México, D.F.: The National Association of Directors of Agricultural Schools, 1964), p. 27; Nicolas Sánchez Durón, "La investigación agrícola y el desarrollo agropecuario de México," *Anales de la Sociedad Mexicana de Historia de la Ciencia y de la Tecnología* 2 (1970): 176; Peter S. Cleaves, *Professions and the State: The Mexican Case* (Tucson: University of Arizona Press, 1987), p. 26; Gómez, *Biografías*, pp. 94, 227–28, 245, 381–82, 439, 450–51; Luís Aboites Aguilar, "Política de irrigación y cambio tecnológico en México, 1920–1940," *Papeles de la Casa Chata* 2, 3 (1987): 8; Gonzalo González, "La hacienda modelo y el agrónomo," November 1922, in Agrónomos; Manuel Carpio, *Una Escuela de Agricultura para agricultores y no para profesionales empleomanos* (México, D.F.: 1923); Ismael Reyes Retana, "Causes principales de la crises henequeñera," September 1922, pp. 3–6; ARFF, Box 105; Luis L. León, "Discurso," April 10, 1926, in DGA, *Concurso del mejor mazorca del maiz* (Tacubaya, D.F.: Imprenta de la Dirección de Estudios Geográficos y Climatológicos, 1926), p. 20.

11. Francisco Bulnes, *Los grandes problemas de México* (México, D.F.: Ediciones de "El Universal," 1926), pp. 138–53; Pettus, *The Rosalie Evans Letters*, p. 187.

12. Eustacio L. Contreras, "Principios de una organización," 1922, pp. 4–16, 46, AGN, Robles papers, Agriculture Section, Box 5, Exp. 69; Fernández y Fernández, *Chapingo*, pp. 66–68; Gonzalo González, "Segundo Congreso Nacional Agronómico," 1922, ARFF, Box 83.

13. Gómez, *Biografías*, pp. 72, 110, 197–98, 368, 509–13.

14. *Agros*, June, August, and October, 1923; Manuel Mesa A., "Un arte desconocido en México," *Agros* 1, 4 (September 1923): 19; Guillermo Gándara, "Modo de estimar la flora del Valle de México," *Memoria y Revista de la Sociedad Científica Antonio Alzate* 45, 1–6 (1926): 93; Gómez, *Biografías*, p. 246.

15. Enrique A. Cervantes, "Monografía del Municipio de Tehuacán," *Memoria y Revista de la Sociedad Científica Antonio Alzate* 45, 7–12 (July–December 1926): 378–402; Cervantes, "Apuntes acerca de la municipalidad de Huauhinango," *Memoria y Revista de la Sociedad Científica Antonio Alzate* 40, 7–9 (July–September 1922): 475.

16. Rufino Monroy, "¿Cual es la raza de gallinas que más conviene explotar?" *Agros* 1, 5 (October 1923): 6.

17. Miguel García Cruz, "El problema legendario de la ENA," p. 3, in Enseñanza.

18. Antonio Rivas Tagle, *El cultivo racional del maiz* (Tacubaya, D.F.: Talleres Gráficos de la SAF, 1929), pp. 8, 283; Romulo Escobar, *El esparago* (México, D.F.: Imprenta de la DGA, 1922), p. 16.

19. Vicente Lombardo Toledano, "El problema de la educación en México," in Lombardo Toledano, *La ciencia y la educación técnica* (México, D.F.: Dirección de Publicaciones del Instituto Politécnico Nacional, 1984), p. 11; Rufino Monroy, "La avicultura en México como se ha vista y como debe verse,"*Agros* 1, 1 (June 1923): 3; Carlos Terrazas Moro, "Agrónomos y hacendados," *Agros* 1, 1 (June 1923): 2; de Dios Bojórquez, "Momento decisivo," p. 1.

20. Augusto Prister, "La química y la agricultura mexicana," *Memoria y revista de la Sociedad Científica Antonio Alzate* 51, 9–10 (1929–1930): 349; Daniel Nugent, *Spent Cartridges of Revolution: An Anthropological History of Namiquipa, Chihuahua* (Chicago: University of Chicago Press, 1993), p. 94.

21. Jorge Guardiola, *El cultivo del frijol* (México, D.F.: Secretaría de Agricultura y Fomento, 1921), pp. 56–57; Contreras, "¿Que es un agrónomo?" *Agros* 1, 2 (July 1923): 5; Mary Kay Vaughan, *Cultural Politics in Revolution: Teachers, Peasants, and Schools in Mexico, 1930–1940* (Tucson: University of Arizona Press, 1997), p. 28.

22. Guardiola, *El cultivo del frijol*, pp. 11–13, 112; Contreras, "Principios," pp. 19, 24–27; Arcadio Sánchez, "Agrónomos y hacendados," *Agros* 1, 4 (September 1923): 8; H. Umberger to Robles, August 11, 1921, and others, AGN, Robles papers, Correspondence Section, Box 41, Exp. 34; A. Cienfuegos y Camus to J. Ballesteros and Gonzalo Robles, April 11, 1925, AGN, Robles papers, Agriculture Section, Box 13, Exp. 112; Gómez, *Biografías*, pp. 364–65; Alfonso E. Bravo, "La agricultura moderna," *Irrigación en México* 1, 3 (July 1930): 50–51; Manuel Alcazar, "Legislaciones sobre plagas," December 1922, pp. 6–8, ARFF, Box 141; Monroy, "La avicultura," p. 3; George C. Booth, *Mexico's School Made Society* (Stanford, Calif.: Stanford University Press, 1941), p. 44; Ramón Eduardo Ruiz, *Mexico: The Challenge of Poverty and Illiteracy* (San Marino, Calif.: The Huntington Library, 1963), pp. 106–8; Estey S. Smith to Secretary of Foreign Relations, Mexico, July 30, 1929, AHSRE, Exp. IV-629-17; SAF, *Memoria 1924–1925*, pp. 37–46; SAF, *Monografía del jitomate* (San Jacinto, D.F.: 1932), pp. 217–18; "Estudios sobre la semilla de algodón 'Maravilla del Sur,' " *El Nacional*, November 19, 1941; "Estudio presentado por la Delegación de Defensa Agrícola de León, Gto.," in Consejo Nacional de Agricultura, *Memoria de la Primera Convención Nacional de Productores de Papa* (México, D.F.: 1939), pp. 24–26; E.W. Eaton, "Chemical Fertilizers Used in Mazatlan Consular District," April 14, 1930; William M. Smale to Secretary of State, June 16, 1930, NA, RG 59, Microfilm Roll m1370, Roll 117, Mexico 1930–1939; "Beneficio obtenido por los Ar-

roceros Sonorenses con la Constitución de su Asociación," *Revista Arrocera* 1, 2 (August 1935): 5; José Antonio Rivera, "El jitomate mexicano de exportación a Estados Unidos" (n.d.), ARFF, Box 117; Los grandes empresas industriales de México, 'Los Mochis,' " *La Hacienda* 20, 5 (May 1925): 135–37; Robert S. Bursley, "Cotton Growing in the Guaymas Consular District," January 26, 1927, NA, RG 166, Box 147; Manuel Ítie, "Las variedades de la caña de azúcar cultivadas en el Potrero, Ver.," *Agricultura* 1, 4 (January–February 1938): 24–25; David Ronfeldt, *Atencingo: The Politics of Agrarian Struggle in a Mexican Ejido* (Stanford, Calif.: Stanford University Press, 1973), pp. 10–11; M.M. Blazquez to K.F. Kellerman, November 30, 1923; J. Garza Zertuche to USDA Sugar Cane Research Department, June 26, 1925; Juan de Noreiga to USDA Bureau of Plant Industry, February 25, 1928, NA, RG 54, Series 656, Box 391.

23. Claude I. Dawson, "Fertilizers in Mexico," September 25, 1923, NA, RG 166, Box 148; Paul R. Kelbaugh, "Pan American Cooperation in Agriculture, *Bulletin of the Pan American Union* 73, 3 (March 1939): 160; José L. Colón, "Pan American Cooperation in Agriculture," *Bulletin of the Pan American Union* 75, 6 (June 1941): 330, who claims 1929; Luis Marín, *El cultivo del algodón* (México, D.F.: Talleres Gráficos de la SAF, 1931), p. 42; Bartley F. Yost and Edward S. Maney, "Cotton Prospects in the Torreon District," April 2, 1924, NA, RG 166, Box 147; Ernest A. Gross to Frank Pace Jr., April 21, 1949, NA, RG 59, Document File 1945–1949, Box 5116; Dario L. Arrieta M., "Comisión Internacional de Plagas" (n.d.), pp. 3–4, RFA, RG 6.13, Series 1.1, Box 10, Folder 121; "Reminiscences of Elvin C. Stakman," June 2, 1970, pp. 896–99, RFA, RG 13, RAC; O.C. Trotman, "Report of Inspection Trip along West Coast of Mexico for the Mexican Fruit Worm," August 1930; R.E. McDonald to C.L. Marlatt, April 9, 1929, NA, RG 7, Boxes 1392 and 1396.

24. Mendoza Vargas, "La enseñanza," pp. 37–38; Edmundo Flores, *Historias de Edmundo Flores, autobiografía 1919–1950* (México, D.F.: Martín Casillas Editores, 1983), p. 139; Gilberto Fábila to Pablo Campos Ortiz, June 29, 1929, AHSRE. Exp.IV-629-7; Fernández y Fernández, *Chapingo*, pp. 104, 128–32; Ernest Gruening, *Mexico and Its Heritage* (New York: The Century Company, 1928), p. 527; Alexander Weddell, "Mexico," June 4, 1926, NA, RG 166, Box 147; SAF, *Memoria 1924–1925*, p. 73; Miguel Álvarez Gleason, "La política de irrigación que más conviene a la agricultura Mexicana," in PNR, *Los problemas*, p. 132; SAF, *Informe de labores 1940–1941* (México, D.F.: Editorial Cultura, 1941), p. 16; Gómez, *Biografías*, pp. 33–34, 126, 229, 255–56; Gortari Rabiela, "Educación," pp. 129–30; Martin H. Greenberg, *Bureaucracy and Development: A Mexican Case Study* (Lexington, Mass.: D.C. Heath, 1970), p. 14.

25. Ruiz, *Mexico*, p. 33; Guillermo Gándara, "La Oficina de Botánica Aplicada y de Plantas de Cultivo del Instituto Agronómico Experimental

de Rusia," *Memoria y Revista de la Sociedad Científica Antonio Alzate* 46, 3–6 (1926): 183–92; Gándara, "La hierba de la leche," *Boletin de la Sociedad Científica Antonio Alzate* 52 (1929–1931): 240; "Necrologia," *Agricultura* 11, 13 (July–August 1939): 37.

26. Hall, *Alvaro Obregón*, pp. 108, 219; Aboites Aguilar, "Política de irrigación," pp. 3–7; Christopher R. Boyer, "Old Loves, New Loyalties: Agrarismo in Michoacán, 1920–1928," *Hispanic American Historical Review* 78, 3 (August 1998): 421; Vaughan, *Cultural Politics*, p. 4.

27. Booth, *Mexico's School Made Society*, p. 146.

28. Robles, "Huertas de niños," December 1, 1922, AGN, Robles papers, Agriculture Section, Box 13, Exp. 117; Fernández y Fernández, *Chapingo*, p. 93; Arrieta M., "La educación," p. 253; SAF, *Memoria 1924–1925*, pp. 65–66.

29. "A los trabajadores del campo, a los profesionistas agrónomos," *El Democrata*, September 18, 1923; Mesa A., *Relatos*, p. 13; de Dios Bojórquez, "Momento decisivo," p. 3; Beltrán, "La Dirección," p. 123; Stepan, *"The Hour of Eugenics,"* pp. 67–73.

30. Robles, "Bases para la fundación de una Granja Regional en Córdoba," December 1922, AGN, Robles papers, Education Section, Box 49, Exp. 23; Marco Antonio Durán, *El agrarismo mexicano* (México, D.F.: Siglo Ventiuno, 1967), p. 117.

31. "Organización actual del Servicio de Agrónomos Regionales," March 20, 1924, AGN, Robles papers, Agriculture Section, Box 10, Exp. 93; SAF, *Memoria 1924–1925*, pp. 66–70; "Lo que debe hacer" (n.d.), AGN, Robles papers, Agriculture Section, Box 5, Exp. 70; Fernández y Fernández, *Chapingo*, pp. 99–100.

32. Silva Herzog, *El agrarismo*, p. 333; "Mercados en el extranjero: Historia del cultivo del jitomate (tomate) y de las legumbres frescas de invierno, en la costa occidental" (n.d.), p. 204, ARFF, Box 117; Justo Lozoya L., "Estudio económico-estadístico sobre el cultivo de la alfalfa en la hacienda del PILAR, Edo. de Coahuila," February 25, 1930, ARFF, Box 11; Roberto Gómez Camacho, "Somero estudio sobre la citricultura en Jalisco," in CNAG, *Memoria de la Primera Convención Nacional de Productores de Naranja* (México, D.F.: CNAG, 1936), p. 2; "Sección Informativa," *Agros* 1, 1 (June 1923): 14; Rivas Tagle, *El cultivo racional*, p. 273; DGA, "Campos Cooperativos de Ensayos," September 7, 1931, ARFF, Box 82.

33. González to Agrónomos Regionales, March 10, 1926, AGN, Robles papers, Agriculture Section, Box 5, Exp. 70; Ernesto Rivas Rivera, "Conferencia sobre la importancia de la selección de la semilla," in DGA, *Concurso*, p. 13.

34. Antonio Rivas Tagle, *La selección de la semilla de maiz* (Guanajuato, México: Talleres Linotipográficos del Estado), 1929; Rivas Tagle, *El cul-*

tivo racional, pp. 35–48, 60–93, 285, 293–311; David S. Ibarra, *Enfermedades que se transmiten a las plantas con su semilla y metodos para prevenirlas* (Guanajuato, México: Gobierno del Estado, 1930); Dirección General de Aduanas, "Circular #31-81-151," April 23, 1930, AHJ, Box AG-27-E, Exp. 118(4), Legajo 1; "Translation, Circular #88, Mexico North Western Railway Company," June 11, 1925, NA, RG 166, Box 62; González Gallardo to Relaciones Exteriores, May 20, 1929, AHSRE, Exp. IV-304-39; Carlos Noriega, "La fruticultura como una transformación nacional de la agricultura," in PNR, *Los problemas*, vol. 2, pp. 611–12.

35. Guardiola, *El cultivo del frijol*, pp. 22–23; Rivas Tagle, *El cultivo racional*, pp. 107–28, 225–27; Arturo I. Aguirre, "El acido fosfórico en los suelos y los diferentes abonos fosfatos," unpublished thesis, ENA, 1922, pp. 12–16; SAF, *Memoria 1924–1925*, pp. 66–69; León Fourtón, *¿Que abonos necesitan las tierras? Medios de que dispone un agricultor para saberlo, campos de experimentación* (México, D.F.: SAF, 1930), p. 6; Julio Blumenkron, *Directorio de la industria azucarera en México* (México, D.F.: 1930), pp. 37–38; Robles, "Proyecto de organización de una Estación Experimental para cultivos tropicales" (n.d.), AGN, Robles papers, Agriculture Section, Box 10, Folder 93; Gómez, *Biografías*, pp. 72, 128–29; Alfonso Dampf, "Como defendernos contra las plagas de la agricultura," September 19, 1931, p. 4, ARFF, Box 139; Genero Arzave, "Estado actual de la producción naranjera en el Edo. De Nuevo León," in CNAG, *Memoria . . . Naranja*, p. 8.

36. George I. Sánchez, *Mexico: A Revolution by Education* (New York: Viking Press, 1936), pp. 67–71; Ruiz, *Mexico: The Challenge*, pp. 43–44; Gruening, *Mexico*, pp. 519–20; Emilio Portes Gil, *The Mexican Schools and the Peasantry* (México, D.F.: Press of the Ministry of Foreign Relations, 1936), p. 20; Nathan Whetten, *Rural Mexico* (Chicago: University of Chicago Press, 1948), pp. 481–82; Callcott, *Liberalism in Mexico*, p. 293; Gómez, *Biografías*, pp. 265, 403; SAF, *Memoria 1924–1925*, p. 66; Bulnes, *Los grandes problemas*, p. 157; SEP, "Proyecto"; Mendoza Vargas, "La enseñanza," p. 5; E. Idem, "Nuevas orientaciones sobre las prácticas agrícola," *Agros*, 1, 3 (August 1923): 18; SAF, *La organización agraria en 1927* (Tacubaya, D.F.: Talleres Gráficos de la SAF, 1928); "La ENA en la actualidad se estan haciendo valiosos trabajos de experimentación," (n.d.), in ENA.

37. Mendoza Vargas, "La enseñanza," p. 6; Dana Markiewicz, *The Mexican Revolution and the Limits of Agrarian Reform 1915–1946* (Boulder, Colo.: Lynne Rienner, 1993), p. 42.

38. "Escuelas Centrales Agrícolas, Programa de Estudios," (n.d.), AHSRE, Exp. IV-662-7; "Acuerdo, SAF y SEP," December 18, 1925, AGN, Robles papers, Agriculture Section, Box 5, Folder 70; "Interview with

Fernando Romero Quintana," in Leobardo Jiménez Sánchez, *Las ciencias agrícolas y sus protagonistas* (Chapingo, México: Colegio de Postgraduados, 1984), p. 333.

39. SEP, "Proyecto de la Escuela Rural: Educación campesina y fomento rural," 1932, pp. 3–6; "Memorandum sobre las Escuelas Centrales Agrícolas" (n.d.), AGN, Robles papers, Education Section, Box 49, Exp. 19 and 23; *Exposición del criterio ideológico del Gremio Agronómico Mexicano* (México, D.F.: Confederación Campesina Emiliano Zapata del Estado de Puebla, 1932), pp. 9–10; Harry M. Miller, "Mexican Log," March 2–18, 1941, p. 7, RFA, RG 6.13, Series 1.1, Box 33, Folder 366, RAC.

40. Merilee Grindle, *Bureaucrats, Politicians, and Peasants in Mexico: A Case Study in Public Policy* (Berkeley: University of California Press, 1977), pp. 52–53; Jesús Silva Herzog, *El agrarismo mexicano y la reforma agraria: Exposición y critica* (México, D.F.: Fondo de Cultura Económica, 1959), p. 375; Fernández y Fernández, *Chapingo*, pp. 86, 117; Silva Herzog, "La generación de 1928" (n.d.), in Agrónomos.

41. Division of Inter-American Educational Relations, "Alumni of United States Colleges of Agriculture"; Lorey, *The University System*, p. 43; Robles, "Educación agrícola, propaganda," September 1921, pp. 3–4, AGN, Robles papers, Education Section, Box 49, Exp. 24.

42. León Fourtón, *Utilidad de los analysis de tierras, reglamento del laboratorio de química, nomenclatura de tierras* (Tacubaya, D.F.: Talleres Gráficos de la SAF, 1929), p. 13.

43. Fernández y Fernández, "Apuntes de las clases de entomología," 1925, ARFF, Box 141; Monroy, "Hagamos la avicultura nacional," *Agros* 1, 2 (July 1923): 5.

44. Contreras, "Principios," p. 14; Rivas Tagle, *El cultivo racional*, p. 265; Raziel Mora Olivera, "Porvenir de la alfalfa en el Mexe, Edo. de Hgo.," unpublished thesis, ENA, 1930, p. 20.

45. Gándara, "Desarrollo del cultivo del maiz en México," *Agricultura* 1, 3 (November–December 1937): 6; Pandurang Khankhoje, *Nuevas variedades de maiz* (Tacubaya, D.F.: Talleres Gráficos de la SAF, 1930); Khankhoje, "Somera descripción de los productos obtenidos por el autor en el campo de la Estación Experimental Agrícola de la ENA, desde 1924 a 1930," *Memoria y Revista de la Sociedad Científica Antonio Alzate* 51, 11–12 (1929–1930): 447–59; Alejandro Brambila Sr., *El analysis químico y la fertilidad de los suelos* (México, D.F.: DGA, 1928), p. 6; Beltrán, "La Dirección," pp. 111–15; Beltrán, "Contribución al estudio del Coccobacillus acridiorium d'Her," *Memorias y Revista de la Sociedad Científica Antonio Alzate* 46, 3–6 (1926): 129–531; Olivero Téllez, "La utilización industrial de la langosta," *La Hacienda* 20, 4 (April 1925): 125; SAF, *Memoria 1924–1925*, pp. 37–49; Moises Herrera, "Los escorpiones de México," *Memoria y Revista de la Sociedad Científica Antonio Alzate* 30, 1–6 (January–June

1921): 137–59; Fernández y Fernández, *Chapingo*, p. 99; Soberón to Oficial Mayor, Secretaría de Relaciones Exteriores, May 10, 1929, AHSRE, Exp. IV-249-2; Alejandro Brambila Jr., "Notas de la Granja Experimental del Sistema Nacional de Riego 'Presdiente Calles,' " Aguascalientes, *Irrigación en México* 1, 6 (October 1930): 81–84; Brambila Jr., "La granja experimental en Rodríguez, N.L.," *Irrigación en México* 1, 1 (May 1930): 44–48; Rivas Tagle, *El cultivo racional*, p. 214.

46. Sánchez Durón, "La investigación," p. 176; Adolfo R. Diéguez to J. Luis Peña, August 6, 1935, AHJ, Box AG-27-F, Exp. 61.33, Legajo 2; Lázaro Cárdenas to Juan Andreu Almazan, November 16, 1935, AGN, Cárdenas papers, Exp. 565.4/70; SAF, *Monografía de la Naranja* (México, D.F.: 1936), p. 1–6; SAF, *Características de los trigos cultivados en la República Mexicana* (México, D.F.: D.A.P.P., 1937), p. 46; SAF, *Memoria 1924–1925*, pp. 14, 66–69; "Organización actual," *Extensión Agrícola* 1, 1 (January 1933): 1; Gómez Camacho, "Somero estudio," p. 2; Noriega, "La fruticultura," pp. 611–12; Manuel Salazar y Arce, "Estudio sobre el cultivo de la papa en la República Mexicana," December 25, 1934, ARFF, Box 136; Gabriel Ítie, "Producción nacional e importación de la papa para semilla," in CNAG, *Memoria . . . papa*, p. 96; Brambila Jr., *El analysis químico*, p. 12; Drew Linard, "Increased Taxation and Agrarian Activities, Piedras Negras District," October 23, 1926, NA, RG 166, Box 148; Raleigh A. Gibson, "Annual Review of Commerce and Industries, Guadalajara," 1930, NA, RG 166, Box 149; R. E. McDonald to C. L. Marlatt, October 22, 1929, NA, RG 7, Box 1398; Gómez, *Biografías*, p. 138; Scott, *The Reluctant Farmer*, p. 169.

47. Enrique Beltrán, "Las investigaciones científicias en México, su ratiquismo actual y manera de promoverlas," *Memoria y Revista de la Sociedad Científica Antonio Alzate* 47, 1–4 (January–April 1927): 111–15.

48. Rivas Tagle, *El cultivo racional*, pp. 285–89; "Proyecto para la creación de un departamento autónomo para fomento de maiz" (n.d.), "Datos generales sobre el establecimiento de una Estación Experimental para el cultivo de la caña y cultivos que entran en su rotación" (n.d.), p. 2; Mesa A., "Memorandum sobre la ENA," August 15, 1933, p. 7, AGN, Robles papers, Agriculture Section, Boxes 13, 10 and 5, Exps. 118, 93 and 70; León, "Discurso," p. 20; Fernández y Fernández, *Chapingo*, pp. 111–18; Enrique Beltrán, *Medio siglo de recuerdos de un biólogo mexicano* (México, D.F.: Sociedad Mexicana de Historia Natural, 1977), p. 132; José Luis Calva Téllez and Carlos Monzón, "La economía agrícola mexicana y el consumo nacional de fertilizantes," in Jesús Gracia Fadrique, ed., *Estado y fertilizantes* (México, D.F.: Fondo de Cultura Económica, 1988), pp. 96–97.

49. Alberto Arce, *Negotiating Agricultural Development: Entanglements of Bureaucrats and Rural Producers in Western Mexico* (Wageningen, the Neth-

erlands: Agricultural University Wageningen, 1993), p. 43; "Directorio de Socios," *Boletín Interior* 1, 1 (August–September 1930): 3–40; Gómez, *Biografías*, pp. 16, 53, 130–31, 193, 328–33, 336–37, 348, 350–51, 362–63, 368, 392, 459, 471–73, 521–22; Greenberg, *Bureaucracy and Development*, p. 52.

50. Scott, *The Reluctant Farmer*, p. 217.

Chapter 3

Revolutionary Agronomy and the Cultural Campaign, 1930–1937

During the late 1920s economic difficulties, bad weather, and rural conflict stimulated debate about Mexico's agricultural policies. Joined by *Jefe Máximo* Calles in 1929, rightists in and outside the Revolutionary Family wanted to end the agrarian reform to give private farmers incentive to innovate, but leftists demanded more land distribution and an intensive campaign to transform peasant agriculture. The *agrónomos* allied with the Left, which triumphed in 1935, when Cárdenas exiled Calles. Five Revolutionary Family members, none of them *agrónomos,* led the SAF: Manuel Pérez Trevino (1930–1932), Francisco S. Elias (1932–1934), Tomás Garrido Canabal (1934–1935), Saturnino Cedillo (1935–1937), and José G. Parres (1937–1940). The *agrónomos'* project for the *campesinos* borrowed heavily from U.S agronomy, as well as from Soviet agricultural policy and science. Contrary to several scholars, it was not a reaction against foreign science that inspired them to study Mexican farming technologies or methods; rather, as geographer Carl O. Sauer observed in 1941: "[T]he little agricultural work that has been done by experiment station people have so far been making that very mistake, by introducing U.S. forms instead of working on the selection of ecologically adapted native items."[1]

Some academics stress the populist goals of Cárdenas' agricultural policy, which used loans, fruit trees, and experimentation stations to win *campesinos'* support. Other scholars, including a few *agrónomos,* praise his "vision of a prosperous rural Mexico" with "peasant communities with access to land, credit, technical assistance, and social services" and compare it to more recent ideas like Integrated Rural Development.[2] Historian Friedrich Schuler claims that *cardenista* bureaucrats viewed the transformation of peasant farming methods and culture as integral steps toward a State-directed economy. Other scholars argue that his programs often alienated *campesinos* by attacking their traditional culture. The *agrónomos*, like most other *cardenistas*, did not come from peasant families; they were outsiders who wanted to transform *campesino* farming and culture to meet the needs of a modernizing capitalist society, while advancing their professional objective of sustaining their alliance with the State.[3] To conduct the cultural campaign, the SAF used new and old projects and methods, including the *Instituto Biotécnico*, a central research institution, regional experimentation stations, extension agents and local offices, publications, farmer and rural youth organizations, and contests and fairs.

THE GREAT DEPRESSION: A PROFESSIONAL CRISIS AND A NEW CULTURAL CAMPAIGN

Calles admired the prosperity of U.S. farmers. A trip to France reinforced his negative opinion of peasant farming, U.S. ambassador Dwight Morrow stimulated his interest in scientific agriculture, and the Great Depression exacerbated Mexico's economic crisis. Joining rightists like the private farmers of the *Unión Agrícola de Jalisco*, Enrique Creel, the *Banco de México*, and the Mexico City daily *Excelsior*, Calles called *agrarismo* a failure and claimed that land tenure insecurity and the *ejidatarios'* poverty and ineptitude caused Mexico's low farm productivity. Other Revolutionary Family members such as Luis Cabrera, President Pascual Ortiz Rubio (1930–1932), and Luis L. León wanted to end the land reform. Some critics denounced the *agrónomos*, like the daily *El Universal*, which called them inept and corrupt. Javier Uranga M., a student of Native American culture, claimed that *agrónomos* felt that farm labor was beneath their social status and that the lack of competent, responsible experts ruined agricultural programs and eroded farmers' faith in the profession and the SAF. Eduardo Villaseñor of the *Banco de México* said that middle-class *agrónomos* did not understand *campesinos'* problems. The SEP admitted that rural teachers did not know how to select seed or grow vegetables, acknowledged that the vast nature of rural problems overwhelmed the Cultural Missions, and claimed that *agrónomos* liked urban office jobs and hated working with peasants in rural areas. In 1930 at least 105 of 296 SAM members lived in Mexico City.[4]

From 1931 to 1937, 95 *agrónomos* graduated from the ENA, and another 107 earned degrees from Antonio Narro. In 1934, 109 EPA *pasantes* (students who finished their course work but not their thesis) were ready to join the school's 189 agronomists who completed the degree. Realizing that attacks on their professional credibility and the end of land reform threatened their alliance with the State, *agrónomos* rebutted their critics. The programs of the 1920s convinced them that *hacendados* did not accept their claim to professional expertise. They broke with Calles and León, lamented the *campesinos'* "tremendous resistance to adopting modern agricultural systems," but supported the *agraristas* by advocating more land reform.[5] To critics who told *agrónomos* to abstain from politics, Mesa A. replied that "to limit our actions to technical works, to studying and solving the scientific problems of agriculture, is tantamount to condemning all prior agricultural policy."[6] In 1933 and 1934 *agrarismo* created jobs for *agrónomos* when President Abelardo Rodríguez (1932–1934) ordered new surveys of rural properties and even more after 1935, when Cárdenas accelerated land distribution. For example, in 1934, 73 agronomists from the EPA worked for the CNA. *Agrónomos* intervened in ejidal politics, such as Augusto Hinojosa, who helped new officials in the *ejido* La Ladera take office, and assisted Cárdenas' project to unify independent peasant leagues, like León García, a member of the Organizing Committee of Peasant Unification who presided at a 1936 *agrarista* convention in Coahuila.[7]

Agrónomos tried to avoid guilt by association with the SAF by calling it an underfunded "asylum of petty politicians" where "generals and colonels" ran ludicrous programs and insisted that the agency stop hiring such "terrible burdens" who diminished the profession's prestige.[8] To replace them, with Lombardo Toledano's support they urged the State to define agronomy as a bureaucratic profession by claiming that the small farmers and *ejidatarios* that it created could not afford their services or understand why they were beneficial. Like Colombia's *técnicos*, *agrónomos* demanded government jobs and claimed that hiring foreign scientists insulted national pride and opened Mexico to exploitation by "charlatans and adventurers." Appealing to the architects of Mexico's racial identity, *agrónomo* Bernardo L. Arrieta M. lamented that "black people in the U.S., always held in disdain," had access to 25 times more extension agents than Mexican farmers.[9] To improve the SAF's programs, in 1933 Rodríguez formed a committee and appointed Beltrán, who had just completed a Ph.D. in zoology at Columbia University, to evaluate research and extension. The Revolutionary Family used his ideas in the *Plan Sexenal* (Six-Year Plan), policy guidelines for the next administration, which ordered the SAF to create a new *campesino*: "[T]he maximum farm, fishery, and forest production means nothing to our national consciousness if it is not accompanied and in some cases preceded by the biological improvement, economic and cultural advance, and social redemption ["*revaluacion*"] of our rural masses."[10]

The SAM supported Beltrán's proposals, recognizing that they meant jobs and a stronger alliance with the State. *Agrónomos* promised that their science would solve the *campesinos'* social, economic, and political problems by establishing intensive agriculture in the *ejido* and proposed research and extension programs for many crops, including bananas, beans, cacao, coffee, corn, cotton, forage grasses, fruit trees, henequen, rice, sugarcane, and wheat. To redeem the land reform, some rightists also urged the government to make the *campesinos* productive farmers, and a few said that scientific agriculture would alleviate rural unrest. *Agrónomos* knew that *hacendados* did not seek or follow their advice, but transforming *campesinos* into scientific farmers would validate their claim to professional expertise and status. For example, ENA professor Emilio Alanis Patiño claimed that "maximum contact with *campesinos*" would establish the "technical authority" of extension agents.[11] Cárdenas accepted their arguments, and the SAF, agricultural banks, and Ministry of Finance and Public Credit hired more *agrónomos*. Responding to their critics, Fernández y Fernández and other *agrónomos* from the ENA tried to tighten professional standards by claiming that EPA and Antonio Narro graduates had an inadequate scientific education, came from affluent urban families, and wanted to be farmers rather than bureaucrats. They urged Cárdenas to forbid them from practicing agronomy and close these schools, but the campaign accomplished little other than to divide the profession and arouse the ire of EPA and Antonio Narro students and *agrónomos*.[12]

Agrónomos and other scientists continued to look abroad to find examples of the cutting edge of agronomy to bring to Mexico. To enhance their scientific knowledge, agricultural students asked Cárdenas to send them to countries that used state-of-the art farming techniques, and *agrónomos* urged Mexican farmers to emulate them, such as the practices of Colombian and Hawaiian coffee growers. Antonio Narro hired agronomist Egidio Gaetano Rebonato, an Italian anti-fascist refugee, to teach. In 1932 Guillermo Gándara claimed that French agronomy reigned supreme during the discipline's taxonomic phase, but U.S. and Soviet scientists assumed leadership with their work on plant genetics. Riquelme Inda thought that the USDA's Bureau of Entomology led the world in pest control but also praised Soviet programs and called France "a great agricultural nation."[13]

As anthropologist John Gledhill notes, Soviet agricultural programs like Stalin's collective farms influenced the *cardenistas'* project for the *campesinos*. Agricultural scientists like Gándara, a member of the Leningrad Institute of Applied Botany who praised the Russians Vavilov, Veronov, Jakushkina, and Pangalo, maintained their connections established in the 1920s. In 1932 the U.S.S.R. hired ENA student Elias Kesselbruner, son of Russian immigrants, to conduct a botanical survey in South America, and he donated some specimens to the school. Believing that small-scale farming could never be productive, some *agrónomos* wanted to create collective *ejidos* to retain an economy of scale permitting the use of machinery and improved methods. For example, the BNCA thought that collectivization would foster standardizing Mexican wheat varieties. Like Soviet agronomists of collective farms, economist Mario Sousa said, "each application of technology brings real improvement to the workers"; he thought that Mexico "must" collectivize and praised Stalin for transforming peasant farmers not "against their will, but rather with organization, credit, tractors, and great experimental fields" and "a technical education policy that changes their material and psychological conditions."[14] Dampf lauded Soviet pest control, particularly aerial spraying, and Portes Gil admired Russian teachers who eradicated "ignorance" by convincing peasants that science explains "all phenomena of life and nature."[15] In 1935 Lombardo

Toledano urged professionals to abandon their bourgeoisie outlook and serve the masses. In 1936 agricultural students asked teachers to discuss "the overthrow of capitalism" and "the great capitalist powers' harmful impact on the development of the national economy," and *agrónomos* interested in Marxism and the U.S.S.R. founded the *Liga de Agrónomos Socialistas* (Socialist Agronomists League, hereafter LAS); ENA student Edmundo Flores joined, read Marx and Lenin, and "converted" as "before, as a child, I had accepted the principles of Catholicism."[16] Prominent *cardenistas* like Ramón Beteta denied that *cardenismo* was socialism, which persuaded historian Victor Alba, but their statements about socializing the land and pictures of Lenin on classroom walls convinced rightists like Cabrera and U.S. journalist Victor F. Calverton that they were Marxists. Similar to Soviet scientists and intellectuals, *cardenistas* thought that science could transform peasant culture, urged scientists to serve the classes redeemed by the Revolution, and advocated scientific autonomy.[17]

To secure raw materials and diplomatic influence, Germany, Italy, and Japan also sought contacts with Mexico's farmers and *agrónomos*. Germany purchased coffee, cotton, *Ixtle* fiber, and rice, and Mexican plow manufacturers and henequen growers bought German machinery. In 1936 and 1937 Mexico imported most of its seed potatoes from Germany. Berlin's *Ibero-Amerikansches Instit* collected agricultural journals like *Jalisco Agrícola y Ganadero*. In 1933 *Agrónomo Regional* Pablo Córdova tested chemical fertilizers in Veracruz with the help of Hans Burbach, a representative of the German-owned Beick, Felix y Cia. Cárdenas asked German engineers about the viability of making fertilizers from the *Gran Canal del Desague*'s black waters. In 1937 Germany's embassy sent a delegate to the National Coffee Growers Convention. Alañis Patino and *agrónomo* Gilberto Loyo studied statistics at the University of Rome. Cárdenas hired two Japanese aquatic biologists to run the Lake Patzcuaro Fisheries Station, and Japanese immigrants established chile cultivation near Ensenada, Baja California.[18] These connections had little impact, but U.S. diplomats worried as international relations deteriorated.

U.S. agronomy continued to influence Mexican agriculture. Farmers grew U.S. rice, cotton, tomato, orange, and potato varieties. When Sonora's governor, Rodolfo Elias Calles, founded the El Yaqui Experimentation Station, he copied research institutes that he visited in Arizona, Texas, and California and obtained seeds from the USDA. With Sonora leading the way, Mexican farmers obtained USDA literature or advice about corn, cotton, forage grasses, wheat, insect pests of apple trees, and growing melons for export. According to Frank Reeves, who had an Iowa State degree, *rancheros* who lived near his hacienda in Durango often asked him for technical advice. The BNCA obtained USDA information on a wide variety of crops, and Dampf secured a report on wheat flag smut in the Midwest. Mexican rice growers praised Iowa hybrid corn and the new rice strains developed at Louisiana's experimentation stations, small-scale industries expert Jesús Echavarrí lauded U.S. beekeepers, and *Jalisco Rural* admired U.S. farmers' prosperity, urging Cárdenas to foster rural peace and land tenure security to bring it to Mexico. Like anthropologist Manuel Gamio, *agrónomos* thought farmworkers deported by the United States during the Great Depression (*repatriados*) could improve agriculture by teaching *campesinos* how to farm like farmers north of the Rio Grande. Discussing northern Mexico's farmers, José Lázcano Romero praised their "more modern" agriculture, a result of "proximity to the U.S.," which "facilitates buying modern machinery cheaply, and because the teachings of the experimentation stations in the south of that great country apply to much of this region," and the Escobar brothers said that the EPA trained better *agrónomos* than the ENA because of its irrigated lands and location near El Paso, Texas.[19]

Contact with U.S. scientists at meetings like the 1930 First Inter-American Conference on Agriculture, the New Jersey Experimentation Station's fiftieth anniversary celebration, which Pascual Gutiérrez Roldan attended, and the 1935 Seventh Inter-American Scientific Congress in Mexico City, to which Garrido Canabal enthusiastically sent SAF and *Instituto Biotécnico* experts, stimulated *agrónomos'* interest in research. In 1930 and 1931 ENA director Juan Manuel Corona Morales established contacts with professors at prestigious U.S. agricultural colleges, and the SAM offered an English course. Juan González brought W.A. Lloyd of the USDA's Extension Service to Mexico and urged the *Agrónomos Regionales*, who told farmers to use the dry farming techniques of the western U.S., to emulate their U.S. counterparts. In 1933, to promote fruit production, Mexico's National Railways hired a U.S. expert. When the J.G. White company's contract expired in 1934, the CNI retained two engineers as consultants. In 1937 nationalists criticized the BNCE for hiring U.S. cotton expert J.H. Harrison, formerly of the Anderson & Clayton Company, but the bank told Cárdenas that he "only gave us his opinions about the *ejidatarios'* farm work."[20]

The government sent SAF employees to the United States for advanced training. From 1930 to 1937, 19 Mexicans studied agriculture at U.S. universities. Seven completed degrees, including *agrónomo* Eduardo Limón Garcia, who earned an Iowa State M.S. in animal husbandry in 1935. Samuel Torres Elorduy, who worked with the USDA during the 1920s, taught plant genetics at the ENA. The SAF and Chihuahua sent *Agrónomo Regional* Manrique González to the United States to study tree nurseries, and he helped the state's growers win first prize at the 1932 Mexico City agricultural fair. That year *agrónomo* Edmundo Taboada R. was Mexico's agricultural attaché in Washington, but his main task was taking classes at Cornell University for his M.A. degree. The SAF sent veterinarian Alfredo Téllez Girón to the United States to study viral livestock diseases, and the ENA paid two students to tour farms in Wisconsin. In 1936 Cárdenas commissioned *agrónomos* Miguel Santillán to discuss plans for a nitrate plant in Las Vigas, Veracruz, with an U.S. engineering firm, and Rene A. Becerra to visit plow factories and the USDA's Agricultural Machinery Experimentation Station in Toledo, Ohio. Dampf won a Guggenheim Fellowship to study insects with the USDA in Washington.[21]

The USDA–SAF pink bollworm and fruit fly campaigns continued. In 1932 the SAF asked the USDA to send bollworm inspectors to the Mexicali Valley, extending the work beyond the Texas border region. On several occasions the SAF ignored farmers who wanted to raise a second cotton crop from their field trash (*zoca*), which facilitated the pest's spread, and accepted the USDA's recommendation to force all growers to clear their fields. The USDA helped the SAF test biological methods to control the pink bollworm. In 1934 Baker searched for fruit flies in Mexico's northwest with Dampf and in Jalisco with Andres Macias of the State's Department of Agriculture. In 1936 four scientists worked at the USDA's Mexico City laboratory, collaborated with SAF pest control specialists, and tested new methods at the Hacienda Santa Engracia in Tamualipas. Thinking that the wind blew the flies across the border, the USDA helped the SAF install traps in Nuevo León. The SAF introduced a parasitic wasp to begin a biological control program but failed to assign enough personnel to raise and distribute them. In 1936 the USDA gave the SAF parasites of the citrus black fly to release in Sinaloa. The USDA wanted to protect U.S. farmers, which sometimes worked against Mexico's interests, such as locating the fruit fly laboratory in Mexico City, far from the orange groves in the north, but Mexican growers and *agrónomos* who worked with USDA employees supported these projects. As a result and in recognition of mutual benefit, in 1934 the United States and Mexico established the International Plagues Commission, which met for the first and only time in October 1935. SAF–USDA contacts led to collaborative studies of selenium poisoning of livestock in Chihuahua and of *chía*, an oilseed plant used to make paints and varnishes.[22]

Henry A. Wallace, Franklin Delano Roosevelt's secretary of agriculture, worried that the Great Depression and international conflict might disrupt U.S. access to rubber and other Asian and African tropical farm goods. He found Latin American agriculture and Mexico's land reform fascinating, and he wanted to promote trade with the region to address both problems. Orange growers and congressmen complained about using taxpayer funds to help Mexico fight fruit flies, and Wallace had to wait for an opportunity to expand the USDA's presence south of the Rio Grande, but private citizens like Carl Alsberg of Stanford's Food Research Institute and L.S. Rowe of the Pan American Union watched the development of agricultural research in Mexico with interest. The San Francisco, Los Angeles, and San Diego Chambers of Commerce gave banana trees to the State of Guerrero, which distributed them to *campesinos*, and promised to buy the fruit. In 1935 a U.S. promoter convinced a bank to test linseed in the Mexicali Valley. U.S. consuls knew that the irrigation districts were markets for farm machinery, and manufacturers' representatives such as the Cleveland Tractor Company helped agricultural banks open new lands to commercial farming. In 1933 SAF chief Elias and Department of Public Health officials reacted favorably when John A. Ferrell of the Rockefeller Foundation's International Health Division (IHD) proposed an agricultural assistance program, but the trustees thought that they already had too many financial commitments. In 1936, seeing "no serious controversial issues," Ferrell suggested an agricultural survey and claimed that Cárdenas would support it.[23] To Diego Rivera, the Rockefellers were enemies of the Revolution, but Cárdenas and *El Universal* appreciated the foundation's contributions to Mexico's public health programs. The trustees rejected Ferrell's idea, in historian Bruce H. Jennings' opinion because of Cárdenas' attacks on U.S. economic interests, but the foundation's records do not mention this concern. The scholarships that the foundation granted to Mexican health care professionals for advanced study in the United States before and after the oil expropriation and Ferrell's 1941 statement that "the Rockefeller name" was "identified throughout Mexico . . . with the Foundation," whose "relationship with Mexico was not strained even when the country had acute issues with the U.S. and its corporations," suggest that financial commitments and the policy of working with elite scientific institutions also influenced this decision.[24]

CREATING A NEW *CAMPESINO*

To construct their project for the *campesinos*, *agrónomos* used U.S. and Soviet agronomy, Porfirian stereotypes, the revolutionary anti-clericism of the SEP's socialist education program, and notions from public health, nutrition, and eugenics. To make a new Mexico with *campesinos* who lived in the twentieth-century world, socialist educators propagated "scientific rather than metaphysical explanations of the universe" to fight "docile acceptance" of "local legends" and "religious beliefs and superstitions" professed by priests.[25] Rafael Ramírez designed the Cultural Missions Program to create "men without religious prejudices," Sub-Secretary of Public Education Gabriel Lucio said that socialist schools used "observation and experimentation" to help students "shed prejudices and fanaticism," and Beltrán believed that teaching *campesinos* biology would replace "anti-scientific prejudices that seek supernatural explanations and laws for individual and collective life" with "a clear and precise concept of biological determinism" and eradicate erroneous folk beliefs that hindered their ability to profit from nature.[26] In Villahermosa a teacher conducted this experiment: "In this first plot, we will plant no seed, but every day we will pray God to give us beans and corn. In the second plot, we will plant seed and tend to it carefully, but we will not pray over it. Then we will see which will do the most for us, God or our own individual efforts."[27] School gardens became battlegrounds in a war on *campesino* culture. As historian George Booth and socialists Nathaniel and Sylvia Weyl recognized, modernizing peasant farming was integral to the *cardenistas'* campaign.[28]

Making an argument that appealed to *agrónomos*, Gamio and Portes Gil said that socialist education would convince *campesinos* to consult professionals to solve their problems instead of priests or *brujos* (sorcerers), like city people who saw medical doctors when they were ill. Reflecting *agrónomos'* disdain for traditional rural culture, many called the *campesinos'* lack of education a significant obstacle for agricultural programs. Like the SEP, the SAF and its *agrónomos* joined the State's campaign to transform the peasants' worldview by using science to break their bonds to the Catholic faith and clergy. The SAM ridiculed the Apostolic Roman Catholic Church, S.A. (in liquidation). In 1931 *El Nacional Revolucionario*, a newspaper run by recent ENA graduate Marco Antonio Durán, even suggested that people might accept fertilizers made from human remains. This anti-clerical campaign peaked when Garrido Canabal ran the SAF. In 1935, claiming that the Church kept peasants in "darkness" (*obscurantismo*), *agrónomo* Jacobo Aragón Aguillón told reporters that the SAF's experimentation stations "show *campesinos* the absurd advice of priests, such as sprinkling holy water on the soil to ensure a good harvest or excommunicating worms that devour crops, are monstrous lies," and Garrido Canabal ordered ENA professors to teach students "the process of the religions and the harm they have brought to workers."[29] He also ordered all male SAF employees under 30 to join his Red Shirts, who covered the floor beneath the *Instituto Biotécnico*'s time clock with religious symbols and forced the staff to step on them when they punched in, threatening to fire anyone who refused. One employee fainted, chemist Pablo Hope y Hope quit, and the discontent that these measures caused made Beltrán resign, notwithstanding his own anti-clerical beliefs. Garrido Canabal fired SAF *agrónomo* Leonel Lemus for refusing to cooperate with the Red Shirts, but the SEP, another anti-clerical agency, hired him to run an agricultural school. In 1935 Cárdenas dismissed Garrido Canabal and appointed the pro-Catholic Cedillo. The SAF's anti-clerical campaign waned, but some *agrónomos* kept the faith, and some agricultural students still wanted to "destroy the clergy."[30] In 1937 the SAM claimed that "it is wrong to think bad luck intervenes decisively in farming; if one looks for the causes of failure, they will surely find lack of effort, intelligence, and careful work, not the negative force of fate," and veterinarian José Figueroa, director of the *Instituto Biotécnico*, wanted to study plant diseases and insect pests to fight "old superstitions" like the belief that they were a type of divine retribution that *brujos* could stop.[31]

The tortilla discourse exerted a strong influence on the *agrónomos'* cultural campaign. Intellectuals like Gamio, medical doctors, and the Ministry of Public Health believed that Mexico's corn-based diet was nutritionally deficient, and Beltrán wanted to change it "to improve the racial condition of our people."[32] *Agrónomo* Alfredo Valle Rivera, an ENA professor, joined the Mexican Eugenics Society. *Agrónomo* Francisco Quintanar denounced the "slavery of corn" that made "Indians negligent" and criticized people who "protect maize, as if it was desirable to keep our Egyptian plows, oxen, burros, adobe houses, etc."[33] Economists with the National Railways and the Ministry of Communications urged peasants to eat bread instead of tortillas and wheat growers to modernize so the poorest Mexicans could afford it. *Instituto Biotécnico* researcher Alfonso Ramírez Paulín claimed that a diet of sunflower seeds made Northern Europeans industrious and told Mexicans to grow and consume them. SAF fruit tree expert Carlos Noriega thought that the *campesinos'* diet caused nutrition-related diseases that eating fruits could prevent. To improve it, the SAF gave peasants vegetable seeds, and experimentation stations and teachers planted gardens to demonstrate cultivation practices. The *Instituto Biotécnico* conducted many nutrition studies such as testing the protein content of milk, eggs, and meat. In 1934 *Agrónomos Regionales* promoted sorghum as a drought-resistant alternative to corn that was suitable for human consumption. In the *ejido* Villa Reyes, San Luís Potosí, the SAF banned tortillas and fed *ejidatarios* bread while testing 32 imported wheat varieties. The agency also urged peasants to plant and eat potatoes. Cárdenas took personal interest in the SAF's sunflower campaign, which began in 1937. Experimentation stations tested their oil content and fed them to poultry, and *Agrónomos Regionales* urged *ejidatarios* to plant them. Reflecting the SAF's interest in domestic hygiene and other issues affecting *campesinos'* health, the *Instituto Biotécnico* wrote and distributed a pamphlet on methods to control pests that infested their homes. *Agrónomo* Alberto A. Gutiérrez gave lectures on personal hygiene in *ejidos* near the La Bartollila Station in Michoacán. The Tlalnepantla Station told *campesinos* to plant other crops to replace maguey, the source of *pulque*, a popular but mildly intoxicating beverage. In a plan that would not pass muster today, to fight alcoholism, Dr. Baltazar Isaguirre Rojo wanted peasants to grow grapes and drink wine instead of *pulque*. In an interesting spin on the tortilla discourse, sugar industry spokesman Julio Blumenkron urged *campesinos* to eat sweets to add energy to their diet.[34]

Emphasizing how foreign capital used science for profit and left Mexicans impoverished, as well as the problems of relying on imported methods to solve agricultural problems, Beltrán wanted the SAF to conduct more research. Many *agrónomos* agreed, and some said that programs must be constructed on a foundation of Mexican research because foreign scientists worked in different ecological and socioeconomic conditions. For example, Gándara claimed that only "rich *hacendados*" benefited from foreign research, but *campesinos* could use the results of studies conducted in Mexico. In 1931, recognizing the job opportunities, Mario Macias V. decried the "error" of using imported science and urged the State to establish 200 experimentation stations, "each run by an *agrónomo*."[35] Taking the argument to its most extreme, in 1934 DGA director *agrónomo* Gumaro García de la Cadena refused to let the *Instituto Biotécnico* publish a journal to exchange with foreign institutions because he thought that its scientists would translate the articles rather than conduct their own research. Pointing to the ecological and financial benefits of traditional technologies like the "*macana*" (digging stick) in comparison to mechanized farming or iron plows on rocky or sloping fields, French immigrant *agrónomo* Gabriel Ítie recognized:

> Any system of cultivation, no matter how rudimentary, is the product of long observations by many generations of farmers, a real adaptation to conditions that gains value unconsciously and eliminates or mollifies many adverse events, producing a state of equilibrium that, if it does not produce the highest yields or the best crops, provides the best guarantee of obtaining harvests year after year.[36]

He urged the SAF to study the *campesinos'* methods before trying to modernize them, but the *agrónomos*, steeped in western science and urban upper- and middle-class attitudes, ignored his suggestions, believing that peasants had nothing to offer their project except cooperation.

Instead, to establish their status as scientific professionals, *agrónomos* tried to invalidate hacienda and peasant agriculture. Attacking *hacendados'* extensive farming, the CNI wanted to eradicate "the erroneous traditions of the Conquistadores," and Gándara claimed that *campesinos* used practices "from the days of Cortez."[37] *Agrónomo* José R. Rosso said that *hacendados* and *rancheros* mistreated their livestock, and his colleagues chastised bean, cacao, coffee, corn, rice, cotton, orange, potato, strawberry, sugarcane, and wheat growers for selecting their seed improperly or allowing introduced varieties to mix. They also criticized native crop strains and urged farmers to purchase certified seed. According to *Campaña en Pro del Maiz* apostle Rivas Tagle, "we use seeds that, without exaggeration, would not even be fed to pigs in other countries."[38] *Agrónomos* called the cultivation practices of farmers in the Guadalajara region and growers of bananas, cacao, coconuts, coffee, corn, oranges, sugarcane, and wheat backward. Armando Martínez Aguilar said that the *campesinos* of Ciudad Guzmán, Jalisco, farmed "with no rational plan" and did not "prepare or try to improve the land or conserve or restore its resources."[39] The SAF claimed that corn-bean intercropping promoted insect infestations and made them impossible to control, *agrónomo* Romulo Delgado Crespo chastised mango cultivators' intercropping system, and Sixto Colina Robles told the orange growers of Coatepec, Veracruz, not to plant coffee and banana trees in their orchards. To *agrónomo* Luis Hermán, "only the poorest or most reactionary *campesinos* still use wooden plows and oxen."[40] Traditional farming practices did not become part of "the new tapestry of folkloric nationalism" woven by the post-Revolutionary State-builders.[41]

Hoping to improve the quality of SAF research, Beltrán became the *Instituto Biotécnico*'s first director, but he had problems finding people whom he considered qualified scientists, and he hired only two *agrónomos*. In 1934 several of the most prominent *agrónomos* urged the SAF to make cereal grain genetics its first research priority, but other agronomists wanted to develop improved cacao, coffee, and orange varieties. Reflecting Cárdenas' populist agenda, in 1935 the SAF published *Lo que significa el Instituto Biotécnico para el campesino* (What the *Instituto Biotécnico* Means for the Peasantry). Beltrán wanted the *Instituto* to direct the SAF's regional experimentation stations and to found only three, but under the DGA's control they proliferated. Garrido Canabal created six, and four were in Tabasco, his home state. In November 1935 Cedillo announced a plan to establish 200 stations by the end of 1936. This was unrealistic, but he founded three before Cárdenas fired him, two in his home State of San Luís Potosí, and he ignored suggestions to close several so that the others could operate efficiently. In 1937 Parres planned to open more stations. The SAF began a few plant breeding projects. In 1933 Khankhoje gathered seeds from *ejidatarios* who grew *maiz granada* but let it cross-pollinate while he was in Europe, and he purified the strain again in 1935. In 1937 the SAF tested his corn at the Oaxaca Station but did not begin an increase program, ignoring Khankhoje's pleas. In 1938 he took a job promoting fruits and vegetables for Mexico's South Pacific Railroad, leaving his apprentice, *agrónomo* Gonzalo Torres Barusta, to breed maize for the SAF. Working with *hacendados* in the Bajío, in 1933 *Agrónomo Regional* David S. Ibarra created a new corn variety. In 1934 ENA professor Juan Rulfo crossbred wheat to show students the procedures. *Agrónomo* Miguel E. Shultz created two wheat strains at the ENA and the Tlalnepantla Station. One gained some acceptance from farmers in Mexico State, but like Khankhoje's corn, the SAF did not increase the seed. In 1937 the *agrónomos* who ran the Pabellón Irrigation District began to crossbreed chile.[42]

Ítie admired the wisdom of *campesino* plant breeders, calling Mexican strains "the product of an unconscious and very old selection" superior in some ways to imported varieties, and he warned his colleagues that foreign hybrids did poorly in Mexico because they needed chemical fertilizers and "very controlled environmental conditions," but the SAF ignored him.[43] Instead, with faith in foreign, especially U.S., plant breeders, the *Agrónomos Regionales*, experimentation stations, CNI, and agricultural banks tried to convince *campesinos* to plant imported corn, forage grasses, grapes, linseed, peanuts, peas, rice, sorghum, strawberries, sugarcane, sweet peas, tomatoes, watermelons, and wheat, and citrus, deciduous, and tropical fruits. U.S. varieties were most popular, but the SAF also got seeds from the U.S.S.R., Germany, Morocco, Canada, the Dutch East Indies, Denmark, Italy, France, China, and Spain. In one program, the SAF's *Agente General* in La Paz, Baja California, supervised farmers who planted a U.S. tomato variety, giving lectures and distributing literature on cultivation practices, pest and disease control, and fertilizer use. *Agrónomo* Carlos Becerril was proud of the high yields of La Laguna cotton growers, who planted seed imported from the United States.[44]

Most *campesinos*, however, had little cash or faith in businessmen and did not buy their seed; they selected it in the field using traditional criteria based on folk knowledge. Invalidating this as backward "superstition," *agrónomos* told peasants to use the *Campaña en Pro del Maiz* to choose their seed and extended it to beans, rice, and wheat. Denying agency to luck, fate, or divine intervention, Rivas Tagle said that "plants, like human beings and animals, inherit their qualities and defects from their ancestors."[45] The SAF distributed bulletins on seed selection, storage, disinfection, and germination testing to *ejido* officers, municipal presidents, state officials, and rural teachers from Chiapas to Baja California, gave *campesinos* seed disinfectant, and ran articles in newspapers and agricultural journals urging them to select seed scientifically. *Agrónomos* told farmers that these methods would lead to significant yield increases with minimum effort and said that *campesinos* could create their own high-yielding varieties like the SAF laborers whom they directed to do the fieldwork in crossbreeding experiments. For example, Arturo Cuellar thought that tomato growers, like European and U.S. plant breeders, could develop improved strains adapted to Mexico's farming conditions. Hybrid seed loses its vigor in later generations, but Luis Marín said that with proper selection in the field, northern Mexico's cotton growers could develop improved varieties and end the need to import them from the United States Similarly, he urged farmers to do their own soil analysis to find appropriate chemical fertilizers.[46] Reflecting their history as transmitters rather than producers of knowledge and technologies, most *agrónomos* did not define plant breeding and soil science as domains accessible only to trained specialists.

The SAF ran an extension service, but in 1933 Elias reassigned González to the Office of Rural Economy to study railroad rates, and by 1937 only six *Agrónomos Regionales* remained. The new

Agentes Generales and agricultural bank employees compensated for the reduction, but some were not *agrónomos*. Extension agents still used *convencimiento* rather than demonstrations to influence farmers. The SAF also distributed written material, ran newspaper articles and created correspondence courses, and designed posters illustrating improved practices for the illiterate. For example, Rivas Tagle published a poem describing a farmer who followed an *agrónomo*'s advice on selecting corn seed, obtained a good harvest, and decided to seek their counsel in the future. The SAF held contests to promote the *Campaña en Pro del Maiz* and the use of iron plows and sponsored agricultural fairs in Mexico City, state capitals, and rural towns. To organize fairs and distribute agricultural propaganda, the SAF enlisted the aid of teachers, municipal presidents, "progressive farmers," and *agraristas*. The agency tried to reach children as well as adults. *Agrónomos Regionales* helped teachers form *Clubes de Fomento Agrícola* in *ejidos* to teach young peasants the methods of the *Campaña en Pro del Maiz*. *Agrónomos* and many farmers realized that many rural teachers knew little about farming, but the former felt that they were allies in a common project for the rural masses and enlisted their aid. For example, in 1931 they helped the SAF's "*Campaña pro Nogal*" (Campaign for Walnuts). *Agrónomo* Julián Rodríguez Adame tested wheat for resistance to *chahuixtle* (a plague that damages wheat and maize when it rains too heavily) at a school in Mexico State. In 1937 Tabasco's *Agrónomo Regional* taught agricultural courses, including the use of iron plows and vegetable cultivation, at Villahermosa's *Escuela Granja Rural* and ran an apiary at the *Escuela Plutarco Elias Calles*. By Cárdenas' order Jalisco's extension agent taught agriculture at the *Escuela Prevocacional Industrial, Comercial y Agrícola* in Jiquilpan, Michoacán, the president's hometown.[47]

To make the cultural campaign effective, *agrónomos* believed that the *campesinos* must be organized. In Gutiérrez Roldán's words, to "emancipate our indigenous rural population" and counteract "several centuries of slavery," along with land reform, the State had to "organize *ejidatarios*, small farmers, and farm laborers to let them receive technical experts' assistance."[48] Historian Arnaldo Córdova claims that this was the primary objective of Cárdenas' agricultural policy. Founded in 1932, the *Consejo Nacional de Agricultura* (National Council on Agriculture, hereafter

CNAG) organized producers and *Consejos Locales de Fomento Agrícola* (Local Agricultural Development Councils) with state governments. Each *Consejo Local* developed a plan to improve agriculture and distributed information on scientific farming. In 1934 Jalisco formed several producer organizations, starting with wheat and peanut growers, and orange and other citrus fruit growers. In 1935, recognizing the problems of organizing corn and bean farmers, the CNAG decided to target crops grown in "a reduced number of zones," like bananas and rice. In 1936, to resolve all of their "economic, technical, and social" problems, the CNAG tried to organize Tabasco's cacao growers. In 1937 the SAF began an intensive campaign to organize farmers in the Mezquital Valley, and General Almazán formed an orange grower cooperative in Nuevo León. The CNAG also sponsored national conventions, the first in 1934 for wheat farmers and later for diverse farm industries like garbanzos, garlic, and horses. The SAF used the conventions to disseminate information about agricultural policy and scientific farming and to provide a forum for technical experts to listen to farmers' concerns.[49]

Building on another 1920s agenda, many *agrónomos* believed that seasonal (*de temporal*) corn and bean farming could not be profitable for *campesinos*. Some thought that small *parcelas* needed high-value crops cultivated intensively with fertilizers and pesticides, while others wanted to shift food production to the northern irrigation districts, where mechanized production was viable. *Agrónomos* identified imported farm products, plants that provided raw materials for industry, or new export crops as potential income sources for peasant producers. The SAF knew that *campesinos* who received land wanted to grow corn and beans, but claimed that after four years they readily switched to commercial crops. To *agrónomo* Salvador Uribe Jasso, olives would end the *campesinos'* "misery." Pointing to unmet U.S. demand after Japan's invasion of Manchuria and China's civil war, as well as to Mexican imports, the SAF promoted sesame cultivation, asserting that it generated two to three times more revenue than corn, and Cárdenas took personal interest in the campaign. To increase the income of Nayarit's *campesinos, agrónomos* taught *ejidatarios* who lived along the Rio Tuxpán to grow cotton. Pharmacists and the CNI thought that *campesinos* could prosper by supplying Mexico's drug and fertilizer industries with castor beans and exporting surpluses to the United States. *Agrónomos* tested them in irrigation districts and published articles in agricultural journals lauding their benefits. The SAF gave many citrus, tropical, and deciduous fruit trees to peasants, such as 10,000 pineapple shoots in Sinaloa and Nayarit and 50,000 papaya saplings in Morelos. The *Agrónomos Regionales*, experimentation stations, and agricultural banks told *ejidatarios* to grow barley, coconuts, grapes, limes, linseed, peanuts, sorghum, tung oil trees, and walnuts. The SAF also studied wild plants to find new crops. Carlos Gajón Sánchez discoverd *lomboy* (*Jatropha*, family Euferbiaceas) in the Baja California desert and claimed that its seed oil could be used to manufacture an anticorrosive lubricant. *Agrónomos* thought that small irrigation works would eliminate the risks of *de temporal* farming, let *campesinos* grow food crops profitably, give them safe drinking water, and even make them responsible and cooperative. Cárdenas asked state and local governments for labor, funds, and suggestions for project locations. In 1937 he allocated 200,000 pesos for small-scale irrigation works, most in his home state, less than 1% of federal expenditures on large projects.[50]

Agrónomos knew that Mexico's soils lacked essential nutrients but also that chemical fertilizers were scarce and expensive, and many urged *campesinos* to use dung or plant green manure crops like clover to keep their fields fertile. To provide a better diet, more income, and enhanced soil fertility, some *agrónomos* advocated small-scale livestock or poultry keeping, but the SAF also tested chemical fertilizers on many crops, including coffee, oranges, sugarcane, vegetables, and wheat, and *agrónomos* urged farmers to use them. Joining *hacendados* who claimed they needed chemical fertilizers to farm profitably with less land, many *agrónomos* and officials such as Parres, Dampf, Marín, and Enrique Sanromán of the DGA urged the State to manufacture them. In 1935 Cárdenas announced plans to establish a Mexican chemical fertilizer industry, and in 1936 he reduced tariffs on imported brands. He considered building a nitrate plant in Las Vigas, Veracruz, and sent geologists to assay several mineral beds. In 1937 he appointed Modesto C. Roland, who wanted to pay government employees commissions for selling State-manufactured agrochemicals, to head the new Commission for the Study and Promotion of the Use of Fertilizers. His work laid the foundation to create the parastatal firm *Guanos y Fertilizantes de México* during the 1940s. Recognizing that many farmers could not afford pesticides, *agrónomos* told them to control pests and diseases by weeding, plowing, and other practices that did not require spending cash. Pointing to ecological benefits, others favored biological pest control. The SAF promoted the cultivation of insecticidal crops and sent expeditions to Tabasco's jungles to find rotenone-bearing plants, but *agrónomos* also knew that insects and diseases caused serious losses and looked for chemicals to fight them. Dampf and ENA professor Francisco García Robledo urged Cárdenas to build pesticide factories and support campaigns to convince farmers to use them. In 1936 the SAF ordered cotton growers in the northern border states to spray with calcium arsenate if they found pink bollworms. In 1937 the Oaxaca Station gave *campesinos* insecticide sprayers.[51]

A few *agrónomos* knew that the cultural campaign was not constructed on a solid foundation of research. Fernández y Fernández called the SAF's experimentation stations "no more than fields where they select seeds" and said that *Agrónomos Regionales* gave "simple advice, without previously conducting experiments that take many years."[52] Ítie recognized that bad advice made farmers lose confidence in the government and its experts. Some *agrónomos* tried to define agronomy as specialized knowledge intelligible only to trained experts. Pablo Tamez González told orange growers that some aspects of plant breeding, in this case the use of mutant buds, were too complicated for farmers to do themselves because they required long-term study by specialists. Sanromán admitted that most farmers lacked the expertise to find the right fertilizer formula for their soil, putting them at the mercy of unscrupulous vendors. Knowing that the sale of inappropriate fertilizers made farmers reluctant to use them, he wanted trained scientists to analyze Mexico's soils. Leonel Robles Gutiérrez called most of his ENA courses "bad," Flores said that "nothing" he learned "helped me improve farming in Guerrero," and Antonio Sandoval, who labeled Mexican research "insignificant" and "insufficient to meet national necessities," belittled his classes in experimental science.[53] Mesa A. urged the ENA to train soil scientists, pest control experts, and plant breeders. In 1937 the SEP said that "technical deficiency" made agricultural research "yield no appreciable benefits," *agrónomo* Armando Esquivel claimed that the ENA "abandons research, diminishing the school's and the nation's prestige," and students went on strike to demand better scientific training.[54]

MEXICAN FARMERS AND THE CULTURAL CAMPAIGN

The government often informed the press that *campesinos* cooperated enthusiastically with the SAF and its *agrónomos*. In January 1936 Cedillo told Cárdenas that the *Campaña en Pro del Maiz* led to productivity gains and claimed that in 1937, 15 to 20% of the *campesinos* would select their corn seed scientifically. Reporters heard Cárdenas and other government spokesmen proclaim that the SAF's campaign reduced corn and wheat imports, *agrónomos* proudly took credit for the *ejido* system's success, and Gándara declared victory: "[O]ur people will not die of hunger" because Mexico attained self-sufficiency in corn.[55]

Anthropologists have described peasants' willingness to change their farming practices to improve their living conditions or economic situation. Some farmers did support and cooperate with government agricultural programs. For example, in Veracruz *ejidatarios* won prizes for adopting the *Campaña en Pro del Maiz*, and in Jalisco farmers asked Inspector Enrique de la Torre for information on castor beans, cotton, rice, and tobacco. In Yucatán, *campesinos* who received henequen plantation land founded the *Instituto Técnico Henequenero*, a research institute. *Agraristas* and labor unions thanked Cárdenas for establishing the Acapulco Station; in Salina Cruz, Oaxaca, they urged him to reverse his decision to close the SAF's local office; farmers in Morelos asked him to open one; and politicians from Campeche, Hidalgo, Puebla, Jalisco, and Michoacán requested advice or programs. The private farmers of the *Unión Regional Agrícola de Chapala* supported the SAF's agents, the conservative *Unión de Vetranos de la Revolución* praised the agency's representative in Mazatlán, and *agrónomo* Salvador Rodríguez M. Marte Gómez claimed that the *ejidatarios* of Tamualipas considered *agrónomo* Carlos Soto, who implemented land reform in the state, "one of their own."[56] With the help of *repatriados* who worked on linseed farms in the United States and eight tons of imported seed, the SAF and Jalisco's Department of Agriculture established the crop near Arandas. The BNCA taught the Pabellón Irrigation District's farmers to grow grapes, which brought some prosperity to the region in the 1940s. In Sinaloa the State's Confederation of Agricultural Associations supported the SAF's sesame campaign, and production increase 10-fold from 1932 to 1937. In Sonora El Yaqui's tests persuaded farmers to plant it, convincing the SAF that the State's cultivators and government supported the agency. Extension agents also succeeded in Oaxaca, but in Nuevo León, Coahuila, and Chihuahua growers rejected the sesame campaign. From 1933 to 1937 Oaxacan farmers tripled acreage planted in castor beans, perhaps due to the SAF's program, but possibly also to a 100% price increase. Peanut tests conducted by the Santa Inés Station during its short life (1934–1935) convinced private promoters and some *ejidatarios*, and production increased in Jalisco and Sinaloa. In 1937 the BNCA persuaded some coffee growers to weed their plantations regularly and use chemical fertilizers. Many farmers asked the SAF to help fight pests and diseases, and other cultivators, such as the cotton growers of Chihuahua's Los Conchos Irrigation District and *ejidatario* coffee producers in Veracruz, tried chemical fertilizers, the latter after a *hacendado* used them successfully on his trees for several years.[57]

In other cases evidence indicates that the cultural campaign did not succeed. *Agrónomos* often complained that farmers refused to cooperate, such as Colina Robles, who said that the growers of Mixquiahuala, Hidalgo, did not join his pest control campaign and reported that the *ejidatarios* of Tonalá refused to participate in his imported wheat tests because the strains that he used the previous year did poorly. SAF bulletins and a grower who raised sesame successfully with the help of *agrónomos* from Santa Inés did not convince his neighbors in Autlán, Jalisco, to plant it. *Agrónomo* J.J. Islas León claimed that the tomato growers of La Paz who read the SAF's pest control publications had no faith in them, and they hired a U.S. expert. According to USDA pink bollworm inspectors, *agrarista* cotton growers in the Rio Grande Valley did not respect SAF employees because their supervisor opposed land reform. Some farmers in La Laguna cultivated *zoca* (a sprout born out of a second sugarcane harvest cut after already being cut once) ignoring the SAF's ban. Some peasant leaders denounced *agrónomos* and the SAF. For example, in 1936 José Flores of Jalisco's *Liga de Comunidades Agrarias* claimed that an SAF employee who organized several meetings fomented *campesino* unrest. In that state recalcitrant *cristeros* threatened agricultural inspectors like Jesús Vanegas, who waited for a military escort before he visited Bolanos, Chimaltitlán, and San Martín. The *ejidatarios* of Miltepec, Veracruz, were outraged when the SAF did not give them the prizes that they won at a fair and refused to hold a fiesta to distribute diplomas that they received in lieu of them. The *ejidatarios* of Las Margaritas, Oaxaca, broke the soil and opened untilled land with the iron plows that Cárdenas donated, but they used wooden ones for rowing, planting, weeding, and all operations on hilly land. When *campesinos* refused to cooperate, some *agrónomos* wanted to use coercion. In 1936 the BNCA began to grant loans in kind in the Pabellón Irrigation District to force farmers to use chemical fertilizers. In 1937, on the USDA's advice, SAF pink bollworm inspectors fined and even jailed growers who did not clean their fields on time, including *ejidatarios*.[58]

The CNAG discovered how hard it would be to organize Mexico's diverse agricultural producers and confronted the complexities of Mexican politics of the 1930s. By 1935 only four governors formed *Consejos Locales*, and many refused to cooperate. In 1933 the Ministry of National Economy claimed that coffee growers in Veracruz and Oaxaca did not trust the government and refused to found producer organizations. In 1936 Cedillo admitted that many *campesinos* would not join them. In Jalisco it was easier to organize commercial farmers like orange and wheat growers. The people of Arandas claimed that "marauding bands" would kill them if they formed a Corn Growers Organization, and in 1934 the state acknowledged "enormous difficulties" organizing corn and bean farmers, such as poverty that made travel to meetings in Guadalajara unaffordable. In 1936 Jalisco's *Consejo Local* stopped giving technical assistance. The CNAG's conventions attracted commercial farmers who could afford to visit Mexico City, and they sometimes made unrealistic demands, like orange growers' request for 194 *agrónomos* to assist their industry. Fernández y Fernández said that the conventions made farmers think that the SAF's employees were "clumsy officials with no spirit of responsibility" or desire to "truly resolve any problem."[59] Planned as a place to show *campesinos* scientific agriculture, the fairs became a forum to entertain city people. In 1936, to prove that correspondence courses "enjoyed great popularity" with *campesinos*, the SAF claimed 6,723 subscribers, but only literate farmers benefited. Many peasants detested socialist education or needed their children's labor in the fields, and in 1936 the SAF lamented that only 634 of 6,000 planned *Clubes de Fomento Agrícola* functioned. As Vaughan notes, rural teachers lacked the training and resources to modernize the *campesinos'* farming practices, and peasants concluded that they knew nothing about farming.[60]

Many problems plagued the SAF, including corruption. Historian Dudley Ankerson claims that Cedillo stole over 1 million pesos while he ran the agency. García de la Cadena put a friend on the payroll at Santa Inés, but he never worked there. Lorenzo Flores Fagoaga, the CNAG's representative in Jalisco, spent most of his time "sick" in a Mexico City hotel. In 1936 the State's dispute with the SAF over funds led to the station's closure. In 1933 peasants from Cuatro Cienegas, Coahuila, denounced a CNA *agrónomo* who resold them seed that they delivered for market, and the SAF fired another after an investigation proved that he spent most of his time on private matters like his wheat business. Army officers in Oaxaca claimed that the personnel with the SAF's campaign against locusts did not report for work, leading to protests from farmers whose crops were devoured. Fernández y Fernández claimed that the ENA suffered from 1934 to 1937 because many *aviadores,* people who did not work there, were on the payroll. Inadequate funding, political turmoil, and bureaucratic turf wars also affected the SAF. In 1936 the agency could not afford effective campaigns against Panama disease of bananas, locusts, fruit flies, and grasshoppers in Sinaloa, and insect pests of cotton in Veracruz but still assigned a few employees to them. Funding shortages delayed the experimentation stations' projects. Cedillo sometimes hired political allies rather than *técnicos.* When he appointed Andres Mendoza, an *agrarista* crony with no degree, to a post in Chihuahua, peasant organizations denounced him, including *ejidatarios* who claimed that he interfered in *ejidal* politics, and the governor and other politicians demanded his dismissal. When Cárdenas fired Cedillo, he purged his allies in the SAF, including Gándara and pest control experts Leopoldo de la Barreda and Pedro Ramírez de Alba, who were not *cedillistas* (followers of agrarian general Saturnino Cedillo), disrupting some programs. Rapid turnover prevailed in the agrobureaucracy, like the case of Nabor Cuervo Martínez, who from 1930 to 1937 changed jobs seven times, and adversely affected some projects. Other agencies sometimes refused to cooperate, delaying programs like pest control campaigns. Department of Forestry, Fish, and Game inspectors thwarted Córdova's effort to increase the productivity of Veracruz's coffee growers by refusing permission to fell their shade trees.[61]

U.S. consuls said that agrarian reform made *hacendados* and *rancheros* refuse to improve their properties or try new farming methods, and these farmers resented or detested the *agrónomos* because of their alliance with the *agraristas* and their role in land distribution. But the SAF believed that they were more efficient and cooperative than *campesinos* and sometimes decided to work with them, such as to increase seed of foreign wheat varieties and test imported corn in the Bajío and chemical fertilizers in Nuevo León, Jalisco, and Veracruz. *Agrónomos* said that they wanted to work with *campesinos*, but with their education, urban origins, and class backgrounds, many had negative opinions of their character. Inspectors with Jalisco's Department of Agriculture thought that the state's farmers had no desire to improve their practices. Colina Robles called the peasants of San Marcos and Techuiltlán, Jalisco, "apathetic," and the CNI said that the "negligence and indolence" of Lake Chapala's farmers, "who refuse to abandon growing corn, wheat, and beans because they are indispensable to their daily diet," thwarted projects to establish chiles, garbanzos, onions, potatoes, tomatoes, and watermelons.[62] Gledhill claims that the class background of *agrónomo* Guillermo Maqueo Castellanos and other BNCE employees made them feel uncomfortable talking to *campesinos*. In 1936, sounding like *hacendados*, "government agronomists" told U.S. Consul Thomas D. Bowman that "the peon, as an independent farmer, has not been successful."[63] Knowing that agricultural schools attracted mostly urban, middle-class students, the SEP wanted more young peasants to enroll but admitted that *campesinos* who graduated felt that their degree raised their status, making them unwilling to return to rural life. In 1929 the ENA adopted the policy but waited until 1937 to implement it. That year Fernández y Fernández urged the SAF to "bring *agrónomos* to the countryside," and *agrónomo* Roberto E. Palacios claimed that they hated working in the fields and suffered from "*empleomania* [the quest for a government job], the fly paper on which all agronomists shamefully fall," supporting Eyler Simpson's assertion that "the Mexican student who actually goes out in the field and looks for facts about his own country is so rare as to be a curiosity."[64]

CONCLUSIONS

Sánchez Durón claims that the research programs of the 1930s were an important advance for Mexican agronomy, but they did not change many *campesinos'* farming practices, lead to significant productivity gains, or establish *agrónomos'* credibility as scientific professionals. Cárdenas created many new *ejidos,* but he also overwhelmed the State's ability to give them technical assistance. Nevertheless, *agrónomos'* participation in the cultural campaign helped solve their employment problem. Further, membership in the Revolutionary Family brought substantial political rewards. In 1933 Rodríguez appointed Marte Gómez to lead the Ministry of Finance and Public Credit. In 1934 Juan de Dios Bojórquez ran his Department of Labor, and Cárdenas appointed him minister of government. In 1938 Carlos Terrazas Moro and Guillermo Liera B. were federal deputies; Ignacio L. Figueroa, Angel Posada, and Ramón Ramos were senators; and Marte Gómez was "elected" governor of Tamualipas. During the Cárdenas administration the Revolution "brought justice" to *agrónomos* as well as *campesinos.* Their alliance with the Revolutionary Family fostered ideological solidarity as they served the *cardenistas'* land reform and cultural programs, but they failed to attain "market closure" in the practice of agronomy. Further, politics remained dangerous business. For example, Calles' ally de Dios Bojórquez resigned his position during Cárdenas' struggle with the *jefe máximo,* left the profession, and became a wire service reporter. Other *agrónomos* abandoned the turbulent world of public employment. For example, in 1932 José Arturo Jiménez Cadet quit his government job to begin a construction materials business.[65]

Why did the peasants resist the cultural campaign? Despite the *agrónomos'* alliance with the *agraristas* and support for *cardenismo,* their class background, prejudices, scientific training, and attempt to obtain professional status by delegitimating peasant farming systems, as well as the cultural campaign's objectives, deterred effective interfaces with the *campesinos.* In 1938 the SAF urged peasants to seek the advice of *agrónomos,* but *campesinos* viewed them through the lenses of past encounters and their antipathy toward urban sophisticates; they were the most recent of many "educated" outsiders who came to their villages to carry out reform programs that often led to negative outcomes. Hence, they saw them as potential exploiters, city people who knew nothing about farming, as the *hacendados* believed, or even agents of godless Communism. Trust is essential for professionals to exercise their authority, but *agrónomos* could not win the *campesinos'* confidence. To protect themselves from market uncertainties and maintain their subsistence autonomy and security, peasants planted corn and beans, used their traditional cultivation practices and crop strains, and rejected foreign hybrids and exotic crops. Nevertheless, like the *campesinos'* encounter with the *cardenistas'* rural teachers or their response to corn-grinding machines, they accepted aspects of the *agrónomos'* modernizing discourse that they found useful.[66] The *cardenista agrónomos* thought that they were "helping" the *campesinos*, but they also demanded that they accept their definition of "good farming." Whether or not the land reform or the cultural campaign led to any productivity gains, in 1938 a serious agricultural crisis engulfed Mexico.

NOTES

1. Bruce H. Jennings, *Foundations of International Agricultural Research: Science and Politics in Mexican Agriculture* (Boulder, Colo.: Westview Press, 1988), pp. 98–99; George I. Sánchez, *Mexico: A Revolution by Education*, (New York: Viking Press, 1936), pp. 145–46; Cynthia Hewitt de Alcantara, *Modernizing Mexican Agriculture: Socioeconomic Implications of Technological Change, 1940–1970* (Geneva: United Nations Research Institute for Social Development, 1976), pp. 19–20; Alberto Arce, *Negotiating Agricultural Development: Entanglements of Bureaucrats and Rural Producers in Western Mexico* (Wageningen, the Netherlands: Agricultural University Wageningen, 1993), p. 40; "Comments by Carl O. Sauer," February 10, 1941, RFA, RG 1.1, Series 323, Box 1, Folder 2, RAC.

2. Arnaldo Córdoba, *La política de mases del cardenismo* (México, D.F.: Ediciones Era, 1974), pp. 106–7; Douglas Bennet and Keith Sharpe, "The State as Banker and Entrepreneur: The Last Resort Character of the Mexican State's Economic Intervention 1917–1970," in Sylvia Ann Hewlitt and Richard S. Weinert, eds., *Brazil and Mexico: Patterns of Late Development* (Philadelphia: Institute for the Study of Human Issues, 1982), pp. 179, 195; Hewitt de Alcantara, *Modernizing*, p. 4; Joe C. Ashby, *Or-*

ganized Labor and the Mexican Revolution under Lázaro Cárdenas (Chapel Hill: University of North Carolina Press, 1967), p. 169; Sergio Reyes Osorio et al., *Estructura agraria y desarrollo agrícola en México: Estudio sobre las relaciones entre la tenencia y el uso de la tierra y el desarrollo agrícola en México* (México, D.F.: Fondo de Cultura Económica, 1974), p. 38; Hewitt de Alcantara, *Modernizing*, pp. 19–20; Marco Antonio Durán, *El agrarismo mexicano*, (México, D.F.: Siglo Ventiuno, 1967).

3. John Gledhill, *Casi Nada: A Study of Agrarian Reform in the Homeland of Cardenismo* (Albany: State University of New York, Center for Mesoamerican Studies, 1991), p. 30; Friedrich E. Schuler, *Mexico between Hitler and Roosevelt: Mexican Foreign Relations in the Age of Lázaro Cárdenas* (Albuquerque: University of New Mexico Press, 1998), pp. 22–23.

4. Alan Knight, "The United States and the Mexican Peasantry, c. 1880–1940," in Daniel Nugent, ed., *Rural Revolt and U.S. Intervention* (La Jolla: U.C. San Diego, Center for U.S.–Mexican Studies, 1988), p. 34; Harold Nicolson, *Dwight Morrow* (New York: Harcourt, Brace and Company, 1935), pp. 310, 321, 332, 351; Córdoba, *La política*, p. 22; Jesús Silva Herzog, *El agrarismo mexicano y la reforma agraria, exposición y crítica* (México, D.F.: Fondo de Cultura Económica, 1959), pp. 325, 385; Valentín Gama, *Consideraciones sobre la producción agrícola e industrial de México* (México, D.F.: Empresa Editorial de Ingenería y Architectura, 1936), pp. 18–19; Charles A. Cunningham, "Criticism of 'Ejido' System," March 28, 1932; Raleigh A. Gibson, "The Union of Agriculturalists of the State of Jalisco," May 7, 1931; J. Ruben Clark to Secretary of State, June 5, 1931, NA, RG 166, Box 149; Enrique C. Creel, "Agricultura y agrarismo," *Memoria y Revista de la Sociedad Científica Antonio Alzate* 51, 5–6 (1929–1930): 157–79; Victoria Lerner, *La educación socialista* (México, D.F.: El Colegio de México, 1979), p. 69; Dana Markiewicz, *The Mexican Revolution and the Limits of Agrarian Reform 1915–1946* (Boulder, Colo.: Lynne Rienner, 1993), pp. 84–85; "Nota editorial: La campaña contra los agrónomos," *Agronómica* 1, 4 (January 1932): 3–4; Javier Uranga M., "El fracaso de las Escuelas Agrícolas y el gran problema de la educación rural," December 1933, AGN, Cárdenas papers, Exp. 534/62; Eduardo Villaseñor, "Las zonas nuevas y el indio," *Revista Mexicana de Economía* 1 (1928–1929): 47; SEP, "Proyecto de la Escuela Rural: Educación Campesina y Fomento Rural," December 1932, pp. 6, 12–13, AGN, Robles papers, Education Section, Box 49, Exp. 19; "Directorio de Socios," *Boletín Interior* 1, 2 (August–September 1930): 3–40.

5. Gilberto Mendoza Vargas, "La enseñanza y extensión agrícola," February 1954, in Enseñanza Agrícola Superior, BCM Bound Volume #145–43; Agrónomos, BCM Bound Volume #168–40 (hereafter Agrónomos); Unión Agronómica Mexicana, *Agrónomos y pasantes de agrónomo de*

la Escuela P. de Agricultura de Ciudad Juárez, Chih. (México, D.F.: 1934); Manuel Mesa A., "Memorandum sobre la ENA," August 15, 1933, AGN, Robles papers, Agriculture Section, Box 5, Exp. 70; *Exposición del criterio ideológico del Gremio Agronómico Mexicano* (Puebla: Confederación Campesina Emiliano Zapata, 1932), p. 6; Luis Hermán, "El cultivo del ajonjolí," *El Nacional*, May 25, 1931.

6. PNR, *Los problemas agrícolas de México* (México, D.F.: PNR, 1934), p. 2; *Exposición*, p. 5; Manuel Mesa A., "La labor oficial de los agrónomos en México," *Agronómica* 1, 3 (December 1931): 34.

7. Rafeal Villareal to Abelardo Rodríguez, December 7, 1933, AGN, Rodríguez papers, Exp. 550/61; Rodríguez to Sebastián Allende, May 2, 1934, AHJ, Box AG-27-E, Exp. 200, Legajo 12; Augusto Hinojosa to Municipal President of Zamora, July 12 and October 19, 1932, AMZ, Box Fomento 1931–1932; Unión Agronómica Mexicana, *Agrónomos*, p. 24; Lyle C. Brown, "General Lazaro Cardenas and Mexican Presidential Politics, a Study in the Acquisition and Manipulation of Political Power" (unpublished Ph.D. dissertation, University of Texas, Austin, 1964), pp. 224–25; Rafeal Lopez Ocampo, "La ley de colonización y el problema de la tierra en México," in PNR, *Los problemas*, p. 239; Dudley Ankerson, *Agrarian Warlord: Saturnino Cedillo and the Mexican Revolution in San Luis Potosí* (De Kalb: Northern Illinois University Press, 1984), p. 173.

8. "Nota editorial: La campaña," p. 4; Bernardo L. Arrieta M., "La educación agrícola como fundamental de reorganización de nuestra economía rural," in PNR, *Los problemas*, p. 255; Ramón Fernández y Fernández, "El fracaso de los agrónomos," *El Universal*, February 4, 1932; Fernández y Fernández, *Los agrónomos* (México, D.F.: Impresa Secretaría de Maquinas, 1933), pp. 2–6; Fernández y Fernández "Los agronómos y la reglamentación de los professiones," March 1934, in Agrónomos; Pascual Gutiérrez Roldán, "Los agrónomos y la Secretaría de Agricultura," *Agronómica* 1, 4 (January 1932):9.

9. Fernández y Fernández, *Los agrónomos*, pp. 9–15; Roberto E. Palacios to Cárdenas, May 3, 1937, AGN, Cárdenas papers, Exp. 506.15/1; Gutiérrez Roldán, "Los agrónomos," p. 8; Vicente Lombardo Toledano, "Professiones liberales y professiones del Estado," *Agronómica* 1, 4 (January 1932): 21; Frank Safford, *The Ideal of the Practical: Columbia's Struggle to Form a Technical Elite* (Austin: University of Texas Press, 1976), p. 234; "Nota editorial: La campaña," p. 5; "Microscopio," *Agronómica* 1, 4 (January 1932): 10; Arrieta M., "La educación agrícola," pp. 257–58; Rosendo de la Peña and Manuel J. Gándara to Fernández y Fernández, July 31, 1936, in Agrónomos.

10. "Total reorganización de la SAF," *El Nacional*, June 5, 1933; Enrique Beltrán, "Instituto Biotécnico (1934–1940) de la SAF," *Anales de la Sociedad*

Mexicana de Historia de la Ciencia y de la Tecnología 1 (1969): 163–83; SAF, *Sugestiones para la elaboración del Plan Sexenal 1934–1940* (México, D.F.: mimeograph, 1933), p. 8; Jennings, *Foundations*, p. 98.

11. Christopher R. Rounds, *From Hacienda to Ejido: Land Reform and Economic Development in Yautepec, Morelos, 1920–1970* (New York: Garland, 1987), p. 127; SAF, *Sugestiones*, p. 155; "Un Congreso de Agrónomos," *El Universal*, January 8, 1934; "Asuntos de vital interés se tratarán en el Próximo Gran Congreso Nacional Agronómico," *El Nacional*, November 7, 1935; "La agricultura y el gobierno de 1934," *Germinal*, September 15, 1933, p. 5; Guillermo Liera B., "El Sexenal y los ejidos," *El Nacional*, November 25, 1934; Alberto Carreño, "La evolución económica de México en los últimos cincuenta años," *Memoira y Revista de la Academía Nacional de Ciencias Antonio Alzate* 54, 1–3 (1934): 65–152; Arturo M. Escandon to Allende, February 25, 1935, AHJ, Box AG-27-F, Exp. 61.33, Legajo 3; SAM, "Boletín #1," January–February 1934, ARFF, Box 180; Marte R. Gómez, *Biografías de agrónomos* (Chapingo, México: Colegio de Postgraduados, Escuela Nacional de Agricultura, 1976), p. 125; Amalio Ruiz Velasco, "Conferencia sustentada en la Exposición de la Feria Primavera en Puebla," *Revista Arrocera* 1, 1 (July 1935): 3–7; Emilio Alanis Patiño, *Diversos aspectos de la situación agrícola de México*, (México, D.F.: Instituto Mexicano de Estudios Agrícolas, 1934), p. 66; "Nota editorial: La campaña," p. 5; "Nota editorial: Los bajos rendimientos," *Agronómica* 1, 3 (December 1931): 7.

12. "Nota editorial: La campaña," p. 4; Escobar Hermanos y Cia to Fernández y Fernández, October 3, 1933, in Enseñanza; "El Congreso Agronómico," *El Nacional*, May 16, 1936; Manuel A. Gándara to members, SAM, November 16, 1940, ARFF, Box 180.

13. "Primer Congreso Nacional de Estudiantes de Agricultura," October 1936, p. 8, in Enseñanza; Heliodoro Monroy, "Discurso," in CNAG, *Memoria de la Primera Convención Nacional de Cafeticultores* (México, D.F.: 1937), p. 2; "Interview with Lorenzo Martínez Medina," in Leobardo Jiménez Sánchez, *Las ciencias agrícolas y sus protagonistas* (Chapingo: Colegio de Postgraduados, 1984), p. 97; Guillermo Gándara, "Las calabazas de México," *Memoria y Revista de la Academía Nacional de Ciencias Antonio Alzate* 53, 6–8 (1932): 209–11; Julio Riquelme Inda, "La próxima guerra universal, la lucha contra los insectos," *Memoria y Revista de la Academía Nacional de Ciencias Antonio Alzate* 51, 3–4 (1931): 133–35.

14. Gledhill, *Casi Nada*, p. 30; "Valiosa aportación a nuestra agricultura," *El Universal*, April 3, 1933; Roberto Trevino Martínez, *El problema agrario de México y la solución revolucionaria (sugestión de una economía agrícola colectivista)* (México, D.F.: Editorial Nuevo Era, 1936), pp. 129–52; Rene A. Becerra to Gonzalo Robles, December 1, 1936, AGN, Robles papers, Agriculture Section, Box 11, Exp. 95; BNCA, "Monografía

Económico-Agrícola del Trigo," unpublished study, 1937, p. 2, ARFF, Box 201; M. Rubenstien, "Relations of Science, Technology, and Economics under Capitalism and in the Soviet Union," in *Science at the Crossroads* (London: Kniga, 1931), p. 57; Mario Sousa, "La explotación agrícola en la U.R.S.S. y en México," in SAM, *Primer ciclo de conferencias* (México, D.F.: D.A.P.P., 1938), p. 14.

15. Alfonso Dampf, "Como defendernos contra las plagas de la agricultura," September 19, 1931, ARFF, Box 139; Emilio Portes Gil, *The Mexican Schools and the Peasantry* (México, D.F.: Press of the Ministry of Foreign Relations, 1936), pp. 21–22.

16. Lombardo Toledano, "Los técnicos sin empleo," *El Universal*, July 3, 1935; "Primer Congreso," p. 6; Edmundo Flores, *Historias de Edmundo Flores, autobiografía 1919–1950* (México, D.F.: Martín Casillas Editores, 1983), pp. 124–34.

17. Ramón Beteta, "Some Economic Aspects of Mexico's Six Year Plan," in Hubert Herring and Herbert Weinstock, eds., *Renascent Mexico* (New York: Couici Friede, 1935), pp. 108–9; Luis Cabrera, "The Key to the Mexican Chaos," in Herring and Weinstock, eds., *Renascent Mexico*, p. 28; Victor Alba, *The Mexicans: The Making of a Nation* (New York: Praeger, 1967, pp. 181–83; Victor F. Calverton, "Red Rule in Mexico's Schools," *Current History* 43 (December 1935): 263–66; Nicolai I. Bukharin, *Economics of the Transformation Period* (New York: Bergman, 1971, trans. from Russian original and authorized German edition, Moscow, 1920), pp. 111–13; Alexander Vucinich, *Empire of Knowledge: The Academy of Sciences in the U.S.S.R. 1917–1970* (Berkeley: University of California Press, 1984), pp. 80–90; J.G. Crowther, *Soviet Science* (London: Kegan Paul, Trench, Trubrer & Company, 1936), p. 19; M. Rubenstien, "Relations," pp. 46–59.

18. Schuler, *Mexico between Hitler and Roosevelt*, p. 51; "Lechugilla" (n.d.), AGN, Robles papers, Agriculture Section, Box 3, Exp. 54; Becerra to Robles, December 1, 1936; "Maquinaria para Yucatán," *El Universal*, March 11, 1937; Hermann B. Hagan to Director, *Jalisco Agrícola y Ganadero*, May 20, 1935, AHJ, Box AG-27-F, Exp. 61.33, Legajo 3; "Estudio presentado por la Delegación de Defensa Agrícola de León, Gto.," in CNAG, *Memoria de la Primera Convención Nacional de Productores de Papa* (México, D.F.: 1939); CNAG, *Memoria . . . Cafeticultores*, p. 1; Cosme Hinojosa to Cárdenas, April 6, 1937, AGN, Cárdenas papers, Exp. 502.1/14; "Interview with Alanis Patiño," in Jiménez Sánchez, *Las ciencias agrícolas*, pp. 358–59; B.F. Osorio Tafall, "Materiales para el estudio del microplanctón del Lago de Patzcuaro," *Sobretiro de los Anales de la Escuela Nacional de Ciencias Biológicas* 2, 2–3 (1941): 1; James A. Noel, "Cultivation of Chiles in the Ensenada Consular District," July 10, 1936, NA, RG 166, Box 153.

19. Howard A. Bowman, "Colonization Project of Colorado River Land Company," December 18, 1934; A.F. Yepis to Secretary of State, August 8 and 23, 1935; Francis H. Styles, "Agriculture Department," January 19, 1933; NA, RG 166, Boxes 149, 151, and 388; "Nuevas especies en Estados Unidos," *Revista Arrocera* 1, 1 (July 1935): 30; Salvador Benevides Jr. to U.S. Department of Agriculture, March 28, 1933, and others, NA, RG 54, Series 656, Box 391; "Beneficio obtenido por los arroceros Sonorenses con la Constitución de su Asociación," *Revista Arrocera* 1, 2 (August 1935): 4–6; Frank Reeves, *Hacienda de Atotonilco* (Yerbanis, Durango: Cia. Ganadero de Atotonilco, 1936); L. Iñíguez de la Torre, "El rancho Americano y el rancho Mexicano," *Jalisco Rural* 17, 7 (July 1937): 177–78; A.J. Pieters to H.E. Allanson, May 10, 1934, NA, RG 54, Series 656, Box 391; "Nota editorial: Los bajos," p. 8; Nathan Whetten, *Rural Mexico* (Chicago: University of Chicago Press, 1948), pp. 270–71; Garrido Canabal, "Memorandum," January 14, 1935, AHSRE, Exp. III-221-4; C.B. Smith, "Como interesar el mayor numero de agricultores," *Extensión Agrícola* 1, 2 (March 1933): 37; DGA, "El cultivo de secano es de gran porvenir en México," November 30, 1931; DGA, "Alto empleado del Servicio de Extensión visita México," August 29, 1931, ARFF, Boxes 5 and 82; "Actividades de la Sociedad," *Boletín Interior* 2, 2 (January 1931): 2; Patiño, *Diversos aspectos*, p. 12; Luis Paredes R. to Luis I. Rodríguez, March 28, 1936, AGN, Cárdenas papers, Exp. 506.1/13; José Lázcano Romero, "¿Hay problema de relocalización de la producción agrícola en México?" in PNR, *Los problemas*, p. 329; "Datos históricos y estadísticos sobre la EPA de Ciudad Juarez, Chihuahua," in *Ecos del aniversario de plata de la EPA de Ciudad Juárez* (Ciudad Juárez, México: EPA, 1931), p. 19; Jesús Echavarrí, "Apicultura," *El Campesino* (Durango), April 1933, p. 17.

20. "El cultivo de la fruta en mayor escala en la República," *El Universal*, October 31, 1933; Victoriano Anguiano to Presidente Municipal de Zamora, June 20, 1933, AMZ, Box Fomento 1933, Exp. 3; Martin H. Greenberg, *Bureaucracy and Development: A Mexican Case Study* (Lexington, Mass.: D.C. Heath, 1970, p. 16; "Numerosas despepitadores de algodón serán adquiridas en el Banco de Crédito Ejidal," *El Nacional*, July 1, 1937; Felix V. Rios to Cárdenas, February 15, 1937; Banco Ejidal to Cárdenas, March 3, 1937, AGN, Cárdenas papers, Exp. 546.6/87.

21. "Interview with Edmundo Taboada Ramírez," in Jiménez Sánchez, *Las ciencias agrícolas*, p. 128; Harry M. Miller, "Mexican Log," March 2–28, 1941, RFA, RG 613, Series 1.1, Box 33, Folder 366, RAC; Division of Inter-American Educational Relations, "Alumni of U.S. Colleges of Agriculture Who Are Citizens of the Other American Republics," April 1944, NA, RG 59, Decimal File 1940–1944, Box 196; "Informe sintético de

las labores que desarrollará la SAF en relación con los postulados del Plan Sexenal," 1936; J. Francisco Andraca and S. Gutiérrez Silva to Cárdenas, May 8, 1936; "Planta de Fertilizantes de Las Vigas, Ver." (n.d.), and others, AGN, Cárdenas papers, Exps. 433/116, 506.1/15, and 506.22/3; Gómez, *Biografías*, pp. 126, 489; Alfredo Téllez Girón, "El Instituto Nacional de Investigaciones Pecuarias y sus antecedentes," *Anales de la Sociedad Mexicana de Historia de la Ciencia y de la Tecnología* 2 (1970): 196.

22. Eduardo Vasconcelos to Jose Mares, October 7, 1932, AHSRE, Exp. IV-428-103; Octavio Gómez del Campo to Director de Fomento Agrícola, April 30, 1934, ARFF, Box 140; "Se intensifica la campaña contra el gusano rosado en los sembradores de algodón," *El Nacional*, October 29, 1935; Dampf, "La importancia de la mosca de la fruta," in CNAG, *Memoria de la Primera Convención Nacional de Productores de Naranja* (México, D.F.: 1936), pp. 2–4; Lee A. Strong to P.A. Hoidale, March 6, 1935, NA, RG 7, Box 1392; "Solo en Tequila existe la plaga," *El Jaliscense* (Guadalajara), August 6, 1934; Dario L. Arrieta M., "Comisión Internacional de Plagas" (n.d.), pp. 1–6, RFA, RG 6.13, Series 1.1, Box 10, Folder 121, RAC; Dampf, "La importancia," p. 5; Lorenzo R. Jiménez to P.A. Hoidale, July 10, 1936, NA, RG 7, Box 1393; R.E. McDonald to Strong, December 7, 1936, NA, RG 7, Box 1400; "Para ayudar a La Laguna," *El Nacional*, January 19, 1935; José Figueroa, "Las intoxicaciones del ganado por el selenio," *Agricultura* 1, 5 (March–April 1938): 25–29; Juan M. Rulfo, "La chía," *Agricultura* 1, 1 (July 1937): 30.

23. Samuel J. Walker, *Henry A. Wallace and American Foreign Policy* (Westport, Conn.: Greenwood Press, 1976); Schuler, *Mexico between Hitler and Roosevelt*, p. 39; "Statement by Earl N. Bressman before the Defense Investigating Committee of the Senate" (n.d.), p. 3, Rockefeller Family Archives, RG 4, Nelson Rockefeller papers, Series 0, Box 10, Folder 78, RAC; Rexford G. Tugwell to Tom Connally, October 14, 1936, NA, RG 7, Box 1389; A.C. Baker to P.A. Hoidale, March 21, 1936, NA, RG 7, Box 1391; "La hacienda de Santa Inés, el exponente de la obra del Estado en pro del campo," *El Jaliscense* (Guadalajara), November 17, 1934; L.S. Rowe to Editors, *Jalisco Agrícola-Ganadero*, November 26, 1934, AHJ, Box AG-27-F, Exp. 61.33, Legajo 3; "Ven mil plantas de plátano en el Edo. de Guerrero," *El Nacional*, December 21, 1935; Karl Wessel to Rodríguez, August 4, 1934, AGN, Rodríguez papers, Exp. 554.1/2; "La linaza en el Valle de Mexicali, Baja California," *El Campesino*, April 1937; Oscar C. Harper, "Mexican Repatriates Take up Farming in the Nuevo Laredo Consular District," May 13, 1931, NA, RG 166, Box 149; Thomas D. Bowman, "Plans for Agricultural Development near Acapulco," April 24, 1934, NA, RG 166, Box 151; Gustavo Esteva, *The Struggle for Rural Mexico* (South Hadley, Mass.: Bergin & Garvey, 1983), p. 42; J.A. Ferrell to F.F.

Russell, September 28, 1933, and others, RFA, RG 1.1, Series 323, Box 1, Folder 2, RAC; Ferrell to Raymond B. Fosdick, October 16, 1936, RFA, RG 1.2, Series 323, Box 10, Folder 63, RAC.

24. Mary Kay Vaughan, *Cultural Politics in Revolution: Teachers, Peasants, and Schools in Mexico, 1930–1940* (Tucson: University of Arizona Press, 1997), p. 38; "Mensaje," January 1, 1937, in *Palabras y documentos públicos de Lázaro Cárdenas, Informes Presidenciales y Mensajes de Año Nuevo, 1938–1940*, vol. 2 (México, D.F.: Siglo Ventiuno Editoriales, 1978), p. 275; "Prominente médico Americano nos visita," *El Universal*, February 13, 1935; Jennings, *Foundations*, pp. 46–47; Luís Vargas y Vargas to Silvestre Guerrero, September 20, 1940, AHSRE, Exp. III-1704-12; Rockefeller Foundation, "Memorandum para el Presidente" (n.d.), AGN, Cárdenas papers, Exp. 534.1/1084; Fosdick to Owen Young, May 26, 1938; Staff Conference, March 9, 1939, RFA, RG 1.2, Series 300, Box 1, Folder 7; Ferrell, "Conference, Vice President Wallace, Fosdick, and Ferrell Regarding Mexico, Its Problems and Remedies," February 3, 1941, RFA, RG 1.1, Series 323, Box 1, Folder 2, RAC.

25. Vaughan, *Cultural Politics*, p. 5; Nancy Leys Stepan, *"The Hour of Eugenics": Race, Gender, and Nation in Latin America* (Ithaca, N.Y.: Cornell University Press, 1991), pp. 33–37; Sánchez, *Mexico*, p. 104; George Booth, *Mexico's School Made Society* (Stanford, Calif.: Stanford University Press, 1941), pp. 19, 55; George A. Kneller, *The Education of the Mexican Nation* (New York: Columbia University Press, 1951), p. 64; William Cameron Townsend, *Lázaro Cárdenas: Mexican Democrat* (Ann Arbor, Mich.: George Wahr Publishing Company, 1952), p. 336; John H. Plenn, *Mexico Marches* (Indianapolis: Bobbs-Merrill, 1939), p. 361.

26. Rafael Ramírez, "The Six Year Plan in Education," in Herring and Weinstock, eds., *Renascent Mexico*, pp. 138–39; Gabriel Lucio, "Trends in Mexican Elementary Education," *Progressive Education* 13, 2 (February 1936): 99; Beltrán, "The Place of the Biological Sciences in Educational Programs," *Progressive Education* 13, 2 (February 1936): 96.

27. S.L.A. Marshall, "Mexico Muddled on New Education," *New York Times*, March 19, 1935.

28. SEP, *Plan de Acción de la Escuela Elemental Agrícola*, (México, D.F.: 1938), pp. 7–49, 50–61; Booth, *Mexico's*, p. 77; Nathaniel Weyl and Sylvia Weyl, *The Reconquest of Mexico: The Years of Lázaro Cárdenas* (New York: Oxford University Press, 1939).

29. Manuel Gamio, "Education and National Integration," *Progressive Education* 13, 2 (February 1936): 86–89; Alanis Patiño, *Diversos*, p. 70; "Proyectos para mejoramiento a la agricultura," *Excelsior*, August 12, 1934; *Exposición*, p. 2; "Microscopio," *Agronómica* 1, 4 (January 1932): 11; "Los cadaveres de los animales," *El Nacional Revolucionario*, May 11, 1931; Jacobo Aragón Aguillón, "Las Estaciones Agrícolas para la experimen-

tación," *El Nacional*, January 19, 1935; Booth, *Mexico's*, p. 65; Robles and Mesa A., "Opinión sobre las sugestiones generales para reorganizar la ENA," January 8, 1935, AGN, Robles papers, Education Section, Exp. 19; Flores, *Historias*, p. 95.

30. Beltrán, *Medio siglo de recuerdos de un biólogo mexicano* (México, D.F.: Sociedad Mexicana de Historia Natural), 1977), p. 152; Gómez, *Biografías*, pp. 320–25; Plenn, *Mexico Marches*, p. 88; "Primer Congreso," p. 9.

31. "Mucho maiz con poco dinero," *Germinal*, January 25, 1937; Figueroa, "Editorial: Discurso pronunciado en la inauguración del Instituto Biotécnico en San Jacinto," *Agricultura* 2, 2 (March–April 1939): 2–3.

32. "La esclavatud del maiz," *El Universal Gráfico*, February 7, 1936; "El maiz no es un alimento completo," *El Nacional*, June 24, 1936; M. Neimann, "La alimentación adecuada como problema nacional," *Medicina y Farmacía*, 3, 35 (October 1936): 5–11; Guillermo Quesada Bravo, "Problemas de salubridad e higiene en las regiones cafeticultores," in CNAG, *Memoria . . . Cafeticultores*, pp. 14–21; SAF, *Sugestiones*, p. 155.

33. Gómez, *Biografías*, p. 491; Francisco Quintanar, "El problema del maiz," *El Mercurio* (Mexicali), June 24, 1931.

34. Fernando Reading and Moises T. de la Peña, "Trigo," January 30, 1936, ARFF, Box 198; Alfonso Ramírez Paulín, "La industrialización del girasol," *Revista del Banco Obrero* 13 (December 1938): 14–17; Carlos Noriega, "La fruticultura como una transformación racional de la agricultura Mexicana," in PNR, *Los problemas*, vol. 2, pp. 605–6; Luis T. Olivera, "Una planta substituya del maiz," *Agricultura* 1, 1 (June 1934): 19; "Trigo en vez de maiz en la alimentación," *El Nacional*, May 15, 1937; Winfield H. Minor, "Factors Relating to Importation of American White Potatoes into Mexico," May 12, 1936, NA, RG 166, Box 153; "El Instituto Experimental de Oaxaca esta dando buenas resultados," *El Nacional*, February 26, 1937; SAF, *Memoria 1937–1938* (México, D.F.: D.A.P.P., 1938), pp. 81, 102–3; Alberto A. Gutiérrez to DGA, September 10, 1937, AGN, Cárdenas papers, Exp. 606.3/164; "Intensificar el cultivo de la vid, reforzaría la campaña contra el alcohol," *El Nacional*, June 25, 1936; Julio Blumenkron, *Directorio de la industria azucarera en México* (México, D.F.: 1930), pp. 28–29.

35. SAF, *Sugestiones*, pp. 8, 38; Alanis Patiño, *Diversos*, pp. 67–68; "Campos experimentales para la resolución del problema agrario," *El Nacional*, February 15, 1935; Gándara, "Desarrollo del cultivo del maiz en México," *Agricultura* 1, 3 (November–December 1937): 11; Mario Macias V., "Pequeñas estaciones experimentales," *El Nacional*, June 10, 1931.

36. Beltrán, "Instituto Biotécnico," p. 175; Gabriel Ítie, "Estaciones agrícolas experimentales," *Chapingo* 1, 2 (September 1935): 22.

37. "Nuestra agricultura, los agrónomos, y la prensa," *Irrigación en México* 6, 2 (February 1933): 102; Gándara, "El desarrollo," pp. 10–11.

38. José R. Rosso, *El alfafero mexicano: El manual de los colonos y campesinos* (México, D.F.: D.A.P.P., 1936), p. 3; Adrián Cordero, "Estudio del Agrónomo Regional," in SAF, *Memoria de la 1o Convención Triguera Nacional* (Tacubaya, D.F.: Talleres Gráficos de la Oficina de Publicaciones, 1934), pp. 146–47; Antonio Rivas Tagle, "Miserissima producción individual del campesino, se impone mejorarla" (n.d.), ARFF, Box 123; DGA, "Use siempre semillas certificadas," October 5, 1931, ARFF, Box 327.

39. "El cultivo del maiz debe modernizarse," *El Universal*, December 17, 1933; Monroy, "Discurso," p. 2; Armando Martínez Aguilar, "Informe de Zona de Ciudad Guzmán, Jalisco," July 10, 1933, p. 8. ARFF, Box 113.

40. "Campaña en pro del mejoramiento de los cultivos de frijol," *El Nacional*, September 26, 1935; Carlos Torres Cordera, "Frijol: Síntesis económico-estadística," 1937, ARFF, Box 90; Romulo Delgado Crespo, "En Veracruz, el principal centro productor de frutas tropicales, la explotación del mango en el Sur del Estado," *El Nacional*, September 7, 1934; Sixto Colina Robles, "Estado actual de la producción de naranja en la region de Coatepec, Ver.," in CNAG, *Memoria . . . Naranja*, p. 2; Hermán, "El cultivo del ajonjolí," *El Nacional*, May 8, 1931.

41. Knight, "Racism, Revolution, and *Indigenismo*: Mexico, 1910–1940," in Richard Graham, ed., *The Idea of Race in Latin America, 1870–1940* (Austin: University of Texas Press, 1990), p. 82.

42. Beltrán, "Instituto Biotécnico," pp. 171–72; "Fragmentos del Acta Constituyente del Instituto Mexicano de Estudios Agrícolas," in Alanis Patiño, *Diversos*, n.p.; Leopoldo de la Barreda, "Enfermedades principales del trigo y la manera de tratarlas," in SAF, *Memoria . . . Triguera Nacional*, pp. 186–87; "Campaña en pro de los cacaoteros," *El Nacional*, April 18, 1937; "Estaciones agrícolas en todo el pais," *El Nacional*, November 29, 1935; SAF, *Sugestiones*, p. 150; José G. Parres to Governor of Jalisco, December 9, 1935, AHJ, Box AG-27-F; "Programa de trabajos que desarrollará la SAF," 1937, p. 11; José de Jesús Urquizo to Subsecretario de Agricultura y Fomento, April 20, 1936, AGN, Cárdenas papers, Exps. 606.3/164 and 534.4/101; Pandurang Khankhoje, *Maiz Granada "Zea Mays Digitata," su origen, evolución, y cultivo* (México, D.F.: Talleres Gráficos de la Oficina de Publicaciones y Propaganda, 1936); Gonzalo Torres Barrusta, "Investigaciones genéticas: Breve platica sobre genética y algunos conclusiones relativos al origen del maiz," *Agricultura* 1, 4 (January–February 1938): 14; "Características del maiz 'Cuatero del Bajío,' " *El Tractor* 2, 10 (March–April 1934): 6; SAF, *Características de los trigos cultivados en la República Mexicana* (México, D.F.: D.A.P.P., 1937), p. 32; "Informe de trabajos realizados sobre mejoramiento de plantas" (n.d.), RFA, RG 6.13, Series 1.1, Box 46, Folder 523; José Rodríguez Vallejo, "Reporte de los estudios para el control del chahuixtle de trigo en México" (n.d), p. 4, RFA, RG 6.13, Series 1.1, Box 45, Folder 520, RAC.

43. Ítie, "Estaciones," p. 22.

44. "Experimentación de trigo extranjero," *El Nacional*, July 7, 1941; "Han iniciado las siembras de tomate en Todos Santos, B.C.," *El Nacional*, November 11, 1937; J.J. Islas León, "Informe de Agencia General, Baja California Sur," June 6, 1935, and others, AGN, Cárdenas papers, Exp. 606.3/101; "Se inicio la convención algodonera," *El Nacional*, October 31, 1935.

45. Rivas Tagle, "Incremento de la producción de maiz: Mediante el escogido de la semilla," *Agronómica* 1, 2 (November 1931): 30–31.

46. Francisco Arias Leñero to Gumaro García de la Cadena, March 27, 1934, AMZ, Box Agricultura y Fomento 1933–1941, Exp. 6; Lorenzo Flores Fagoaga to Andres Macias, March 25, 1935, AHJ, Box AG-27-F, Exp. Folleto sobre el Maiz; Urquizo to Campesino, March 26, 1937; "Programa de labores que desarrollará la SAF," September–December 1937, AGN, Cárdenas papers, Exp. 606.3/164; Arturo R. Cueller, "El cruzamiento del tomate," *El Nacional*, December 27, 1935; Luis Marín, *El cultivo del algodón* (México, D.F.: Talleres Gráficos de la SAF, 1931), p. 40; Luis Marín, "Los abonos del cafeto," in CNAG, *Memoria . . . Cafeticultores*, p. 6.

47. SAF, *Memoria de trabajos de las Direccciones de Agricultura y Ganadería dependiente de la SAF durante el periodo presidencial de Lázaro Cárdenas* (México, D.F.: D.A.P.P., 1940), pp. 18–19; Juan A. González, "Influencia de las tarifas de transporte sobre la producción y venta de la naranja," in CNAG, *Memoria . . . Naranja*; "El cultivo de café se intensifica," *El Nacional*, May 9, 1937; Rivas Tagle, "Selecciona tu semilla," *San Luis Potosí Agrícola-Ganadero* 1, 4–5 (April–May 1937): 3; Figueroa to Presidente Municipal de Zamora, September 1933, AMZ, Box Fomento 1933, Exp. 3; DGA, "Concurso de mejores trabajos con arados," February 9, 1932; DGA, "Resultados de la campaña 'Cultivo del Nogal,' " February 18, 1932, AHSRE, Exp. IV-615–61; "Organizaciones agrícolas," *Extensión Agrícola* 1, 1 (February 1933): 21–22; Gutiérrez Roldán, "Los agrónomos," p. 9; Arrieta M., "La educación," p. 251; BNCA to Cárdenas, January 24, 1936; Ramón Sánchez to Cárdenas, January 6. 1936, AGN, Cárdenas papers, Exps. 530/24 and 534.6/309; "Las problemas que presenta el trigo en nuestro país," *El Nacional*, May 27, 1935; Vaughan, *Cultural Politics*, pp. 58, 64; SAF, *Memoria 1936–1937* (México, D.F.: D.A.P.P, 1937), p. 85.

48. Gutiérrez Roldán, "Los agrónomos," p. 9; "Nota editorial: Los bajos," p. 4; Mesa A., "La labor oficial," pp. 34–35.

49. Córdova, *La política*, p. 49; "Consejos Locales de Fomento Agrícola en todos los Estados," *El Porvenir* (Monterrey), August 3, 1934; "Instructivo para orientar el criterio del personal foraneo de la SAF sobre el CNAG" (n.d.), AGN, Cárdenas papers, Exp. 500/25; Enrique Acosta Raalu to Flores Fagoaga, February 7, 1935, AHJ, Box AG-27-F, Exp. 61.23, Legajo 1; José Carenga L. to productores de cacao, November 25, 1936,

ARFF, Box 26; "Programa de labores que desarrollará" and "Exportación de la naranja de Montemorelos," *El Universal*, October 9, 1937; "La plenaría de productores de naranja va a inaugurarse," *El Nacional*, August 4, 1936; Antonio Martínez Barragan to Presidente Municipal de Zamora, January 9, 1937, AMZ, Box Agricultura y Fomento 1936–1940, Exp. 1.

50. Becerra, "El capital en la agricultura mexicana," in PNR, *Los problemas*, p. 45; Gilberto Fábila, "La producción ejidal frente a la producción agrícola privada," in PNR, *Los problemas*, p. 388; Hermán, "El cultivo del manzano y los cuidados de el," *El Nacional*, May 4, 1931; Alfonso Alatorre to Secretario de Relaciones Exteriores, July 5, 1930, AHSRE, Exp. IV-287–11; Jesús Ceja Barajas to Presidente Municipal de Zamora, March 6, 1936, AMZ, Box Agricultura y Fomento 1936, Exp. 3; "Producción y consumo de higuerilla," *El Nacional*, February 4, 1937; Steven E. Sanderson, *The Transformation of Mexican Agriculture: International Structure and the Politics of Rural Change* (Princeton, N.J.: Princeton University Press, 1986), pp. 235–36; "Preferente lugar del maiz en la agricultura," *El Nacional*, January 3, 1936; DGA, "Oportunidad para nuestros agricultores, el ajonjolí," December 11, 1931, ARFF, Box 10; "Campaña de Agricultura," *El Nacional*, April 24, 1935; "El cultivo del algodón en Nayarit," *El Nacional*, March 28, 1935; Salvador Uribe Jasso, "El cultivo del olivo," *Agricultura* 1, 2 (September–October 1937): 53; "El mercado de la semilla de higuerilla en los Estados Unidos," *La Farmacía* 8, 8 (August 15, 1936), in AGN, Cárdenas papers, Exp. 564.1/262; Marín, "La higuerilla o ricino," *Agricultura* 1, 2 (September–October 1937): 14–27; SAF, *Memoria 1937–1938*, p. 112; "Lucha contra los insectos," *El Nacional*, October 6, 1937; Carlos Gajón Sánchez, "Flora útil del territorio sur de la Baja California," *Agricultura* 1, 3 (December 1937): 24; Francisco Vázquez del Mercado, "Irrigation in Mexico," *Bulletin of the Pan American Union* 71, 3 (March 1937): 264; Vázquez del Mercado to Presidente Municipal de Zamora, March 30, 1935, AMZ, Box Agricultura 1935, Exp. 2; "Nuevas Orientaciones," *Irrigación en México* 8, 1 (January 1934): 3; "El Gobierno del Estado de Jalisco," *El Nacional* (special edition), November 1935; Patiño, *Diversos*, p. 61; Rafael Prieto Sousa, "Labor social de la Comisión Nacional de Irrigación," *Irrigación en México* 15, 1–6 (July–December 1937): 3–4; "Mensaje," pp. 272–73.

51. Rafael F. Lazo, "Conservación y aplicación del estiércol," *El Campesino* (Durango) (April 1933): 13; Ernesto Montolongo, "El uso de los abonos y mejoradores," *San Luis Potosí Agrícola-Ganadero* 1, 4–5 (April–May 1937): 5–7; Sindicato Nacional de Agricultores to Consejo Nacional de Economía, "Es urgente la fabricación de nitratos sintéticos en México para fertilizar nuestras tierras agrícolas," January 14, 1931; Eduardo Turcott B., "Informe sobre el estado actual en que se encuentra la industria cafetera en el Estado de Oaxaca," June 27, 1933, ARFF, Boxes 1 and 29;

"Discurso pronunciado por el Secretario de Agricultura," *Agricultura* 1, 3 (December 1937): 1–4; "Mensaje de año nuevo," January 1, 1935, in *Palabras*, p. 199; Presidential Decree 311.1(011)/12925, *Diario Oficial*, April 2, 1936; "Seis milliones de pesos para la fundación de plantas nacionales de fertilizantes," *El Nacional*, January 5, 1937; "Un estudio sobre la industria del guano," *El Universal*, July 9, 1937; "Acuerdo que crea la Comisión Nacional para el Estudio y Fomento de la Utilización de Fertilizantes Nacionales," November 19, 1937, AGN, Robles papers, Agriculture Section, Box 1, Exp. 5; Jesús Gracia Fadrique, "El desarrollo de la industria química y la industria de las fertilizantes en México, 1759–1948," in Jesús Gracia Fadrique, ed., *Estado y fertilizantes* (México, D.F.: Fondo de Cultura Económica, 1988), p. 175; Macias, "El pulgón: Una de las plagas que ataca mas los cultivos de Jalisco," *Jalisco Agrícola y Ganadero* 1, 11 (October 1934): 3–10; Alfonso Madriaga, "Plagas de hemipteros en las naranjas," in CNAG, *Primera . . . Naranja*; Dampf, "Como defendernos," p. 11; SAF, *Memoria 1936–1937*, pp. 85, 105–6; Francisco García Robledo, "Las perdidas de la producción agrícola por plagas y enfermedades," in PNR, *Las problemas*, pp. 291–92; Lee A. Blohm, "Agricultural Items from Chihuahua," December 15, 1936, NA, RG 166, Box 146.

52. Fernández y Fernández to Secretaría de Hacienda y Crédito Público, September 7, 1936, ARFF, Box 131; Fernández y Fernández to Fernando Romero Quintana, October 11, 1937, in Enseñanza.

53. Ítie, "Estaciones," p. 22; Pablo Tamez González, "El porque de las deficiencias fisiológicas y del polimorfismo en una población de especies citricas," in CNAG, *Memoria . . . Naranja*, p. 4; Enrique Sanromán, "Los abonos en México," *Agricultura* 1, 2 (September–October 1937): 28–31; William C. Cobb, "Interview with Leonel Robles Gutiérrez" (n.d.), p. 5, RFA, RG 13, RAC; Flores, *Historias*, pp. 138, 222; Antonio C. Sandoval, "La ENA y el Consejo Nacional de Educación," *Chapingo* 1, 2 (September 1935): 8–9.

54. Enrique Arreguín Jr., "Proyecto para actividades de investigación científica y de enseñanza técnica, a cargo de sabios alemanes, austriacos, y españoles," June 28, 1937, AGN, Cárdenas papers, Exp. 534/100; Armando Esquivel, "Consideraciones que se hacen acerca de las reformas que se solicitan respecto del presupuesto de egresos, 1937," November 26, 1936, pp. 2–3, in ENA, BCM Bound Volume #168–2; Plenn, *Mexico Marches*, p. 93; Fernández y Fernández, *Chapingo hace 50 años* (Chapingo, México: ENA, Colegio de Postgraduados, 1976), pp. 115–16; García Cruz, "El problema legendario"; 1939, Mesa A., "Memorandum," p. 11; "Interview with José Rodréguez Vallejo," in Jiménez Sánchez, *Las ciencias agrícolas*, p. 379.

55. "Buenos resultados de las estaciones para agricultura," *El Nacional*,

October 5, 1936; Gándara, "El desarrollo," p. 10; "Importación de cereales extranjeros," *El Nacional*, November 17, 1937; "Informe sintético"; Cárdenas, "Informe," September 1, 1937, in *Palabras*, p. 107; "Editorial: Los Servicios de Extensión Agrícola en el pais," *Agricultura* 1, 2 (October 1937): 2.

56. Alan R. Sandstrom, *Corn Is Our Blood: Culture and Ethnic Identity in a Contemporary Aztec Indian Village* (Norman: University of Oklahoma Press, 1991), pp. 374–75; "Reparto de utilidad," *El Nacional*, February 14, 1937; Manuel Navarro Jr. to Rodríguez, November 19, 1933, AGN, Rodríguez papers, Exp. 334.1/831; Celso Muñoz to Cárdenas, April 10, 1935; Urquizo to Subsecretario de Agricultura y Fomento, April 20, 1936; Dionisio Garcia Leal to Cárdenas, August 11, 1936, AGN, Cárdenas papers, Exps. 500/24 and 565.4/70; Gómez, *Biografías*, pp. 482–83.

57. "El cultivo de lino se estimulará," *El Nacional*, June 26, 1936; "Aguascalientes, fuente de riqueza que se establece en esta entidad," *El Nacional*, March 30, 1936; Emilio López Zamora, Emilio, *El agua, la tierra, los hombres de México* (México, D.F.: Fondo de Cultura Económica, 1977), p. 118; "El ajonjolí, planta de porvenir en Nuevo Leon," (n.d.); Córdova to BNCE, January 25, 1938, ARFF, Boxes 10 and 28; "Cultivo y venta de ajonjolí en el Edo. de Sinaloa," *El Nacional*, December 3, 1936; "Los trabajos de las estaciones experimentales," *El Nacional*, September 12, 1936; Gonzalo González H., *La producción del ajonjolí en México* (México, D.F.: BNCA, 1941), cuadro 18; Confederación de Asociaciones Agrícolas de Sinaloa to Adolfo Diéguez, August 11, 1934, AHJ, Box AG-27-E, Exp. sin numero; "El cultivo de cacahuate," *El Nacional*, September 14, 1937; "El cultivo del café se intensifica," *El Nacional*, May 9, 1937; Bernardo Ávilez, *Cultivo y comercio de algodón en México* (México, D.F.: SAF, Dirección de Economía Rural, 1939).

58. "La hacienda de Santa Inés el exponente de la obra del Estado en pro del campo," *El Jaliscense* (Guadalajara), November 17, 1934; González H., *La producción*, cuadro 4; Islas León, "Informe, Agencia General, Baja California Sur," June 6, 1935; and others; Alberto Escribano to Cárdenas, May 27, 1935; José Flores to Cárdenas, July 30, 1936, AGN, Cárdenas papers, Exps. 505.2/3, 606.3/101 and 703.4/187; "La industria tomatera en California," *El Nacional*, December 10, 1935; "Weekly Report, Pink Bollworm and Thurberia Weevil Project," October 11 and 25, 1937, NA, RG 7, Box 1400; "No se permite cultivar zoca," *El Siglo de Torreón*, February 18, 1937; Sixto Colina Robles to Director de Fomento Agrícola, June 2 and December 5, 1934, AHJ, Box AG-27-E, Exp. 022, Legajo 2; Jesús Vanegas to Departamento de Control Agrícola, October 14, 1935, AHJ, Box AG-27-F, Exp. Informe de la Inspección a la 8a Zona Agrícola; Abraham Iszacvich, *Modernización campesina* (México, D.F.: Editorial Edicol,

1980), pp. 40–41, 125; SAF, *Informe de Labores 1941–1942,* (México, D.F.: Editorial Cultura, 1942), p. 44.

59. "Programa de labores que desarrollará"; Secretaría de Economía Nacional, *El café: Aspectos económicos de su producción y distribución en México y en el extranjero* (México, D.F.: Editorial Cultura, 1933), pp. 104–7; Cedillo to Governor of Jalisco, March 9, 1936, and others, AHJ, Box AG-27-F, Exp. 61.20, Legajo; People of Arandas to Departamento de Control Agrícola, March 15, 1936, AHJ, Box AG-27-G, Exp. 61.25, Legajo 7; "Acuerdos resolutivos de la Primera Convención Nacional de Productores de Naranja," August 15, 1936; Fernández y Fernández to BNCA, September 7, 1936, ARFF, Box 131.

60. Urquizo to Cárdenas, March 28, 1936; "Programa de trabajos que desarrollará la SAF," 1937, 24–29, AGN, Cárdenas papers, Exps. 534.4 and 606.3/164; "Informe sintético"; Vanegas to Departamento de Control Agrícola, October 31 and November 10, 1935, AHJ, Box AG-27-F, Exp. Informe de la Inspección de la 8a Zona; Lerner, *La educación,* pp. 129–131; Vaughan, *Cultural Politics,* pp. 85–86.

61. Ankerson, *Agrarian Warlord,* p. 136; Frank Tannenbaum, *Mexico: The Struggle for Peace and Bread* (New York: Alfred A. Knopf, 1950), p. 75; González y González, *Los dias del Presidente Cárdenas,* (México, D.F.: El Colegio de México, 1981), p. 155; "Asuntos tratados por el Consejo de Agricultura del Estado," June 12, 1934, AHJ, Box AG-27-E, Exp. 100, Legajo 6; Flores Fagoaga to Governor of Jalisco, November 16, 1935, AHJ, Box AG-27-F, Exp. Santa Ines 1935; Urquizo to Cedillo, April 20, 1936, AGN, Cárdenas papers, Exp. 534.4/101; Casamiro Rosada and Jorge Castillo to Rodríguez, September 18, 1933; José G. Parres to Luís Echeagarary, September 6, 1933, AGN, Rodríguez papers, Exps. 139.73/4 and 139.3/3; Pablo E. Macias to Cárdenas, October 17, 1936, AGN, Cárdenas papers, Exp. 506.17.6; Fernández y Fernández, *Chapingo,* p. 140; "Informe sintético"; Gutiérrez to Cárdenas, February 20, 1937; "Plaga en las plantaciones," *El Nacional,* October 19, 1934; Parres to Cárdenas, July 17, 1935; Gándara to Cárdenas, August 17, 1937, and others; Rodrigo M. Quevedo to Cárdenas, January 12, 1936, and others; Luís Segura Garza to Cárdenas, June 26, 1935, AGN, Cárdenas papers, Exps. 506.17/6, 703.2/642, 702.1/84, and 506.17/7; SAF, *Memoria 1936–1937,* pp. 103–4; Miguel Llamas y Gómez to Rodríguez, March 5, 1934, AGN, Rodríguez papers, Exp. 561.32/59; McDonald to Strong, March 25, 1935, NA, RG 7, Box 1401; Baker to Strong, February 6, 1935, NA, RG 7, Box 1392; Cordova to BNCE, January 25, 1938; Gómez, *Biografías,* p. 138; Ann L. Craig, *The First Agraristas: An Oral History of a Mexican Agrarian Reform Movement* (Berkeley: University of California Press, 1983), p. 127.

62. George H. Winers, "Review of Commerce and Industry, Guada-

lajara," 1934, NA, RG 166, Box 148; Adrian A. Bantjes, *As If Jesus Walked on Earth: Cardenismo, Sonora, and the Mexican Revolution* (Wilmington, Del.: Scholarly Resources, 1998), p. 50; Luis González y González, *San José de Gracia, Mexican Village in Transition*, John Upton, trans. (Austin: University of Texas Press, 1974), p. 189; "Se establecerá un campo de abastecimiento de semillas de trigo," *El Tractor* 2, 13 (October 1934): 4; "Campaña para la explotación racional del suelo," *Extensión Agrícola* 1, 4 (May 1933): 18; A. Macias to Governor of Jalisco, December 30, 1933, AHJ, Box AG-27-E, Exp. 22, Legajo 2; Enrique González C. to Departamento de Control Agrícola, September 1, 1937, AHJ, Box AG-27-I, Exp. Agricultura y Ganadería, 1939; *Comisión Nacional de Irrigación, Estudio Agrológico de la Cienega de Chapala, Proyecto de Los Corrales, Michoacán-Jalisco* (México, D.F.: Editorial Cultura, 1932), pp. 143–64; Colina Robles to Director de Fomento Agrícola, October 2, 1934, AHJ, Box AG-27-E, Exp. 022, Legajo 2.

63. Bowman, "Agriculture and Agrarianism in Mexico," November 12, 1936, p. 5, NA, RG 166, Box 146; Gledhill, *Casi Nada*, p. 99.

64. "Función de las Escuelas Dependientes del Departamento de Enseñanza Agrícola y Normal Rural," (n.d.), AGN, Cárdenas papers, Exp. 534/62; *Ley que creo el Consejo Nacional de la Educación Superior y la Investigacion Científica* (México, D.F.: 1936), p. 8; Palacios to Cárdenas, May 3, 1937; Palacios to Governor of Chiapas, March 10, 1937, AGN, Cárdenas papers, Exp. 506.15/1; Miguel García Cruz, "El problema legendario de la ENA," 1939, p. 2, in Enseñanza; Fernández y Fernández to Gutiérrez Roldán, October 8, 1937, in Enseñanza; Eyler N. Simpson, *The Ejido: Mexico's Way Out* (Chapel Hill: University of North Carolina Press, 1937), pp. 579–80; Lerner, *La educación*, p. 147.

65. Nicolas Sánchez Durón, "La investigación agrícola y el desarrollo agropecuario de México," *Anales de la Sociedad Mexicana de Historia de la Ciencia y de la Tecnología* 2 (1970): 177; Alan Knight, "Cardenismo: Juggernaut or Jalopy," *Journal of Latin American Studies* 26, 1 (February 1994): 73–107; "Se necesita un organo de gobierno que haga justicia al campesino," *El Popular*, October 17, 1938; Randall Collins, "Market Closure and the Conflict Theory of the Professions," in Michael Burrage and Rolf Torstendahl, eds., *Professions in Theory and History* (London: Sage Publications, 1990), pp. 24–25; Gómez, *Biografías*, pp. 154–55, 247, 312.

66. "Campesino ejidatario," *La Tierra* 1, 2 (May–June 1938): 3; González y González, *San José de Gracia*, pp. 49–50, 184–85; Gledhill, *Casi Nada*, p. 52; Paul Friedrich, *Agrarian Revolt in a Mexican Village* (Englewood Cliffs, N.J.: Prentice-Hall, 1970), p. 81; Roy V. Scott, *The Reluctant Farmer: The Rise of Agricultural Extension to 1914* (Urbana: University of Illinois Press, 1971), p. 217; Svante Beckman, "Professionalization: Borderline

Authority and Autonomy in Work," in Burrage and Torstendahl, eds., *Professions*, p. 128; Vaughan, *Cultural Politics*, p. 23, 192; Pilcher, *¡Que vivan!*, p. 111; John Tutino, *From Insurrection to Revolution in Mexico: Social Bases of Agrarian Violence* (Princeton, N.J.: Princeton University Press, 1986), pp. 26–28.

Chapter 4

Agronomy under Attack: The Demise of the Cultural Campaign and the Roots of a New Agronomy, 1938–1943

Although some scholars claim that Cárdenas' land reform stimulated farm productivity, including in the *ejidos*, fostering Mexico's post-1940 industrial revolution, from 1937 to 1940 pests and diseases, insufficient production, and inadequate distribution of food, commercial, and export crops forced him to spend scarce foreign exchange on imports, caused a crisis that led to criticism of his agricultural policy, and gave credence to the enemies of agrarian reform. Corn shortages occurred in Jalisco, Yucatán, Aguascalientes, La Laguna, and Ciudad Juárez, and in 1940 Mexico spent about 1 million pesos to import it. *Chahuixtle* ravaged Mexico's wheat fields and in 1940 destroyed nearly 80% of Sonora's wheat crop, requiring $2 million for imports during the first three months of 1941. In 1938 scarcity caused the price of beans to quadruple in Mexico City. Rice production could not satisfy demand in Veracruz. Supplies of both products remained short in 1939, and popular discontent and denunciations of merchants and the government ensued. During the 1930s the soap industry boomed, leading to a fourfold increase in coconut oil imports. In 1939 war in Asia threatened to disrupt access to supplies, and some Mexicans urged the SAF to promote oilseed crops. Mexico exported many Roatan bananas, which brought some prosperity to Tabasco, but in September 1937 *chamusco* (known elsewhere as Sigatoka disease) appeared. The spore already struck plantations in Central America, forcing the United Fruit Company (UFCO) to invest several million dollars to control it. By 1940 all of Mexico's banana-growing regions were heavily infested. Cacao cultivation dated to pre-Columbian times, but during the 1920s production declined when Tabasco's growers switched to the more profitable banana, and nearly half of consumption was imported. Many Mexicans saw a serious farm crisis. The head of the conservative *Unión Nacional de Vetranos de la Revolución* urged Cárdenas to send the army to the fields to increase production.[1]

These economic and political problems were the "proximate cause" of the Green Revolution. Many scholars call 1940 a crucial turning point in Mexican history. Regarding agricultural policy, they claim that President Manuel Avila Camacho (1940–1946) began a period when the State focused on urban problems and neglected the *ejidos,* and some say that the Rockefeller Foundation shaped this agenda. By 1940 many *agrónomos* were already tired of working with *campesinos;* they belonged to or allied with the "urban elite" who supported the foundation; they were not defenders and students of peasant farming "little interested in importing technologies from abroad," as one author describes them.[2] As political leaders and peasants redefined or renegotiated the Revolution, the crisis began the process that transformed the *agrónomos* from *agraristas* to *técnicos.* Their cultural campaign and *agrarismo* tied them to the State but also alienated many Mexicans. Critics of the land reform found easy scapegoats, and they continued their attacks after Avila Camacho took office and World War II began to affect Mexican agriculture. While the foundation contemplated expanding its activities in Mexico, *agrónomos* worried about their professional status. Dissatisfied with the ineffective and unpopular cultural campaign, they considered other ways to establish their credibility as experts. The MAP and other U.S. wartime technical assistance projects provided them.

THE CULTURAL CAMPAIGN AND THE CRISIS

The Ministry of National Economy said that success promoting more remunerative crops, increased *campesino* food consumption, the scarcity of fertilizers, and the loss of peasant laborers to public works projects caused the crisis, and the SAF also blamed the weather. To address the situation, the State intensified the cultural campaign. Cárdenas asked governors to spend more money on agricultural programs, and governors in Mexico, Colima, and Jalisco cooperated. Parres planned to "organize, control, and direct animal and plant production in all its aspects" and to use "ideology" as well as "competence" as criteria for selecting new SAF employees. In 1940 he agreed to "supervise ejidal politics, land distribution, communal property, contracts, and protection of women's and children's rights."[3] Copying the SEP, the SAF continued its cultural campaign, in Cárdenas' words "to prepare young *campesinos* to improve the crops of our lands," in new primary and vocational agricultural schools. The school in Juchitán, Oaxaca, had 28 students and gave 184 adults "lessons in agriculture."[4] The *agrónomo* in charge formed producer associations and pest control brigades, distributed fruit and mulberry trees, and raised livestock, corn, cotton, alfalfa, and vegetables. On Sundays the director of Tekax, Yucatán's Vocational Agricultural School, organized sporting events and lectured on seed selection and disinfection. Expressing concern about the *campesinos'* health, *agrónomo* Salvador Ramírez M. campaigned against alcohol in *ejidos* near the Tuxpán, Veracruz, Experimentation Station.[5]

Disregarding an internal report questioning the effectiveness of the *Campaña en Pro del Maiz,* to solve the corn crisis the SAF expanded it and told reporters that it would increase national production by 50%. Newspapers urged *ejidatarios* to plant corn instead of more remunerative crops. Cárdenas appealed to nationalism, taking advantage of the surge in patriotism after the oil expropriation. The SAF sent teams of *agrónomos* to Michoacán, Mexico State, Guerrero, and the Bajío to promote the *Campaña* and ordered the *Agrónomos Regionales* in Hidalgo, Nuevo León, Tamualipas, Puebla, and Morelos to give demonstrations and distribute disinfected seed. The SAF made the *Campaña* mandatory for all farmers under its influence, distributed literature, and ran newspaper and magazine articles to convince other cultivators to change their practices. The agricultural banks claimed to favor loan applications from farmers who selected, disinfected, and tested their seed, and the CNI gave them first access to water. The SEP and the Department of Indian Affairs also promised to contribute. A new agricultural school in Tlaxcala provided free seed disinfection. In Jalisco the BNCE distributed disinfectant to treat 17,000 hectares of corn. In June 1938 the SAF claimed that 10% of maize growers used "modern practices" to obtain yields nearly two times higher than those of traditional farmers, but in 1939 the crisis deepened. *Agrónomos Regionales* in Tamualipas, Jalisco, Hidalgo, and Puebla lectured *ejidatarios* on the *Campaña*. The SAF worked with rural teachers, the Department of Indian Affairs, and Jalisco's agricultural inspectors, giving them seed disinfectant and instructions on its use, and distributed more *Campaña* literature. The ENA held a National Seed Selection Contest, awarding prizes for the best examples of corn, wheat, beans, and barley. The *Instituto Biotécnico* issued a new pamphlet claiming that *campesinos* could earn "profits they never dreamed of" by cooperating with the *Campaña*. In April 1940 the CNAG finally held a National Corn Growers Convention, but the SAF cut funding for *Campaña* literature and disinfectant.[6]

Despite the crash program, other goals prevented a total commitment to maize farming. Reiterating the tortilla discourse, *agrónomo* Roberto Galvez wanted peasants to grow and eat peanuts, and General Rubén García urged Cárdenas "to fight so that less corn and more wheat, fruit, meat, etc. is consumed."[7] The SAF thought that potatoes were more nutritious than corn and told *campesinos* to grow them, establishing demonstration plots in six states, and continued to promote sunflowers, giving seed to *ejidatarios* in Morelos, Sonora, Hidalgo, Mexico State, Jalisco, and Querétaro, growing them in Nuevo León, Tamualipas, Guanajuato, and Michoacán, and testing them as forage at the Briseñas Station. Cooperating with the Ministry of Public Health, the SAF planted soybeans in three states "to accustom the *campesinos* to consuming" them, and ran articles and distributed vegetable seeds to convince peasants to grow them "to provide sustenance with the nutritional elements and vitamins needed for a proper diet."[8]

Ignoring Cárdenas and *agrónomos* who said that the switch to more profitable crops caused the corn shortage, as well as the editors of *El Mundo* and *El Dictamen*, who ridiculed the idea that peasants should not grow maize, the State continued to urge *campesinos* to try them. With the support of the Ministry of the National Economy, Parres told reporters who asked about the crisis: "Mexico must abandon corn and bean cultivation to produce products more profitable and suited to our agrological conditions."[9] To improve the *campesinos'* diet and increase their income, the government distributed more fruit trees, including to the most marginalized rural Mexicans like the Cora Indians of Nayarit, who received 300 peach saplings. In 1938 the SAF proudly announced that over 120,000 fruit trees were distributed "to replace corn, beans, and other less remunerative crops."[10] With similar motivations the SAF tested and promoted alfalfa, sorghum, vanilla, pineapples, and other crops. Inspired by a visit from a specialist from New Orleans and a project developed in Pennsylvania, the *Instituto Biotécnico* prepared a pamphlet and built ponds to convince *ejidatarios*, especially Native Americans, to raise U.S. edible frogs. The SAF also began several new projects to address economic problems caused by the war, including promoting coconut palms to supply copra, as well as cacao, quinine, rubber, and other oilseed crops. In 1939 *agrónomo* Conrado Rodríguez y Rodríguez proposed an intensive oilseed campaign as a "substitute for cereals, whose production does not make economic sense in many regions."[11] An example of *indigenismo*, Gándara wanted *campesinos* to grow "*xochinactle*," an herb that the Aztecs used to flavor chocolate. The SAF continued to organize farmers and build small-scale irrigation works, especially in Michoacán.[12]

Urging farmers to make seed selection "an on-going practice in their fields," SAF corn breeder Gonzalo Torres Barusta claimed that "genetic intervention to improve crop varieties" was "now accessible to all people."[13] Inspired by the tortilla discourse and research at the University of Illinois, Rivas Tagle urged corn growers to create high-protein varieties and potato farmers to organize private plant-breeding firms. The SEP told *campesinos* that they could develop their own improved strains, claiming that children could do it in school gardens. Reflecting *agrónomos'* rising interest in plant genetics, the SAF expanded its corn-breeding program and began or evaluated new projects with wheat, beans, sesame, cotton, and bananas. The SAF distributed seed of a few new strains, like "Chalco" corn and "Glorieta de México" beans, to *ejidatarios*, and the BNCA purified chile varieties at Pabellón, but *agrónomos* still told *campesinos* to plant foreign seeds, and the SAF tested or distributed imported wheat, corn, potatoes, sugarcane, peanuts, barley, sesame, rice, and cotton. In 1939 the SAF ordered La Laguna's cotton growers to plant only "Marvel of the South," in the Juarez Valley and "Acala" and "Mebane" in Matamoros. The SAF hoped that foreign corn would improve the *campesinos'* varieties by random hybridization.[14]

The SAF claimed that farmers' failure to use chemical fertilizers contributed to the crisis. Some *agrónomos* still supported organic methods, but in 1938 *El Nacional* proclaimed that "the imperious necessity of using another method of fertilization, known as chemical fertilizers . . . in a widespread manner," and León's potato farmers, Senator Gonzalo Bautista of Puebla, Modesto, Rolando of the Ministry of National Economy, and the LAS wanted the State to manufacture them, in the latter case to build an industrial economy to foster the transition to socialism.[15] The SAF tested them on corn, alfalfa, coffee, oranges, and potatoes; the Ministry of National Economy experimented in Mexico State and the Federal District; and the BNCA tried them on wheat and chile at Pabellón. The agricultural banks granted loans to help farmers in Mexico State, La Laguna, and the Pabellón Irrigation District purchase chemical fertilizers, and the *Confederación Nacional Campesino* (National Peasant Confederation, the "official" peasant league, hereafter CNC) urged *ejidatarios* to try them. Similarly, SAF experts like Gándara and José Figueroa favored biological pest controls and natural insecticides like rotenone, believing that chemical pesticides harmed farmers' health and land, but the BNCE promoted them in La Laguna, and the Oaxaca Station showed *ejidatarios* how to use them to control ants.[16]

Responding to politicians' and *agraristas'* pleas, the SAF sent *Instituto Biotécnico* and DGA experts to assess the *chamusco* outbreak. After destroying the sick trees that they found, they said that the spore was under control and did not threaten the banana industry. In early 1938, when the plague spread and appeared in other regions, the SAF organized an eradication campaign, but the slow arrival of funds delayed it for several months. Parres sent Jeannot Stern of the *Instituto Biotécnico* to Honduras to study UFCO's efforts to control *chamusco*, but to save money the SAF chose a cheaper but less effective powdered fungicide. With rejection rates soaring and export revenue falling, in 1939 the SAF admitted that its campaign was ineffective and the situation desperate, but Cárdenas allocated less money and released it slowly. Growers ran out of funds, spraying stopped, and the spore spread like wildfire through banana-growing regions. *El Universal* said that farmers in Tuxtepec, Oaxaca, left the area after the disease destroyed their trees. In Tabasco *chamusco* damage was devastating. State revenue declined dramatically, teachers were not paid, trade unions protested, and U.S. buyers suspended operations. By

1941 Tabasco exported no bananas. The spore also infested plantations in Chiapas, Oaxaca, and Veracruz.[17]

Other problems plagued the SAF. Poor record keeping ruined data from tests of wheat, sesame, and beans. Overwhelmed by the *chamusco* crisis, the government could do little to fight locusts in Yucatán, the "*chapulin*" (a type of cricket) in Sinaloa and Tabasco, or *chahuixtle* in the wheat fields of Sonora and the Bajío. In 1939 worms infested the new linseed plantations in Arandas, arresting the industry's development. Ineffective pest and disease control programs eroded farmers' faith in the government when the blights damaged or destroyed their crops. Farmers who participated in the sesame campaign were angry when the SAF had no seeds to distribute. The agricultural schools reported poor attendance, which the SAF attributed to the *campesinos'* poverty and need for child labor. Jalisco's municipal presidents often hired relatives, or *compadres,* as agricultural inspectors, but they did not know how to apply seed disinfectant and sometimes overcharged farmers for the service. The State could not afford to build experimentation stations and a laboratory or begin other agricultural programs. Antonio Martínez Barragan, head of the CNAG, admitted that the vastness of the problem overwhelmed his attempt to organize Mexican farmers and introduce new technologies. To address the situation, the SAF planned to use "persuasive but also coercive methods."[18]

AGRONOMY UNDER ATTACK: THE *AGRÓNOMOS* REDEFINE THEIR PROFESSION

Despite the crisis, some Mexicans supported the SAF and the *agrónomos.* With the outburst of nationalism after the oil expropriation, government employees, *agraristas,* trade unions, and even conservative groups like the *Unión Agrícola Regional de Chapala* praised Cárdenas' effort to increase farm production, and *ejidatarios* from Guerrero, Jalisco, and Oaxaca, municipal presidents, and other members of the Revolutionary Family asked him to bring them modern agricultural science. Some *agrónomos* had local support, such as José O. Chávez, whose friends included the governor and *Diputados Federales* of Morelos; Ramírez M., praised by the *agraristas* of Alamo, Veracruz; and Manuel de Regolos, backed by the governor of San Luís Potosí. The SAF often told the press that its programs attained their goals and received enthusiastic cooperation from the *campesinos*. In 1938 the agency lauded the soybean campaign because farmers grew them in "regions where it was unknown" and claimed that production of wheat and corn in *ejidos* increased, the shortages notwithstanding.[19] In 1940 Parres said that the Bajío's farmers always selected, disinfected, and tested their seed and claimed that the SAF "taught the *campesinos* the necessity of relying on the technical advice of *agrónomos*."[20] To assist and benefit from the State's commitment to economic development, more young Mexicans joined the profession. From 1938 to 1943 the ENA granted 136 degrees, 53 in 1943, and Antonio Narro issued 98. The agrobureaucracy continued to grow, and new agencies like the *Comité Reguladora de Subsistencias*, which bought and sold farm goods, hired *agrónomos*. In 1938 and 1939 the ENA's directors, especially the conservative Soberón, refused to enroll more *campesinos*, and most ENA students came from the urban middle class. With the "social importance of the white collar" they favored jobs in Mexico City over fieldwork in rural areas.[21]

Some of the SAF's projects succeeded. With the help of El Yaqui's *agrónomos*, by 1939 Sonoran farmers grew foreign wheats tested for *chahuixtle* resistance. In 1940 Alfonso Díaz del Pino, director of the Tlalnepantla Station, supervised the successful cultivation of alfalfa in four *ejidos*, which earned more revenue than the corn, beans, and barley that the *ejidatarios* usually grew. The SAF convinced some *de temporal* corn growers in the Valley of Mexico to plant Burbank wheat and popularized the "Marquis" strain in the Bajío. By 1941 imported varieties brought some prosperity to Nuevo León's orange growers, and those in Atotonilco el Alto and Ayo el Chico, Jalisco, used the organization that the State formed to improve the quality of their trees, pest control efforts, and cultivation practices. An SAF and South Pacific Railroad lime and lemon campaign led to a significant increase in production. Sonora's vegetable cultivators' cooperative "main-tain[ed] close contact with the agricultural agents of the Federal and State governments and follow[ed] their instructions."[22]

However, the crisis unleashed an outburst of criticism of Cárdenas' agricultural policies. In 1938 the National Chamber of Commerce claimed that Mexico did not produce enough corn because the State gave land owned by "the industrious" to people who, "not being legal owners, do not care about making it productive."[23] Ex-president Calles and the *Unión Nacional Sinarquista*, a quasi-fascist peasant league, agreed. Cattle ranchers in Jalisco and coffee growers in the south refused to innovate without exemptions from the Agrarian Codes. In 1938, fearing imminent expropriation, Sonora's tomato growers left their fields fallow. Even the pro-government *El Universal* admitted that land reform contributed to the crisis. The conservative *Excelsior* claimed that wheat imports proved that the *ejido* had failed and said that Parres had mismanaged the SAF, which forced Mexico to import corn, wheat, and beans. Civil engineer Jorge Tamayo chastised him for not establishing an experimentation station in La Laguna. Praising the practices and prosperity of U.S. cultivators, Jalisco's wheat growers said that the State did nothing to help them.

Cárdenas' enemies also attacked *agrónomos*. Denouncing several who sold mules to *ejidatarios* in La Laguna, *Excelsior* said that corrupt bureaucrats ran the State's agricultural programs. In Chiapas the *Comité de Orientación Social* lambasted the SAF's *Agente General*. To the right-wing *Confederación de Partidos Independientes*, "communized, irresponsible, and inefficient" *agrónomos* wanted "absolute control" over *ejidatarios*. The *Unión Agrícola Regional de Chapala* claimed that the CNA's *agrónomos* treated *pequeños propietarios* (small owners or proprietors) unfairly and told president-elect Avila Camacho that *campesinos* "do not understand the importance of selecting their seed and applying disinfectant."[24] Even *agraristas* impugned their integrity. Bruno Jiménez of Tabasco's Ejidal Banana Growers Union claimed that the personnel assigned to the *chamusco* campaign spent most of their time drinking in Villahermosa's bars and urged Cárdenas to investigate and punish the guilty. Farmers' complaints led to *Agente General* J. Jesús Islas León's transfer from Baja California to Veracruz, where he replaced an *agrónomo* denounced by the *Sociedad Local de Crédito Agrícola* for "divisionist labors," but trade unions said that he "slandered and defamed workers" and demanded his dismissal.[25] SAF employee Guillermo Rodríguez G. told Cárdenas that *agrónomos* had no credibility with *campesinos* because they knew the theory but not the practice of farming. To some critics, at the ENA *agrónomos* learned to enjoy city life, became enraptured by *empleo-*

mania, and lost their desire to solve farmers' problems; even the SEP said that Mexican experts knew little about agricultural conditions or how to improve them.[26]

Evidence indicates that many farmers had no faith in *agrónomos* or the SAF and resisted their programs. The *campesinos* of La Bartolilla, Michoacán, did not appreciate the services offered by the SAF's experimentation station, which included lectures on the *Campaña en Pro del Maiz*, growing wheat, sunflowers, sorghum, linseed, vegetables, and fruit trees, beekeeping, silk production, and free seed disinfection and livestock vaccination, for in 1938 they convinced Cárdenas to turn its lands into an *ejido*. Pabellón's farmers got Cárdenas to reverse his decision to give the BNCA's experimentation station to the SAF. In 1940 the SAF abandoned the sunflower campaign, partly because Mexico had no factory that made oil from the seeds and also "due to the difficulties in suddenly introducing a new plant into the people's diet."[27] *Agrónomo* Joaquín de Zayas claimed that potato growers refused to use the pest and disease control practices that he recommended. ENA professor Leopoldo de la Barreda said that farmers felt that quarantines were "unjustified sanctions against the people."[28]

Stung by the criticism and valuing their role in the State, the *agrónomos* reacted strongly to the crisis. In February 1939 the SAM announced an alliance with the CNC, lauded the profession's *agrarista* heritage, including its struggle to sustain the agrarian reform during the crisis of 1930, advocated more land distribution, and denounced its critics. The CNC backed the SAM's demand that agronomy be defined as a profession of government service. Refuting critics who said that they stayed indoors, *agrónomos* claimed that their "direct contact with the peasant masses" made them the only professionals who understood the *campesinos'* problems.[29] They helped peasants get land. For example, Ramírez A. organized a new *ejido* for longshoremen who lost their jobs because of *chamusco*. At the millworkers' request, the *Agrónomo Regional* in Zamora studied "the economic and social conditions" of a sugar plantation to "determine the convenience of its expropriation."[30] Reflecting the cost of membership in the Revolutionary Family, the SAM supported candidate Avila Camacho but also expelled two *agrónomos* who backed General Almazán, champion of devout Catholics and the Right. Further, "political enemies" murdered *agrónomo* Manuel Maldonado Moreno, who actively campaigned for him.[31]

In small concessions to traditional agriculture, in 1938 the SAF asked *campesinos* to share their knowledge of farming, and *agrónomo* Quintín Ochoa deemed the disease control practices of potato farmers in Chihuahua's mountains rudimentary but effective, but *agrónomos* still blamed *campesinos* when their programs failed. Calling the peasants of Tuxpán "indolent," Ramírez A. said that his station made little progress because of "the inefficiency of the people" who "do not need to work."[32] A commission that Cárdenas sent to the coast of Guerrero denounced the peasants for doing nothing to improve their farming methods. Gilberto Mendoza Vargas said that the "the apathy and opposition of banana growers" was a tremendous obstacle to the *chamusco* campaign, and he claimed that a grower threatened to shoot SAF personnel if they entered his farm to look for infected trees.[33] Diseases killed many fruit trees and grape vines that the SAF gave to Pabellón's farmers, who did not use the cultivation practices that *agrónomos* recommended. Reflecting hostility toward peasant farming, with the help of the agricultural banks, the SAF gave *campesinos* iron plows to replace traditional implements, but farmers in Puebla, many of them Native Americans, refused to cooperate and used digging sticks to sow their fields. *Agrónomo* Manuel Santuario decried *campesinos'* "marked resistance" to new crops, and his colleagues lambasted coconut, corn, rice, pineapple, and wheat growers for not selecting their seed properly or for using poor cultivation practices. Reflecting a negative opinion of *campesino* corn, the SAF wanted to prevent "the hybridizations, morphological and ecological changes it keeps suffering with the worst results for farmers."[34]

Many *agrónomos* wanted to force *campesinos* to follow their advice. The 129 members of the *Frente Revolucionario de Agrónomos Mexicanos* (Revolutionary Front of Mexican Agronomists, hereafter FRAM) told agricultural banks to lend *campesinos* machinery, seeds, and fertilizer rather than cash and thought that land in new irrigation districts offered "optimal conditions of control" that meant that it should be given only to farmers who agreed to follow their directions, while the LAS wanted all cultivators who participated in agricultural programs to sign contracts to work "under the direction of the Ministry's technical personnel."[35] In 1938 the BNCA considered using credit to "force" the "evolution of agricultural technique"; in 1939 the bank refused to give loans to Veracruz coffee growers who did not farm "scientifically," *agrónomo* Pablo Córdova urged the BNCE to adopt the same policy, and Parres, who lamented "the deficient preparation of our peasant masses, especially in Indian communities and many of the *ejidos*," wanted to eradicate the "primitive technologies most of the nation's farmers use."[36]

Many *agrónomos* believed that the cultural campaign's failure diminished their prestige. They urged the SAF to abandon correspondence courses, fairs, and the distribution of written material because *campesinos* did not benefit from them. Fernández y Fernández and Mesa A. thought that internal turmoil destroyed the SAF's credibility, especially "under the heavy boots of [the] ignorant provincial" Cedillo.[37] With the support of the *Confederación de Trabajadores Mexicanos* (Confederation of Mexican Workers, CTM), *agrónomos* urged the SAF to forget "the economic aspects of the organization of production" and "the social improvement of farmers."[38] Miguel García Cruz claimed that politics dominated the agency, detracting from research and extension programs and eroding farmers' confidence. By 1940 disillusion pervaded the SAF, and some *agrónomos* wanted to reject the *campesinos*. To increase production, they urged Cárdenas to repair relations with *pequeños propietarios*, "which is not hostile to *ejidatarios* or the spirit of the Mexican Revolution."[39] The CNI urged him to grant lands irrigated by the new La Angostura Dam to Yaqui Valley farmers whose farms were expropriated. In 1938 *agrónomo* Aaron Merino Fernández helped the Avila Camacho family protect sugar baron William O. Jenkins' lands from the actions of *zapatista* peasants in Puebla. Luis León, a *Callista* who lost his job as director of *El Nacional* during Cárdenas' struggle with the *jefe máximo*, decided to work with private farmers in Sonora.[40]

However, with land reform under siege, as *agraristas*, most *agrónomos* wanted to save it: if the first phase of the agrarian revolution was land distribution, the second was to turn the land into productive farms. In 1938 Cárdenas doubled the ENA's budget, which improved student services and the quality of the faculty, but not enough to impress Romulo Escobar or the Rockefeller Foundation's survey team. The school's directors refused to introduce specialized studies, but *agrónomos* knew that State-supported science would sustain their membership in the Revolutionary Family, and more of them began to define agronomy as a domain accessible only to trained specialists. They asked for training in experimental methods, especially plant breeding. Backed by the CNC, the LAS urged Cárdenas to provide more funds for research, "the only way to make our agriculture progress," including experimentation stations where *agrónomos* could produce certified seed and improved corn, wheat, bean, potato, pineapple, banana, sugarcane, and cotton varieties.[41] To develop a research methodology, the socialist Durán tried to stimulate his colleagues' interest in plant nutrition and soil science, and Edmundo Taboada used the methods that he recommended to test chemical fertilizers on chiles. Mesa A. wanted *agrónomos* to be "true men of science," and José Figueroa claimed that Mexico needed scientists like those in "developed countries."[42] Backed by the CNI, *agrónomos* told Parres to hire skilled professionals rather than political favorites to end the SAF's role as a refuge for "unproductive soldiers from the Revolution," limit projects to manageable number, and provide job security to give them time to produce results.[43] *Agrónomos* admitted that scientific farming worked best on irrigated land while *de temporal* agriculture offered only the prospect of limited gains with the *Campaña en Pro del Maiz*. In 1940 the SAF agreed that *campesinos* could not improve or standardize their corn.[44]

A WINDOW OF OPPORTUNITY: WAR READINESS, ENEMY AGRONOMY, AND U.S. AGRICULTURAL ASSISTANCE

The oil companies' privileged treatment of foreign managers and *técnicos* contributed to their expropriation, and some Mexicans, like chemist Andres Acosta, thought that they should not be allowed to work in Mexico unless there was no alternative. In August 1939 the State of San Luís Potosí codified this idea, but instead, confronted by an agricultural crisis, Mexico opened the door to foreign science and scientists, supporting historian Nora Hamilton's claim that there was "a strong element of continuity in Mexico's dependence on foreign capital even during a period when Mexican nationalism achieved its strongest expression."[45] In 1937 the SEP suggested hiring European refugee scientists, but because few were in the desperately needed fields of botany, plant genetics and pathology, entomology, and animal diseases, the agency wanted to recruit these experts in the United States. The 1939 Organic Law for Education allocated funds to bring foreign *técnicos* to Mexico to teach, and the Ministry of Finance and Public Credit also wanted to find scientists who agree to work in Mexico. Spanish Civil War refugees helped improve several branches of Mexican science, including the 12 agronomists hired by the ENA, the SAF, and the State of Baja California. In 1940 José Luis de la Loma of the University of Madrid began to teach plant genetics and experimental methods at the ENA, but there were not enough refugees to transform Mexican agronomy. To improve their scientific skills, the FRAM, the LAS, and Fernández y Fernández, who wanted to send an *agrónomo* to the Philippines to study coconut plantations and run them in Mexico for the SAF, urged Cárdenas to fund study abroad. With Mexican research "in its infancy," the LAS wanted to "use the results of foreign science."[46]

Some Mexican scientists still admired the U.S.S.R., even after the Hitler–Stalin Pact and the invasion of Finland alienated many Marxists. The *Instituto Biotécnico* corresponded with the Leningrad Institute of Industrial Plants. The SEP praised USDA advances in entomology but said that the U.S.S.R. "puts science to the people's service in an unprecedented manner."[47] The CNC's leaders and Marxists like Lombardo Toledano joined leftist *agrónomos* who believed that small-scale farms could not be modernized and wanted to collectivize all *ejidos* to introduce better technologies. In 1938, like Mexico's Communist Party, even the more conservative FRAM advocated collectivization, albeit with the *ejidatarios'* consent. In 1939 Gonzalo Robles ordered a complete set of Marx's and Engel's works. In 1940 the LAS urged Cárdenas to copy Stalin's industrialization plan. Enrique Beltrán praised Stalin's "policy of peace," and Mesa A. became treasurer of the Society of Friends of the Soviet Union, which called the U.S.S.R. "the only country where people do not exploit other people" and criticized the Hitler–Stalin Pact as "imperialist propaganda."[48] The Axis nations also had some influence on Mexican agronomy. To obtain copra for industry, German businessmen followed the SAF's coconut palm campaign. In 1938 Alanis Patiño continued his graduate studies in Italy. In 1940 Fernández y Fernández visited Japan, and the *Instituto Biotecnico*'s scientists took Tokio Kaboraki, a Japanese pest control expert, on a tour of Mexico City's research institutes.[49]

During the crisis, like many enemies of land reform, some *agrónomos* urged farmers to imitate U.S. practices and claimed that the SAF should be more like the USDA. LAS member Lázcano Romero praised its cotton industry and experimentation stations. *Agrónomo* Eduardo Turcott B. called Sonora's rice growers "more advanced" because they planted U.S. strains. Sugar workers and chicken farmers wanted to study U.S. and European farm technologies. Farmers from Jalisco asked the Pan American Union for technical publications, and in northern Mexico they obtained pest control advice from R.F. Crawford of the University of New Mexico. The USDA–SAF pink bollworm and fruit fly projects continued to bring U.S. agricultural science across the Rio Grande. In 1938 R.E. McDonald wrote an article on the pest for the SAF's journal *Agricultura*. Responding to the petitions of drought-stricken farmers, the SAF allowed *zoca* cultivation in La Laguna but in 1939 reinstated the ban with the growers' support. The USDA donated predators for a biological control campaign and took SAF employees on autogiro flights to locate fields with cotton trash. Mexican inspectors fined farmers who ignored sanitary regulations, and the SAF rejected requests to cultivate *zoca*. In 1940 the BNCE suspended the credit of *ejidatarios* whose fields sprouted cotton trash. Expressing the bonds between Mexican and U.S. pest control experts, in 1938 UNAM opened its new "Leland Ossian Howard Laboratory." The USDA's fruit fly lab still operated, and the agency assisted another SAF biological control campaign against the citrus black fly, but Wallace wanted to expand its role in Latin America beyond pest control. To study the origins of corn, he sent USDA scientists to Yucatán, Durango, and Coahuila. In December 1938 Undersecretary of Agriculture M.L. Wilson, Lee A. Strong of the pink bollworm project, and two other USDA employees met SAF *agrónomos* in Mexico City, who "received them warmly," including Parres, who wanted more collaboration, but Cárdenas postponed these discussions because of the upcoming election.[50]

Conflict in Europe and Asia created U.S. government interest in providing technical assistance to Latin American farmers. To avoid congressional resistance, Wallace and other USDA officials planned to promote "complementary crops" that could not be grown in the United States to provide strategic raw materials for war. Japan's occupation of the East Indies, which would disrupt access to rubber, was their greatest concern, but they also worried about medicinal plants like quinine, as well as other tropical farm products, and claimed that these goods would compensate for lost European and Asian purchases of surplus foodstuffs. To enhance U.S. influence in Latin America, the State Department wanted to foster academic exchanges and bring scientists and intellectuals to the United States for advanced course work. Advocates claimed that these programs would combat Axis influence, including the activities of businessmen and technicians, more effectively than propaganda campaigns. They knew that Latin Americans who attended U.S. universities often exerted considerable influence and occupied important positions at home and became admirers of U.S. culture, economic might, and political institutions.[51]

In December 1937 Roosevelt established an interdepartmental committee to advance these agendas. In June 1938 he signed Act 545, which gave him authority to send government employees to work in other nations, starting the Cultural Relations Program for Latin America. The State Department had reservations about agricultural assistance, arguing that U.S. economic policy opposed creating industries that could not compete in international markets in normal conditions. Congress also resisted, forcing Wallace to spend Agricultural Adjustment Act funds to send the first USDA survey teams to Haiti, Ecuador, and Paraguay. The Inter-American Development Bank promised to promote only "products for sale in the U.S. which are non-competitive with American industry and agriculture."[52] To appease the State Department, in 1940 Wallace agreed to concentrate on crops that could compete with Old World producers in peacetime. However, in addition to winning the support of elites who appreciate music, literature, and the arts, to reach Latin America's masses Wallace thought the Cultural Relations Program should help subsistence food producers.[53]

Mexico wanted U.S. assistance to end the agricultural crisis. When Bolivia invited CNI specialists to study irrigation projects, *agrónomos* learned that foreign training enhanced their prestige and brought international recognition. In 1938 the SEP sent two professors and students from the National School of Biological Sciences to U.S. universities to study genetic research on corn, and an SAF employee visited Wallace in Washington, D.C., to foster more USDA collaboration. To fight *chahuixtle*, the SAF consulted Stakman and in 1939 sent *agrónomo* Jesus Patiño N. to Java, Hawaii, and the Philippines to study tropical agriculture, especially sugar industry research. The BNCE paid an *agrónomo* and five *ejidatarios* from La Laguna to learn cotton grading at Texas A&M and 20 to visit General Electric and International Harvester. From 1938 to 1940 11 Mexicans studied agriculture at U.S. universities. Cárdenas obtained a bid for a chemical fertilizer plant from the American Chemical Construction Company. The CNI hired U.S. engineers for its large projects in the north. The SAF asked A.C. Baker to study a potato pest, and he helped the SAF obtain wasps to kill cottony cushion scales, which infested Mexico City's gardens. In 1940 Cárdenas let Tabasco's governor, Francisco Trujillo Gurria, seek U.S. aid to fight *chamusco*.[54]

Wallace wanted Mexico to participate in the Complementary Crops Program because he believed that rubber could be grown in the southern states, but land reform and the oil expropriation led to State Department resistance. In July 1938, considering a request for a railroad expert and a farm economist, Laurence Duggan approved the former but said that allowing the latter "to operate agricultural properties taken from American citizens" in La Laguna was "not only ambiguous but uncomfortable," and Ellis O. Briggs suggested sending "a few boll weevils . . . of vigorous stock and guaranteed to multiply with astounding rapidity" instead.[55] In March 1939, thinking that Mexico might obtain German equipment by bartering embargoed expropriated oil, Secretary of State Cordell Hull rejected a U.S. engineer's request for specifications on American machinery for a CNI dam. In May the USDA refused another inquiry about cotton graders, but the State Department provided a list of private citizens whom the Mexican government could contact directly. Resistance faded when Mexico asked for fisheries experts. To remove Japanese scientists who worked for the Mexican government at Lake Patzcuaro and derail the expansionist agenda of the firm Nippon Suishan Kaisha, which harvested shrimp in the Gulf of California, in the spring of 1939 the Bureau of Fisheries sent Milton J. Linder to Mexico and established a permanent assistance program in March 1940. Reflecting the spirit of cooperation after Mexico settled Sinclair Oil's claim, Mesa A. attended the Convention of Coffee Producing

Nations in Washington, D.C., and the SAF and USDA reestablished the *Comisión Internacional de Plagas,* holding a meeting in Monterrey. In May Mexico joined the Inter-American Development Bank, which wanted to encourage agricultural research in Latin America. At the State Department's request, Congress funded a USDA survey of Mexican agriculture. Aware of Mexico's own demand to establish *Hevea brasilensis* rubber plantations, Cárdenas accepted USDA financial and technical support.[56]

Many academics supported the Cultural Relations Program, and diplomats like George Messersmith knew that private organizations such as the Rockefeller Foundation could assist it. When Roosevelt formed the Interdepartmental Advisory Council on Technical Cooperation, he invited representatives of the foundation and other philanthropies to meetings. The foundation's officers had misgivings about participating in the government's propaganda campaign against fascism, partly because they thought that Latin Americans saw the Cultural Relations Program in this light, and wanted to expand their presence in the region cautiously. They also disliked the State Department's frequent requests to fund projects. In 1938 Donald H. Stevens of the Social Science Division warned that "conditions . . . today call for special checks on political or commercial implications in the operation or in the indirect outcomes of a project."[57] In 1940, to avoid asking Congress for funds, the State Department invited the foundation and other philanthropies to meetings to plan an Institute for Tropical Agriculture, but the officers felt that it was not suitable for a Natural Sciences (NS) Division program. As the Cultural Relations Program grew, foundation employees in Latin America communicated regularly with U.S. officials, and the IHD helped the Ministry of Public Health promote cinchona, a source of quinine, but the officers still disliked the State Department's methods. By the end of 1940 the trustees wanted to expand the foundation's presence in Latin America, including the NS.[58]

In December 1940 Vice President-elect Wallace attended Avila Camacho's inauguration. The trip convinced him that Mexico wanted a Rockefeller Foundation agricultural program, and he shared his opinion with its representatives, and Ambassador Josephus Daniels made similar suggestions to George C. Payne of the IHD. The officers knew that more funds would be available for NS projects in Latin America as Axis forces advanced in Europe and Asia, recognized Mexico's importance to the United States, and asked several U.S. Latin American specialists for advice on an agricultural program. In February 1941 George I. Sánchez, who studied the *Centrales,* told the foundation about the failure of earlier attempts to introduce U.S. farm technology and emphasized the importance of solving the problems of *ejidal* agriculture. In March Sauer told Joseph Willits of the Social Sciences Division that some traditional agricultural practices were very efficient in Mexico's environment and warned that U.S. scientists lacked "the life-time of learning that native practices represent real solutions of local problems, partly because he comes as an alien to a strange situation."[59] Henry M. Miller met Marte Gómez, Mexico's new minister of agriculture, and other SAF officials, who told him that Avila Camacho supported an agricultural assistance program, leading to the trustees' approval of a survey in April. In July Stakman, soil scientist Bradfield, and plant geneticist Mangelsdorf visited Mexico. In May the USDA sent its own survey team, Harry T. Edwards, a tropical crops expert, and James H. Kempton, a botanist from the Division of Cereal Crops and Diseases who studied corn in Yucatán in 1938.[60]

Their approach to Mexico's agricultural problems was quite similar to that of the *agrónomos.* Wallace compared Mexican agriculture to farming in biblical times, lamented the prevalence of rural poverty and hunger, thought increased farm productivity would alleviate it, stressed the importance of nutrition and public health, and claimed that the SAF's research lacked planning and continuity that U.S. scientists' assistance and U.S.-trained *agrónomos* could correct. Ruben Brigham, the USDA's assistant director, proposed cooperative projects to improve the Mexican diet. The agency's survey concentrated on complementary crops but also recommended more funding for the ENA and for research, including breeding improved food crop strains. Wallace's interest in poor farmers influenced wartime agricultural programs. In 1942 the United States sent sociologist Nathan Whetten to Mexico, and the State Department planned to give agricultural anthropology scholarships to learn "the underlying reasons for the types and systems of culture which exist," such as "why is corn grown as it is in the various regions of Mexico," and understand "people's reactions to social reforms, especially as related to agriculture" because "plans are conceived which we consider desirable but the end results of which cannot be foreseen, for we cannot forecast just how people will respond to them."[61]

Wallace's ideas also influenced the foundation. In March 1941 Miller wrote that "the worthwhileness of research in agriculture and any attempt to better the economy of Mexico through improving the lot of the Indian agriculturalist depends to a large degree on a combination of scientific agriculture and an approach through extension division (county agent) workers (and their wives), who understand the Indians."[62] The survey team told the SAF to give researchers more job security and advocated work in plant breeding and soil conservation. They claimed that *agrónomos* knew political, economic, and scientific theory, but not practical farming or research methods, and proposed NS–IHD collaboration with objectives that resembled the *cardenistas'* cultural campaign because "the improvement of agriculture and rural life involves not only problems in technics of crop and animal production but also amelioration of living and health conditions."[63] A report to foundation president Raymond B. Fosdick noted:

> To anyone who has traveled in Mexico and has examined the extensive and rapidly-growing literature on the country, it is at once apparent that the great majority of Mexican people are poorly fed, poorly clad and poorly housed. Sanitary conditions are much below the standards to which Americans are accustomed; as a consequence both pests and pestilences flourish. Illiteracy is common and adequate education is relatively rare. . . . In short, the . . . standard of living of the Mexican people is pitifully low.[64]

The officers wanted to tie the NS agricultural assistance program closely to the IHD.

The foundation responded to the advice. Officer A.R. Mann warned that "efforts to improve the agricultural economy must be indigenous and arise out of native abilities, native plant and animal stocks, and the cultural characteristics of the people."[65] Evaluating the survey team's report, the officers noted that "it would be wholly impossible, even if desirable, to impose modern American culture upon Mexico; any improvement that is to be brought about must come within the framework of Mexican culture."[66] The MAP's scientists, however, observed peasant farming and decided to reject it. Mangelsdorf found modern agriculture on the farms of "a few progressive farmers" "train[ed] in the U.S. or elsewhere" and criticized the practices of "uneducated peons."[67] The survey team considered the role of *ejidatarios* in the modernization of Mexican agriculture. Acknowledging the "latent talent among the humblest," who "appear to be educable to a certain point," they admitted that "any improvement in agriculture must come as the result of a recognition of the opportunities offered and the difficulties imposed by the [*ejido*] system" but concluded that "it would probably be wholly impossible to introduce hybrid corn on a significant scale to an unorganized, largely ignorant peasantry," although "it might be possible . . . through the ejidal banks," and decided that "improvement will be most rapid if the best talent [members of "the old landed aristocracy" who "had attended an agricultural college in the United States"] is put to work on the problem."[68] Stating the foundation's definition of modernity, Mann said that agricultural aid "release[d] . . . greater resources for other social needs and institutions."[69]

The contacts, converging interpretations, and success of the pink bollworm and fruit fly campaigns reinforced *agrónomos'* interest in U.S. agricultural science, and they appreciated the opportunity for assistance. Carlos Ayala Origel began a book about Wallace's Pioneer Hybrid Seed Company. With the FRAM's support, in 1941 Eduardo Morilla Zafa, head of the DGA, proposed a Pan American conference to foster cooperation among the Western Hemisphere's agricultural scientists. The *Bloque de Agrónomos Revolucionarios* (Revolutionary Agronomists Block, hereafter BAR) urged Mexico "to open its doors" to foreign scientists "kindred and assimilable to our population, devoid of any humiliating superiority, without racist complex, nor coveting any anti-social privileges."[70] *Carta Seminal* was skeptical, but *Excelsior* praised the visit of the foundation's survey team. In October Lester D. Mallory, U.S. agricultural attaché in Mexico City, received several spontaneous requests for USDA complimentary crops aid, including from Tabasco's banana growers, who asked for rubber trees to replant their *chamusco*-infested plantations. In November Marte Gómez told the press that he agreed with Wallace's and the foundation's recommendations, especially assistance in plant genetics. Trujillo Gurria obtained advice on treatment procedures for *chamusco* from a U.S. firm.[71]

To cooperate with the rubber program, Avila Camacho let the

USDA run the El Palmar Station in Veracruz to grow and distribute *Hevea* trees. The United States said that synthetic rubber could not compete in the foreseeable future and claimed that Latin American *Hevea* would be cheaper than the East Asian product if producers used scientific techniques, but some Mexicans demanded trade concessions, claiming that these sources would destroy Mexico's industry when the war ended. To allay their fears, at the January 1942 Rio de Janeiro meeting of Ministers of Foreign Affairs the United States agreed to avoid "in so far as possible the establishment or expansion of production of substitute or synthetic commodities which is economically artificial and might displace the consumption of natural products available in other American Republics," excepting "only those industries which are indispensable for national defense provided that such defense needs cannot be met with natural products."[72] The SAF and rubber industry workers supported the program. Marte Gómez and Luis León became vocal advocates, reassuring the public that rubber producers could compete with Asian and synthetic rubber because they used U.S. agricultural science, but they also urged the United States to give Latin American growers tariff protection after the war. Mexico also cooperated with the USDA to develop the production of rubber from guayule and *cryptostegia grandiflora*. In 1942 the SAF planted *cryptostegia* in Tamualipas, worked to improve the extraction process at a USDA laboratory in Ciudad Victoria, and studied the production of guayule rubber. To 1943, despite advances in synthetic rubber, officials continued to tell Mexico and other Latin American nations that the United States wanted to develop a permanent plantation rubber industry for national defense, believing that the war justified the deception.[73]

To establish new farm industries, Avila Camacho let many U.S. scientists work in Mexico. The lack of State Department funds derailed an SAF–USDA extension project, but in 1942 Mexico offered a site for the Inter-American Institute for Tropical Agriculture, and Mexico City hosted the Second Inter-American Conference on Agriculture, which *El Nacional*'s editors praised. By 1943 government agencies like the USDA and wartime organizations like the Office of Economic Warfare managed many projects in Mexico, including insecticidal plants like pyrethrum, Japanese mint, and derris, medicinals like henbane, belladonna, quinine, and Red Squill, exotics like "*monilla*," and rubber-bearers like *cryp-*

tostegia. Promising that demand would remain high after the war, at the request of the United States the SAF, the States of Chiapas and San Luís Potosí, and the agricultural banks promoted castor beans. Contacts with the USDA led to more biological pest control campaigns. U.S. experts toured the Yaqui Valley, Sinaloa, and Nayarit to find land for rubber-bearing plants and visited fruit orchards in the northeast and along the Pacific Coast and Yucatán to study the henequen industry. Gómez asked experts from the Hawaiian Pineapple Company to establish the crop in Mexico. Some farmers, such as the sheep ranchers who hired Louis L. Madsen, obtained USDA assistance on their own. In 1942 USDA bollworm inspectors said that Mexican farmers cooperated more enthusiastically than at any time in the past. *Agrónomo* Pascual Gutiérrez Roldán invested in a firm that planned to produce *cryptostegia* for the Rubber Development Corporation. These projects generated some resistance. The many visiting experts and the U.S. government's failure to coordinate its programs irritated Gómez and other SAF officials. El Palmar's Mexican employees did not keep good records or follow their USDA advisers' suggestions.[74]

The MAP's scientists were another group of U.S. experts who went to Mexico during the war, but as employees of a private organization they were not under direct government control. Knowing that the many U.S. officials and programs were a source of irritation, the foundation used this independence to its advantage. After the survey, to sustain goodwill, the foundation gave the Ministry of Public Health soybean seeds, and Stakman analyzed races of *chahuixtle* for the SAF. In October 1942 Gómez sent a formal request for foundation assistance, then waited anxiously until an agreement was signed in February 1943. *Chahuixtle* was his primary concern, and he asked the officers to give it top priority. The contract also specified corn breeding, soil management, and livestock disease control projects. Stakman claimed that "their basic concepts are exactly the same as ours and even many of the details are identical or very similar," especially their appreciation of "the need for basic research."[75] To address the SAF's other concerns, the MAP conducted preliminary studies of diseases of citrus fruits, coconut palms, vanilla, potatoes, and beans and visited Yucatán and Tabasco to study the henequen and cacao industries. To assist the Ministry of Public Health, the MAP planned to test soybeans. The MAP's scientists wanted to study soil management to find "methods of maintaining fertility without the expense of chemical fertilizers," but the foundation also funded UNAM's program to train chemical engineers for new fertilizer factories.[76] The foundation told the SAF that the MAP was a long-term program, not an emergency wartime project like those of the U.S. government.[77] Officials such as Ambassador Messersmith, R.E. Moore, the assistant director of the Office of Foreign Agricultural Relations (OFAR), the USDA's Earl N. Bressman, and Nelson Rockefeller followed the MAP's progress with great interest. Foundation officers and MAP scientists communicated regularly with U.S. government representatives. The MAP assisted some wartime projects, for example, by testing castor beans and studying quinine, but rejected invitations to join others, like the pyrethrum campaign. The foundation wanted to bring Mexican students to U.S. universities, recognizing that "the Latin American returns as a leader in his country and as a friend of the U.S." but in 1941 refused to do this for the Cultural Relations Program.[78] However, the foundation did have a foreign policy agenda. Before the MAP began to operate, some of the trustees, especially John D. Rockefeller III, wanted to start these programs in other nations if the MAP succeeded. Stakman claimed that agricultural assistance would improve U.S.–Mexican relations, and the foundation asked the survey team to evaluate the prospects for establishing these programs in other Latin American nations. To enhance the *agrónomos'* credibility and prestige, the officers and the MAP's scientists wanted the SAF to issue all publicity about the project. They reported that SAF *agrónomos* such as Augusto Pérez Toro of the *Instituto Tecnico Henequeñero* welcomed them warmly, recognized that they benefited from U.S. scientists' advice and assistance, cooperated enthusiastically, including Limón and Taboada, who later offered some resistance, and claimed that the SAF's directors felt that the MAP would exert "a stabilizing influence."[79]

To promote academic exchanges, the United States sent E.J. Kyle, dean of Texas A&M's School of Agriculture, to similar institutions in Mexico and seven other Latin American nations. *Agrónomos* appreciated the opportunity to obtain foreign training. In June 1942, after touring agricultural colleges in Texas, *agrónomo* Hector Lazos said that the ENA could not train "true specialists" and told the Second Inter-American Conference on Agriculture that Mexico "should utilize the present current of sympathy [in the United States] to establish scientific relations and technical training programs involving the exchange of students between them."[80] In 1942 and 1943 Mexico took advantage of wartime scholarships by sending *agrónomos* to the United States to study agricultural economics, soil science, animal husbandry, agronomy, plant breeding, and the lime industry. Eduardo Suárez, minister of finance and public credit, used connections to enroll his sons in U.S. agricultural colleges in spite of their insufficient qualifications. In 1943, agreeing with Bradfield's opinion of an ex-student from Portugal who did "as much in an unofficial capacity to promote, among his countrymen, understanding and appreciation of America as any salaried official from our country," the MAP awarded its first fellowship to *agrónomo* José Rodríguez.[81] With the U.S.–Soviet alliance, LAS member Edmundo Flores applied to the University of Wisconsin, and Fernández y Fernández, Durán, and former ambassador to the U.S.S.R. Jesús Silva Herzog recommended him. From 1941 to 1943 at least 19 Mexicans studied agriculture at U.S. universities, including 6 who received no financial assistance from either government.[82]

Before Pearl Harbor some USDA officials claimed that food shortages in Latin America threatened national security because they provided ammunition for Axis propagandists. In 1941 the State Department considered continuing cultural relations programs, including aid for food crop producers, after the war and extending them to other underdeveloped nations, believing that they fostered good relations, alleviated poverty, and promoted democracy. Nelson Rockefeller, the Coordinator of Inter-American Affairs (CIAA), wanted to increase Latin American food production to support the war effort. In 1942 U.S. scientists ran CIAA Food Supply Programs in eight countries, and Rockefeller claimed that they demonstrated the U.S. commitment to regional development. Hearing about food shortages in 1942 and 1943, on several occasions he asked the State Department for permission to begin one in Mexico. In October 1943 he suggested that the USDA assist the MAP. Thinking of postwar conditions, the State Department studied British, Free French, and Soviet cultural relations projects in Latin America. By the end of 1943 many members of the foreign policy community believed that helping Mexico grow more food, raise standards of living, and industrialize served U.S. postwar diplomatic and economic interests.[83]

GÓMEZ'S REFORMS: INFUSING SCIENCE IN THE SAF AND MEXICAN AGRONOMY, 1941–1943

Marte Gómez's appointment pleased senior *agrónomos* who thought that a member of their profession should run the SAF. To address problems caused by the war and advance his industrialization program, in July 1941 Avila Camacho exhorted the SAM to make farmers more productive. As members of the Revolutionary Family, both sides in a disputed SAM election supported his agricultural policies, as did Graciano Sánchez, head of the CNC. Historian Peter Smith claims that Avila Camacho purged bureaucrats to foster more conservative policies, but many *cardenista agrónomos* had important jobs in the SAF, the agricultural banks, and the CNA. In 1941 the SAM claimed that its influence on agricultural policy reached the highest point in its history, and new president Cesar Martino proclaimed his colleagues' duty to make the *ejido* productive to fulfill Zapata's vision of a prosperous and autonomous peasantry. With a more conservative Revolutionary Family, he attacked LAS members who still had government jobs. *Agrónomos* continued to define the profession as one of government service and demand positions in the bureaucracy. Emilio Gutiérrez Roldán called *agrónomos* who became private farmers traitors.[84]

Reflecting their desire to be credible scientific professionals, in 1941 the SAM cheered when León told *agrónomos* to be serious *técnicos* in addition to *agraristas*. Damién Correu and Tabaoda admitted that extension campaigns conducted without prior research led to negative results that eroded farmers' faith in their profession. To "enhance the prestige of agronomy," Durán urged *agrónomos* "to adopt a completely scientific outlook."[85] The BAR resolved:

> The primary goal of the agronomists of our society is to use their knowledge to solve Mexico's agricultural problems, divested completely of political interests, because we believe the most harmful thing for the development of agriculture, the life of our profession, and the activities of all agronomists is the overlap that might occur in our function as technical professionals with political ups and downs ["*arribismos*"].[86]

Chemical engineer Santos Amaro urged Avila Camacho to found a committee to promote the use and manufacture of chemical fertilizers and warned that "as a technical commission, it should be totally removed from politics and its members selected exclusively for their competence."[87] A group of recent ENA graduates told Avila Camacho that they supported his agricultural policy but were "embarrassed" that "professional politicians" like Gómez and Martino rather than "prestigious technical specialists" represented the profession.[88] *Agrónomos* continued to search for a scientific language to establish their credibility as experts with specialized knowledge, and some recognized that genetics provided one that could improve their tarnished public image. Durán said that Mexican agronomy had no experimental tradition, lamented that *agrónomos* only transmitted foreign knowledge rather than develop their own, and claimed that plant breeding "cannot be left to the peasant, who does not have the resources or the theoretical knowledge to do it." The SAM urged the ENA "to develop agricultural research in its genetic and technical aspects."[89] Morilla Safa claimed, "Mexican agriculture cannot develop fully unless it is based on genetics," the SAM advocated "research and the application of genetic principles" to improve the quality and yield of crop plants, and Miguel E. Schultz, who bred wheat during the 1930s, told Avila Camacho that "to select or improve seed, the use of special techniques that are beyond the ability and intellectual development of our *campesinos* is indispensable."[90]

Agrónomos continued to denounce *campesino* farming, and many admitted that the cultural campaign had failed. In 1941 the BAR lamented that "the agricultural evolution of our country has been imperceptible and frustratingly slow; we live in a technical backwardness far from the modern era," and *agrónomo* Rodrigo Olvera confessed that "we have not even taught *campesinos* to select and disinfect their seed."[91] Marte Gómez called Cárdenas' SAF ineffective and the practices used by experimentation stations often inferior to local methods, and Mesa A. claimed that they led to no gains in productivity. Responding to U.S. visitors' suggestions, Gómez closed several stations. In 1942, to avoid giving bad advice by basing extension on research, he disbanded the *Agrónomos Regionales* and gave the SAF's schools to the SEP. He rejected several requests for new stations and projects to modernize several branches of agriculture. The ENA put more emphasis on practical farming and experimental science, but the directors thought that the school needed a stronger research program to complement teaching. Using his USDA contacts, Gómez began several research projects, including a new corn-breeding program, and hired U.S.-trained *agronomos* like Jesús Garza Morales, Taboada, and Limón to run it. In 1943 plant breeding and other research at the *Insituto Técnico Henequeñero* impressed visiting USDA scientists; the SAF distributed Limón's and Taboada's self-fertilizing "Celaya" and "Argentino" corn to farmers in the Bajío and gave growers new strains of chiles with effective dosages of chemical fertilizers thanks to research at Pabellón. Many *agrónomos* supported Gómez's emphasis on science, and they proposed plant-breeding programs to create high-yield wheat, cotton, cacao, sugarcane, and pineapple varieties. Carlos Anzures urged the SAF to begin an intensive program to raise the productivity of *de temporal* corn growers. In 1943 Gómez announced that "the agricultural research that develops in diverse areas will demand the contributions of many specialists."[92]

The scientific successes pleased Gómez, but the legacies of the cultural campaign still affected the SAF. In 1941 and 1942 he opened new stations in Gutiérrez Zamora, Veracruz, to study vanilla, in Zacatepec, Morelos, for sugarcane, in Llera, Tamualipas, his home state, for grapes and citrus fruits, and in La Paz, Coahuila, for cotton and other crops that could be grown in La Laguna. The SAF urged *campesinos* to adopt the *Campaña en Pro del Maiz*, including its most prominent corn breeder Limón. Rivas Tagle told peasant wheat growers to try foreign *chahuixtle*-resistant strains. SAF projects with the governors of Campeche, Chiapas, San Luís Potosí, and Jalisco used *cardenista* methods *sans* anti-Catholicism, but most employees were not *agrónomos*. The tortilla discourse led to soybean, family vegetable garden, and fruit orchard campaigns and proposals in favor of garbanzos, tomatoes, bananas, and sweet potatoes. To increase *campesinos'* income, the SAF promoted olive trees, in one case working closely with the *ejidatarios* of Maravatio, Michoacán, as well as vanilla, sesame, and sunflowers. Gómez wanted to test black and Jamaican pepper, Chinese tea, cloves, edible mushrooms, Roselle hemp, and flowers. To end the shortage of seed oils, the SAF studied tung oil trees and planted coconut palms. In 1943 San Luís Potosí planned to eradicate corn cultivation on irrigated lands. Many *agrónomos* lobbied for small-scale irrigation works, and from 1940 to 1941 the

CNI increased funding from 1,508,876 to 3,971,000 pesos. Avila Camacho also spent more money than Cárdenas to donate iron plows to *campesinos*. Some *agrónomos* wanted to collectivize more *ejidos*, ignoring Avila Camacho's and Gómez's opposition, but the SAF admitted its lack of interest in the problems of most *campesinos*: "In promoting secure and scientific agriculture in Mexico, one should think primarily about farming irrigated lands."[93]

As historian Lane Simonian notes, Avila Camacho was interested in soil conservation and established an agency to study and fight erosion. Some *agrónomos* supported this agenda and got jobs in new conservation districts, but Avila Camacho also wanted farmers to use agrochemicals to support his industrialization program. The SAF tested chemical fertilizers on sugarcane and chiles and in 1942 began to pay half their cost for farmers who would try them, spending 350,000 pesos on the program by 1944. In San Luís Potosí the BNCE gave them to *ejidatarios* in lieu of cash loans, and the CNC and *El Nacional* wanted the government to manufacture them. In 1943 Avila Camacho founded the parastatal firm *Guanos y Fertilizantes de México* and asked the United States for machinery for new fertilizer factories. To promote pesticide use, in 1943 he exempted several products from import duties.[94]

The SAF had many problems before the MAP began. Poor record keeping and field plot techniques ruined 1941 tests of sesame and wheat varieties. There were no Mexican rubber experts to staff El Palmar. In Chiapas an *agrónomo* from the National Export Commerce Bank convinced owners of *chamusco*-infested banana plantations to grow sesame, but the climate produced "a lot of plant and no seed," and they lost money. *Agrónomo* Manuel Castaños claimed his colleagues in La Laguna had no interest in research. Locust swarms and other pest control problems drained the SAF's budget. Avila Camacho allocated more funds to fight the former, but the foundation's survey team said that some were diverted, and Brondo Whitt claimed that the "indolence of the *campesinos*" and the "lack of responsibility of some employees" caused its failure. After visiting UFCO's Central American plantations and considering several proposals for expensive campaigns, the SAF ignored pleas for help and forced farmers to fight *chamusco* on their own, providing only technical advice. In 1942 Tabasco's growers used their organization to borrow money from the National Export Commerce Bank, bought sprayers and fungicides, treated their plantations, and in 1943 shipped fruit to the United States, where it earned high prices. In Chiapas only wealthy farmers got these credits, and they profited from wartime demand while *ejidatarios* could not export their damaged fruit. Tehuantepec's growers got no help, and the boom passed them by. In 1943 they called *chamusco* "a matter of life and death."[95]

In 1941 sugar was scarce in northern Mexico, and supply shortages continued into 1943. The loss of access to copra from Oceania threatened the soap industry, and in 1942 the price of vegetable oils and fats soared. That year local maize shortages occurred, and by September 1943 bad weather, low productivity, and the diversion of land to crops for the U.S. war effort led to unrest in several regions, including Yucatán, Chiapas, and the Bajío. In Durango angry consumers rioted and stole corn from a National Railways warehouse. *Chahuixtle* still ravaged wheat fields. In cities the price of food soared, and Mexico spent nearly $21 million to import over 50% of the wheat that its citizens ate. *El Universal* tried to diffuse the situation by advocating trading more profitable export crops for corn, while others told Mexicans to eat more wheat. The MAP began at a fortuitous moment, as the SAF was under pressure to avert another food crisis. Calling 1943 "the worst [agricultural] year of the century," Gómez admitted that he could do little to solve it.[96]

Some *agraristas* and farmers backed the *agrónomos* and the SAF. *Agraristas, pequeños propietarios,* army officers, and governors applauded Avila Camacho's call to increase production and asked the State to bring them experimentation stations, extension agents, and new technologies. In 1943 the CNC praised Gómez's support for small-scale irrigation and offered to help *ejidatarios* ask the CNI for new projects, but food shortages and wartime inflation led to more criticism of the land reform, led by the right-wing *Partido de Acción Nacional,* as well as more attacks on the *agrónomos*. *Excelsior* said that "Communist" BNCA *agrónomo* Hipolito Cardenas instigated student demonstrations in Ayotzinapa, Guerrero, ousted "Communist infiltrators in the BNCE," claiming that 90% of its employees were sympathizers, and lambasted *agrónomos'* urban outlook: "[I]n their life none of them has grown a humble tomato; the only thing they know about rural areas is what they see from the highway on their way to bathe in the sea in Acapulco."[97] A Mexico City doctor called Gómez and Martino "charlatans of national politics," Puerto Vallarta businessmen denounced CNA *agrónomos* for telling *ejidatarios* to hoard their crops to drive prices higher and renege on their debts, Coronel Ricardo G. de la Llave said that *ejidatarios* did not know how to farm, despite the efforts to teach them, and Federal Deputy José Ch. Ramírez told the SAF to concentrate "on scientific matters to improve cultivation."[98] In 1943 *Excelsior* claimed that *agrónomos* lied about the cause of the corn shortage, in the editors' opinion, land tenure insecurity. *Hoy* called Gómez "ignorant" and urged him to resign.[99] *Sinarquista* peasants in Michoacán denounced the "unfair" actions of BNCE *agrónomos* who "enslave *campesinos* like in the detestable Soviet Union," and in Guanajuato they told Avila Camacho that *agrónomo* Salvador del Castillo behaved like "a brazen octopus with despicable instincts."[100] The *Sindicato Pequeños Agricultores* (Small Farmers Union) asked him for an audience because "it is better to listen to small farmers than *agrónomos burocráticos*," and an angry poultry farmer told the press that *agrónomos* knew nothing about science because the ENA taught them "to exploit the national treasury, not the land."[101] Banana growers in Chiapas complained when the government forced them to pay for *chamusco* control. In 1943 cotton farmers denounced the SAF for forcing them to plant 10% of their land in corn. In Veracruz *agraristas* demanded that *agrónomos* "totally refrain from all political matters and dedicate themselves only to their technical labors," and *ejidatarios* in La Laguna said that the BNCE's representatives exploited them.[102]

Many *agrónomos* complained that farmers refused to follow their advice. ENA student José Zálvidar Flores said that the *ejidatarios* of Lake Chapala ignored bureaucrats' suggestions, and *agrónomo* Herminio Corral found that they often sold the BNCE's selected seed and planted maize chosen by traditional criteria. Whetten claimed that *campesinos* sold their best corn ears because they earned higher prices and sowed kernels from the scrawniest cobs. During the early 1940s the MAP's scientists thought that *agrónomos* had no credibility with farmers, neither *hacendados* who saw them as a threat to their property nor *ejidatarios* who benefited from their politics. Borlaug claimed that "there was no confidence in his ability to change agriculture, and he was a worthless social parasite, who lived from the taxpayer's contributions."[103] Discussing Limón's work in the Bajío, Mallory reported that "local farmers poke fun at the plant geneticist" and said that "the only thing the technically trained men can do is make corn grow smaller," peasants worried that "this new seed would be forced upon them by the new experimenters," and some conducted "actual sabotage," "destroying test plots" and "removing the paper bags from the plants used for pollenation."[104] *Agrónomos* resented this resistance and professed negative stereotypes about "stubborn" and "spiteful" *campesinos*, especially Native Americans. By 1941 some wanted to reorganize *ejidos* like haciendas with *agrónomos* as *mayordomos* or to *"disciplinar a los ejidatarios"* to overcome their resistance. Zálvidar Flores, Anzures, and Jesús Valenzuela Rivera urged *agrónomos* to supervise *campesinos* in irrigation districts and *ejidos* closely. Even Limón, who wanted to breed better corn for illiterate farmers, told banks to force *campesinos* to disinfect their seed before lending them money. After working as a BNCE agent for nine months, Flores "hated the people, the job, and all things connected with them," felt that he "was not born to be a poorly paid rural bureaucrat," and admitted "the difficulty of establishing any contact between *campesinos* and bureaucrats."[105]

CONCLUSIONS

Although the *Cardenista agrónomos'* intentions appear admirable, most of them lacked the scientific training to develop new technologies on their own. Further, the sincerity of their desire to emancipate the *campesinos* by modernizing their farming practices is questionable. Vaughan claims that some of Cárdenas' rural teachers earned the *campesinos'* respect and trust by abandoning the anti-Catholic rhetoric of socialist education, losing "their condescension toward the peasantry as sick, poor, profligate, and savage," accepting "a host of other daily practices" like giving *pulque* to children because local water was contaminated, and learning why their farming practices might be superior to those taught in agricultural schools.[106] Mrs. Charles M. Fox, an English expatriate who worked in the Ministry of Public Health during the Cárdenas administration, found that community development projects in Native American villages worked best when women helped plan and implement them, which won the confidence of female villagers who then convinced their recalcitrant husbands to cooperate. Similarly, a Young Men's Christian Association (YMCA) agricultural program in Camohmila, Morelos, in which the wives of the project directors, a village representative, and the local teacher jointly planned projects won the confidence of women and schoolchildren and introduced the cultivation of fruits, vegetables and herbs, making preserves, beekeeping, and improvements in pest control and domestic hygiene. A village leader claimed that the project realized Zapata's vision of a better future for the *campesinos*.[107] Having less direct contact with *campesinos* and more scientific training than the teachers and as urban middle- and upper-middle-class males with many prejudices about peasants, Indians, and women, *agrónomos* could not make this accommodation. The peasants' negotiations led the *agrónomos* to seek new clients or advocate coercive methods to force *campesinos* to cooperate. The *Cardenista agrónomos* did not want to learn from the *ejidatarios*; they tried to control them to establish their position as scientific professionals.

Mexico independently decided to seek aid to increase farm productivity; it was not a response to U.S. pressure. Some leftist *agrónomos* initially viewed U.S. overtures as Yankee imperialism, but they changed their opinion because they thought that assistance would help their profession. The Rockefeller Foundation's agenda for the development of Mexican agronomy was similar to the objectives of *agrónomos* of all political orientations, who wanted to resurrect their professional credibility by learning experimental science. The MAP's scientists thought that Mexican agronomy was too political, but many *agrónomos* also wanted to de-politicize their profession before they arrived. Contrary to some critics of the Green Revolution, U.S. or Rockefeller family economic interests were secondary considerations when the foundation decided to provide scientific aid to Mexican agriculture, but by 1943 the U.S. government and the foundation believed that agricultural assistance projects like the MAP or the Food Supply Program would advance economic and foreign policy interests, as Harry M. Cleaver and Andrew Pearse assert. In January 1942 the United States began a Cultural Relations Program in China and started another in the Near East in 1943. These World War II programs were the foundation of U.S. Cold War foreign aid policy.[108]

NOTES

1. Guillermo Padilla to Presidente Municipal, Tuxpán, Jalisco, March 15, 1937, and others, AHJ, Box AG-27-I; "Gran escasez de maiz en Aguascalientes," *El Universal*, February 2, 1937; Aaron S. Brown, "Ec-

onomic Effects of the Mexican Agrarian Reform," November 1, 1938, pp. 7–9; William P. Blocker to Secretary of State, "Rise in Price of Essential Food Commodities," March 7, 1939, NA, RG 166, Boxes 149 and 146; SAF, *Estudio agro-económico del maiz* (México, D.F.: 1940), p. 142; "No pague a mas de $0.22 kg. de frijol," *El Siglo de Torreón*, March 13, 1939; "Frijol Japoneses para remediar la escasez local," *El Siglo de Torreón*, April 28, 1939; "Acuerdo que señala el precio máximo de venta, en toda la república, del frijol Chileño 'Red Kidney,' " *Mercado Agrícola-Ganadero*, April 8, 1939; "Importación de seis mil toneladas de frijol," *El Nacional*, May 9, 1939; "Se importa trigo de los Estados Unidos," *El Universal*, November 16, 1937; "El pan crecio y bajo de peso," *El Universal*, March 3, 1939; "Tremenda merma en la producción de trigo," *El Universal*, April 30, 1940; Norman Borlaug, "The Impact of Agricultural Research on Mexican Wheat Production," *Transactions of the New York Academy of Science* 20, 3 (January 1958): 279–80; "Estan por agotarse las existencias de arroz en Veracruz," *El Universal*, November 1, 1938; "En Veracruz el arroz cuesta un peso el kilogramo," *El Universal*, October 1, 1939; "Es angustiosa la escasez de arroz," *El Universal*, September 3, 1939; "El Comité de Subsistencias y su trabajo," *El Nacional*, February 11, 1939; Conrado Rodríguez y Rodríguez, "Principales zonas productores de oleaginosas y producción," in CNAG, *Memoria de la Primera Convención Nacional de Productores de Oleaginosas* (México, D.F.: 1939), pp. 105–6; Ramón Fernández y Fernández, *El autoabastecimiento nacional de copra* (México, D.F.: BNCA, 1941); "Exportación del plátano en el país," *El Nacional*, September 3, 1935; Stacey May and Galo Plaza, *The United Fruit Company in Latin America* (Washington, D.C.: National Planning Association, 1958), p. 154; John Kobler, "Sam the Banana Man," *Life*, February 19, 1951; Gonzalo González H., *La producción de cacao en México* (México, D.F.: BNCA, 1939), p. 1; Frente Revolucionario de Agrónomos Mexicanos (hereafter FRAM), *Programa de acción para la SAM, 1939–1940 (una contribución inicial de los agrónomos para la elaboración del Segundo Plan Sexenal)* (México, D.F.: Gráficos, S.A., 1938), p. 46; Franciso Orozco González to Cárdenas, September 3, 1938, AGN, Cárdenas papers, Exp. 437.1/556.

2. Susan R. Walsh Sanderson, *Land Reform in Mexico: 1910–1980* (Orlando, Fla.: Academic Press, 1984), p. 57; Alain de Janvry, *The Agrarian Question and Reformism in Latin America* (Baltimore: Johns Hopkins University Press, 1981), p. 123; Cynthia Hewitt de Alcantara, *Modernizing Mexican Agriculture: Socioeconomic Implications of Technological Change, 1940–1970* (Geneva: United Nations Research Institute for Social Development, 1976), pp. 12, 19–20, 101; Gustavo Esteva, *The Struggle for Rural Mexico* (South Hadley, Mass.: Bergin & Garvey, 1983), pp. 8; Victoria Lerner, *La educación socialista* (México, D.F.: El Colegio de México, 1979), p. 175; Howard F. Cline, *The United States and Mexico* (Cambridge, Mass.:

Harvard University Press, 1953), p. 225; Charles N. Myers, *Education and National Development in Mexico* (Princeton, N.J.: Princeton University Press, 1965), p. 36; Blanca Torres, *Hacia la utopía industrial* (México, D.F.: El Colegio de México, 1984), pp. 40–41; Luis Medina, *Del cardenismo al avilacamachismo* (México, D.F.: El Colegio de México, 1978), p. 19.

3. *Carta Seminal*, September 24, 1938, p. 24; "Relación de los productos cuya demanda es afectada por la Guerra Europea" (n.d.); Comité Estatal Pro-Mejoramiento del la Producción Agrícola, Colima, "Circular Num. 1," July 15, 1938; "Memorandum para acuerdo con el Presidente" (n.d.), AGN, Cárdenas papers, Exps. 564.1/2016, 437.1/556, and 606.3/164; "Fomento a la agricultura de la nación," *El Nacional*, June 28, 1938; "La labor del Departamento de Agricultura," *Jalisco Rural* 19, 6 (September 1939): 95–98; SAF, *Programa de labores, 1940* (México, D.F.: mimeograph, 1940), p. 9.

4. "Informe," September 1, 1940, in *Palabras y documentos públicos de Lázaro Cárdenas, Informes Presidenciales y Mensajes de Año Nuevo, 1938–1940*, vol. 2 (México, D.F.: Siglo Ventiuno Editoriales, 1978), p. 186; SAF, *Memoria 1939–1940* (México, D.F.: D.A.P.P., 1940, pp. 97–98, 147–53.

5. SAF, *Memoria 1938–1939* (México, D.F.: D.A.P.P., 1939), pp. 56–61; Salvador Ramírez M., "Memorandum," February 16, 1939, AGN, Cárdenas papers, Exp. 506.1/29.

6. Guillermo Rodríguez G., "Estudio formulado en la Oficina de Geografía Económico Agrícola," May 2, 1938, pp. 8–9, AGN, Cárdenas papers, Exp. 437.1/556; "Selección de las semillas para el maiz," *El Nacional*, February 28, 1938; "Impulso a una zona maicera," *El Nacional*, June 12, 1938; "Introductory Statement," *La Tierra* 1, 1 (April 1938): 2; "Protección a agricultores y campesinos," *El Nacional*, March 3, 1938; "Existencias de frijol en la República," *El Nacional*, April 1, 1938; SAF, *Memoria de trabajos de las Direccciones de Agricultura y Ganadería dependiente de la SAF durante el periodo presidencial de Lázaro Cárdenas* (México, D.F.: D.A.P.P., 1940), pp. 12–13; "Debe crecer la cosecha en el maiz," *El Nacional*, April 11, 1938; "Los cultivos del maiz en la República," *El Nacional*, March 1, 1938; "Hay que intensificar nuestra producción agrícola," *El Universal*, June 17, 1938; SAF, *Memoria 1938–1939*, pp. 24–26; ENA, "Concurso Regional de Semillas," 1939, in ENA, BCM Bound Volume #168-2; Daniel Moreno Sanchez to Departamento de Agricultura y Ganaderia, December 1, 1939, AHJ, Box AG-27-I, Exp. Informes de los Inspectores; Instituto Biotécnico, *Campaña Pro-Maiz, el primer paso* (México, D.F.: D.A.P.P., 1939), p. 4; SAF, *Estudio agro-económico*.

7. Roberto Galvez, "Acietes, el cacahuate," *Revista de Industria* (November 1938): 28-29; "Hambre, maiz, hambre," *El Universal*, January 30, 1939.

8. "Minucioso estudio acerca del cultivo de la patata," *El Nacional*,

March 28, 1938; "La cosecha de papa será muy abundante este año," *El Universal*, November 30, 1938; SAF, *Memoria 1939–1940*, pp. 108, 116; Roberto Mejia Morales, "Impuestos que afectan a la producción y venta de papa," in CNA, *Memoria de la Primera Convención Nacional de Productores de Papa* (México, D.F.: 1939), pp. 157–59; SAF, *Memoria 1937–1938* (México, D.F.: 1938), pp. 109–25; "Los girasoles son plantas muy utiles," *El Sol* (Monterrey), December 3, 1938.

9. Miguel A. Quintana, "El emplazamiento de la producción de cereales," *El Nacional*, February 16, 1939; "El maiz debe ser desechado como indice económico," *Novedades*, March 11, 1940.

10. "La entrevista de hoy," *El Nacional*, September 29, 1938; Norberto Aguirre, *Necesidades y posibilidades agrarias de México* (México, D.F.: Talleres Gráficos de la Penitenciaria, 1938), p. 64; Enrique Sanromán, "Producciones y precios del maiz," 1939, ARFF, Box 122; "Ya no cultivemos maiz," *El Mundo*, January 4, 1939; "A sembrar maiz," *El Dictamen* (Veracruz), February 18, 1939; "Informe," September 1, 1938, in *Palabras*, p. 140; "Plantas de durazno se distribuyen a los Coras de Nayarit," *El Popular*, March 2, 1939; "128,052 arboles de frutas varias se han regalado en este año," *La Prensa*, July 8, 1938.

11. "Alfalfa rendidora," *Jalisco Rural* 14, 4 (July 1939): 60; SAF, *Memoria 1930–1940*, pp. 138, 256–59; "Relacion de los productos"; Roberto Durán M., "Es necesario fomentar algunas industrias agrícolas nacionales," *Agricultura* 2, 13 (July–August 1939): 65–67; Rodríguez y Rodríguez, "Principales zonas," p. 106.

12. Guillermo Gándara, "Nota acerca del Xochinactle o Teunacaxtle," *Agricultura* 1, 8 (September–October 1938): 40–43; Ramírez M., "Informe," March 27, 1939, AGN, Cárdenas papers, Exp. 506.1/29; SAF, *Programa de labores, 1939* (México, D.F.: 1939), pp. 7–8.

13. Gonzalo Torres Barusta, "Aplicación de la genética para obtener lineas puras con caracteres determinados," *Agricultura* 1, 8 (September–October 1938): 3.

14. Antonio Rivas Tagle, "Importante problema genético para mejorar la alimentación de las masas," *Agricultura* 1, 9 (November–December 1938): 4–8; Rivas Tagle, "Rendamientos y mejoramiento de los metodos de cultivo," in CNA, *Memoria . . . Papa*, p. 104; Antonio Soto Solis, "Instructivo para realizar la selección racional de la semilla del maiz," *El Maestro Rural* (August 1939): 16–19; Ernesto Montelongo, "Estudio a los tipos de maiz de Chalco," *Agricultura* 2, 2 (March–April 1939): 30–34; SAF, *Memoria 1939–1940*, pp. 90, 116–17, 259, 470; José Iñíguez to Cárdenas, March 25, 1940, AGN, Cárdenas papers, Exp. 506.12/20; Gabriel Ítie, "El mejoramiento de platano," *Agricultura* 1, 7 (July–August 1938): 3–7; "Campo modelo de fomento agrícola e industrial," *El Universal*, No-

vember 8, 1940; BNCA, *Informe del Consejo de Administración a la Asamblea de Accionistas, 1938* (México, D.F.: 1939), pp. 94–95; Horatio Mooers, "Wheat and Cotton Notes, Mexicali, B.C.," December 9, 1938, NA, RG 166, Box 388; "Se regulariza la calidad de los algodones en la Nación," *El Nacional*, March 4, 1939.

15. Edmundo Sterling and Jesus Rulfo, "Estudio sobre la posibilidad y costeabilidad del cultivo de la vid y frutales en la Comarca Lagunera," November 28, 1939, ARFF, Box 202; "La escasez de maiz en la República," *El Nacional*, February 26, 1938; Modesto C. Roland, "El problema de los fertilizantes," *Hoy*, July 30, 1938; Estudio presentado por la Delegación de Defensa Agrícola de León, Gto., in CNA, *Memoria . . . Papa*, pp. 69–73; LAS, "Una revisión a nuestra política económica," May 1940, ARFF, Box 180.

16. CNC, *Bases generales que en materia agraria fueron aprobados por el Pleno de la CNC para el Plan Sexenal y que sostendrán en la Convención que convocará el PRM* (México, D.F.: 1939), p. 9; SAF, *Memoria 1939–1940*, pp. 82–84; SAF, *Memoria 1938–1939*, pp. 82–89; SAF, *Memoria de trabajos*, p. 22; José Figueroa, "Editorial: Discurso pronunciado en la inauguración del Instituto Biotécnico en San Jacinto," *Agricultura* 2, 2 (March–April 1939): 3; Gándara, "La rotenona, nuevo insecticida," *Agricultura* 2, 1 (January–February 1939): 21–23; James G. Byington, "Increasing Acreage of Ejidal Land Being Devoted to Wheat," April 1, 1939, NA, RG 166, Box 146; "Contra las plagas de los algodoneros," *El Siglo de Torreón*, July 27, 1938.

17. Tito Livio Calcaneo to Cárdenas, October 13, 1937, and others, AGN, Cárdenas papers, Exp. 506.17/22; "La invasión de la plaga ya contenida," *El Nacional*, November 5, 1937; Jeannot Stern, "Nuevos aspectos del problema del chamusco de plátano en México," *Anales de la Escuela Nacional de Ciencias Biológicas* 1, 1 (October–December 1938): 161–67; "Valiosa cosecha perdida por falta de transportes," *El Universal*, February 25, 1944; Gilberto Mendoza Vargas, "Informe de los acuerdos tomados en la Primera Convención Resolutiva del Problema Platanero de Villahermosa, Tab.," in SAF, *Memoria de la Segunda Convención Resolutiva del Problema Platanero* (México, D.F.: 1939), pp. 23–25; SAF, *Memoria 1937–1938*, p. 131; SAF, *Memoria 1939–1940*, pp. 78–80; Joseph F. Burt, "Information for the Bureau of Entomology and Plant Quarantine," December 13, 1938, NA, RG 166, Box 391; "La enfermedad del chamusco del plátano ha invadido las Zonas de Defensa establecidas en Nautla, Tuxtepec y Coatzacoalcos," *Boletín Platanero y Agrícola* 1, 1 (September 1938): 5; "Cunde el 'chamusco' en los plataneres de Chiapas," *El Universal*, September 25, 1939; "La Southern Banana se retira de México," *El Universal*, October 16, 1939; Douglas M. Crawford, "Bananas—Mexico, Production

and Export Situation," May 25, 1942, NA, RG 166, Box 312; Thomas T. Poleman, *The Papaloapan Project: Agricultural Development in the Mexican Tropics* (Stanford, Calif.: Stanford University Press, 1964), p. 70.

18. Fernández y Fernández, "La experimentación agrícola," *Germinal*, June 28, 1939; SAF, *Informe de labores 1940–1941* (México, D.F.: Editorial Cultura, 1941), p. 27; SAF, *Informe de labores 1941–1942* (México, D.F.: Editorial Cultura, 1942), p. 46; Humberto Cantu Echeverría to Cárdenas, June 17, 1940; Flavio Ramírez Álvarez to Cárdenas, October 2, 1939, AGN, Cárdenas papers, Exps. 506.17/6 and 506.17/27; Manuel Santuario, "Producción de ajonjolí en la región de Veracruz," in CNA, *Memoria . . . Oleaginosas*, pp. 107–10; Wales W. Signor to Secretary of State, October 3, 1940, NA, RG 59, Document File 1940–1944; Fernando Alatorre, "La enfermedad de la cercospora en las hojas del plátano," March 10, 1939, ARFF, Box 145; Daniel Moreno to Departamento de Control Agrícola, June 9, 1939; Ramón Casillas González to Departamento de Control Agrícola, November 1, 1939, AHJ, Box AG-27-I, Exp. 1939, Informes de los Inspectores; Alfonso Martínez Barragan, "Informe del Secretario del CNA," in CNA, *Memoria . . . Papa*, p. 5; Parres, "Memorandum sobre los puntos fundamentales que en concepto del titular de la SAF deben tratarse de manera preferente en el Consejo de Coordinación de Economía Nacional," October 11, 1939, AGN, Cárdenas papers, Exp. 564.1/2016.

19. Banjidal Rioverde, San Luís Potosí, May 6, 1938, and others; Jesús Ramírez and others to Cárdenas, April 8, 1940; Luís Rivas López to Secretario Particular, April 3, 1939; Jeronimo Herrera to Cárdenas, September 28, 1939, and others; Evaristo R. García to Cárdenas, February 20, 1939; Reynaldo Pérez Gallardo to Cárdenas, June 13, 1940, AGN, Cárdenas papers, Exps. 506.11/103, 702.1/84, 506.1/32, 506.17/6, and 506.1/29; Simón Gutiérrez to Cárdenas, May 15, 1939, AGN, Avila Camacho papers, Exp. 485.1–10; LAS, "Acta de la Sesión del Comité Ejecutivo," January 18, 1940, ARFF, Box 85; "La campaña contra el 'chamusco' ha tenido un éxito," *El Nacional*, January 15, 1938; "Adelantados obtenidos a la fecha en el campo de fomento agrícola en Tlalnepantla, México," *El Nacional*, April 2, 1938; SAF, *Memoria 1937–1938*, p. 112; "Informe," September 1, 1938, in *Palabras*, p. 140.

20. SAF, *Memoria de trabajos*, 2, 22–23.

21. Gilberto Mendoza Vargas, "La enseñanza y la extensión agrícola," February 1954, in Enseñanza Agrícola Superior, BCM Bound Volume #145–43; Parres, "Nueva orientación," p. 52; LAS, *Boletín de Informaciones* (July–August 1939): 44; Flores, *Historias*, p. 201; Miguel García Cruz, "El problema legendario de la ENA," p. 3; Emilio López Zamora to LAS, April 14, 1939; Miguel G. Stoupignan, "El Articulo 3o, letra muerta en Chapingo," June 10, 1940; in ENA, BCM Bound Volume #168–2; David E. Lorey, *The University System and Economic Development*

in Mexico since 1929 (Stanford, Calif.: Stanford University Press, 1993), p. 64; Gómez, *Biografías*, p. 348.

22. Gabriel Serrano, "El cultivo del trigo en el estado de Sonora," *Agricultura* 2, 5 (September–October 1939): 25; SAF, *Memoria de trabajos*, pp. 22, 38–39; *Memoria 1939–1940*, pp. 103–104; SAF, *Memoria 1939–1940*, p. 99; Roberto J. Muñoz, "Los frutales en Jalisco," *Jalisco Rural* 19, 3 (June 1939): 40–42; Copy of paper on Agriculture read at Saltillo, Coahuila, Mexico, on April 2, 1941 by Engineer Rigoberto Villareal; Henry T. Dwyer, "Vegetable Crop Outlook—Southern Sonora," November 14, 1940, NA, RG, 166, Boxes 146 and 152.

23. Lester D. Mallory, "Mexico—Limes and Lime Oil," January 10, 1942, NA, RG 166, Box 312; Pandurang Khankhoje, "El limón Mexicano: Su importancia, cultivo, selección, mejoramiento, empaque, mercado, etc.," May 1939, ARFF, Box 121; SAF, *Informe 1940–1941*, p. 33; Manuel Avila Camacho, *Address to the Mexican Agronomists* (Mexico, D.F.: Department of State for Foreign Affairs and the International Press Service Bureau, 1941), p. 10; William Galbraith, "Scarcity of Corn in Mexico," February 23, 1938, NA, RG 166, Box 148; "Causes a que se atribuye la escasez del maiz en la República," *El Universal*, February 21, 1938; Sergio Reyes Osorio et al., *Estructura agraria y desarrollo agrícola en México: Estudio sobre las relaciones entre la tenencia y el uso de la tierra y el desarrollo agrícola en México* (México, D.F.: Fondo de Cultura Económica, 1974), p. 40.

24. Fidel Saenz F. and Pascual García R. to Avila Camacho, December 31, 1940, AGN, Avila Camacho papers, Exp. 133.2/6; Medina, *Del cardenismo*, p. 19; Unión Agrícola Regional de Chapala to Cárdenas, January 20, 1938; Moreno to Departamento de Agricultura y Ganadería, Jalisco, October 21, 1939, AHJ, Box AG-27-I, Exps. Informes de los Inspectores and 51.50, Legajo 9; Enrique Galván y López, "Sugestiones generales sobre el café," October 1938, ARFF, Box 29; "Disminuyó la cosecha de tomate en Sonora," *El Universal*, October 28, 1938; "El mejor aprovechamiento de la tierra," *Excelsior*, January 9, 1939; "La insuficiente producción de maiz," *El Universal*, February 21, 1938; "Perifonemas," *Ultimas Noticias de Excelsior*, October 4, 1940; "Editorial," *Excelsior*, July 13, 1939; Alfonso M. Grajales Gómez to Cárdenas, April 11, 1939; J. Rosario Heras Uriate to Cárdenas, September 7, 1938, AGN, Exps. 703.4/371 and 437.1/556; "El gobierno debe proteger la producción nacional de trigo," *Jalisco Rural*, 16, 2 (May 1939): 21; "La Confederación de Partidos Independientes opina sobre el Plan Sexenal," *El Universal*, November 19, 1939; Jorge L. Tamayo, "Transformación de la Comarca Lagunera: Sus perspectives y problemas," *Apendice al Tomo 55 de las Memorias y Revista de la Academía de Ciencias Nacionales Antonio Alzate* (México, D.F.: 1941), p. 48; Rafael Dormantes to Avila Camacho, December 6, 1940, AGN, Avila Camacho papers, Exp. 151/3.25.

25. Jesús Adargas to Cárdenas, June 5, 1938; Juana Hernández G. to Cárdenas, February 27, 1939; Lázaro Martínez Soto to Cárdenas, September 23, 1938, AGN, Cárdenas papers, Exps. 606.3/101 and 703.4/324.

26. Rodriguez G., "Estudio," pp. 9–11; SEP, *Instituto Politécnico Nacional, Anuario 1939* (México, D.F.: 1939), p. 51.

27. Alberto A. Gutiérrez to Director General de Agricultura, September 10, 1937; Godofredo F. Beltrán to SAF, August 9, 1938, AGN, Cárdenas papers, Exps. 606.3/164 and 506.1/30; "Pasan a ser dependencias de Agricultura," *El Nacional*, April 30, 1938; SAF, *Memoria de trabajos*, p. 40, 177; SAF, *Memoria 1939–1940*, p. 116.

28. Joaquín de Zayas, "Cultivo de la papa," in CNA, *Memoria . . . Papa*, p. 81; Leopoldo de la Barreda, "Cuarentinas (vigentes y aconsejables), in CNA, *Memoria . . . Papa*, p. 139.

29. SAM, "Manifesto a la nación," *El Universal*, February 5, 1939; "La SAM contesta al General Amaro," *El Universal*, March 9, 1939; Emilio López Zamora to R.H. Llergo, September 12, 1938, ARFF, Box 124; FRAM, *Programa*, p. 16; CNC, *Bases*, pp. 18–19; "Plan de Acción Agrícola," *El Nacional*, June 22, 1938.

30. Ramírez A., "Memorandum"; SAF, *Memoria 1949–1940*, p. 124.

31. BAR to agrónomos de la República, October 20, 1940; Plutarco Sánchez Herrera and Emilio Gutiérrez Roldán to members, SAM, September 28, 1939, ARFF, Boxes 181 and 180; Marte R. Gómez, *Biografías de agrónomos* (Chapingo, México: Colegio de Postgraduados, Escuela Nacional de Agricultura, 1976), p. 353.

32. "Al comenzar," *La Tierra* 1, 1 (April 1938): 1; Quintín Ochoa, "Enfermedades causantes de la perdida de las cosechas de papa en el Distrito de Jiménez, Chih.," *Agricultura* 2, 2 (March–April 1939): 35–36; Ramírez A. to Parres, December 1, 1938, AGN, Cárdenas papers, Exp. 506.1/29.

33. Comisión Intersecretarial de Estudios, "Informe sobre la jira Acapulco, Gro. a Salina Cruz, Oax.," May 1940, AGN, Cárdenas papers, Exp. 506.1/40; Gilberto Mendoza Vargas, "Informe general sobre los trabajos de la campaña experimental contra el chamusco," in SAF, *Memoria de la Segunda*, pp. 10–11.

34. J. Raul Rodríguez, "Memorandum con datos sobre las plantaciones de arboles frutales y vides en el Distrito de Riego de Pabellón, Ags.," December 7, 1938, ARFF, Box 202; BNCA, *Informe . . . 1939*, p. 40; Manuel Santuario, "Agricultura y crédito rural," *Agricultura* 2, 2 (March–April 1939): 97; Roberto Muñoz, *El arroz: Monografía* (México, D.F.: D.A.P.P, 1939), pp. 16–27; Manuel Mesa A., "La agricultura (estudio para el Plan Sexenal)," January 2, 1939, p. 8, ARFF, Box 153; "Los cultivos del maiz en la República," *El Nacional*, March 16, 1938; Eduardo Chávez, "Ventejas de las labores con arado," *Jalisco Rural*, 20, 4 (September 1940):

63–65; SAF, *Memoria 1939–1940*, p. 134; BNCA, *Informe del Consejo de Administracion a la Asamblea de Accionistas, 1939* (México, D.F.: 1940), p. 24.

35. FRAM, *Programa*, pp. 19, 38–39; *Boletín de Informaciones* (July–August 1939): 45.

36. BNCA, *Informe . . . 1938*, p. 20; BNCA, *Informe . . . 1939*, p. 48; Parres, "Nueva orientación que se propone para la SAF," August 7, 1939, pp. 18–19, 46, in Política Agrícola, BCM Bound Volume #179-53; Pablo Córdova to BNCE, January 25, 1938, ARFF, Box 28; SAF, *Memoria 1939–1940*, pp. 95–96.

37. Mesa A., "La agricultura," pp. 46–48; Manuel Mesa A., "La política agrícola," *Futuro*, December 1940, p. 13–15; Ramírez M. to Parres, July 29, 1938, AGN, Cardenas papers, Exp. 506.1/29; LAS, *Boletín de Informaciones*, July–August 1939, p. 45; Fernández y Fernández, "La situación actual de los agrónomos," pp. 3–4, in Agrónomos, BCM Bound Volume #168-40 (hereafter Agrónomos).

38. Mesa A., "La agricultura," pp. 72–73, 93; Fernández y Fernández, "La Ley de Secretarios del Estado y la coordinación agrícola" (n.d.), p. 8; CTM, "Bases generales para el Segundo Plan Sexenal," February 20, 1938, ARFF, Box 142.

39. García Cruz, "Ley de Servicios Agrícolas Nacionales" (n.d.), ARFF, Box 142; Edmundo Taboada's Business Card, ARFF, Box 82; William C. Cobb, "Interview with Richard Bradfield," p. 27, RFA, RG 13, RAC; Marco Antonio Durán, *Apuntes de agrobiología* (México, D.F.: Instituto Mexicano de Estudios Agrícolas, 1940), pp. 5–6; David S. Ibarra, "Un problema nacional: La erosión del suelo," April 1941, p. 3, in SAM Segundo Consejo Nacional Directivo, BCM Bound Volume #186–5 (hereafter SAM Segundo); Aurelio Leal Trevino to SAF, July 18, 1938, AGN, Cárdenas papers, Exp. 437.1/556; "Segunda colaboración de los agrónomos para la elaboración del Segundo Plan Sexenal," (n.d.). ARFF, Box 142.

40. "Sugestiones adicionales de la Comisión Nacional de Irrigación en relación con la incrementación de la producción," September 21, 1939, AGN, Cárdenas papers, Exp. 564.1/2016; David Ronfeldt, *Atencingo: The Politics of Agrarian Struggle in a Mexican Ejido* (Stanford, Calif.: Stanford University Press, 1973), pp. 34–37; Gómez, *Biografías*, pp. 328–33.

41. García Cruz, "El problema legendario," p. 3; López Zamora to LAS; Stoupignan, "El Articulo 3o"; LAS, *Boletín de Informaciones* (July–August 1939): 30; "Metodos de cultivo," *Germinal*, October 28, 1938; CNC, *Bases generales*, p. 10.

42. Durán, *Apuntes*, p. 7; Emilio López Zamora, *La situación del Distrito de Riego de El Mante* (México, D.F.: Editorial Ramírez Alonso, 1939), pp. 74–75; Durán, "Al margen de la agrobiologia," *Agricultura* 1, 8 (Sep-

tember–October 1938): 9–10; Durán, *Apuntes*, pp. 8–9, 12–15; Edmundo Taboada R., *Algunos datos relativos al metodo de Hoffer para investigar deficiencias de elementos fertilizantes en los suelos* (México, D.F.: 1941); Mesa A., "La política," p. 15; Figueroa, "Editorial: discurso," p. 1.

43. "Puntos generales que sugiere la Comisión Nacional de Irrigación en relación con la incrementación del la producción," September 21, 1939, AGN, Cárdenas papers, Exp. 564.1/2016; García Cruz, "El fracaso de la experimentación agrícola en México," *El Popular*, May 23, 1939; García Cruz, "Ley de Servicios," *El Popular*, May 23, 1939, pp. 3–4; Augusto Pérez Toro, "Algunas problemas del henequen," *Agricultura* 2, 1 (January–February 1939): 18–20; FRAM, *Programa*, pp. 14–15; Mesa A., "La agricultura," p. 46.

44. FRAM, *Programa*, pp. 9–10, 40–42; C. Herminio Pérez y Pérez, "Ponencia," in CNA, *Memoria . . . Papa*, pp. 244–277; SAF, *Estudio agroeconómico*, pp. 181–82;

45. Andres Acosta L. to Cárdenas, April 18, 1938, AGN, Cárdenas papers, Exp. 546.6/87; *Decreto No. 33, Que reglamenta el Articulo Constitucional sobre el ejercicío de los profesiones técnicos-científicos* (San Luis Potosí: 1939); Nora Hamilton, *The Limits of State Autonomy: Post Revolutionary Mexico* (Princeton, N.J.: Princeton University Press, 1982), p. 200.

46. Enrique Arreguín Jr., "Proyecto para la organización de un Instituto Nacional de Investigaciones en Ciencias y Letras," September 1937, p. 8, AGN, Cárdenas papers, Exp. 534/100; Luis González y González, *Los dias del Presidente Cárdenas* (México, D.F.: El Colegio de México, 1981), pp. 280–91; José Sarukhán, "The Status of Mexican Science and Technology Research: Potential Avenues for Collaborative Programs with the U.S.," *Mexican Studies* 5, 2 (Summer 1989): 266; "Los grandes cultivos nacionales, el olivo base de porvenir de la agricultura e industria de Baja California," *El Nacional*, September 16, 1941; Fernández y Fernández, *Chapingo hace 50 años* (Chapingo, México: ENA, Colegio de Postgraduados, 1976), p. 136; "Mexico's New Education Law," *Bulletin of the Pan American Union* 74, 6 (June 1940): 480; Secretaría de Hacienda y Crédito Público, *Presupuesto general de egresos de la Federación 1938* (México, D.F.: 1938), p. 4; "Interview with José Luís de la Loma y de Oteyza," in Leobardo Jiménez Sánchez, *Las ciencias agrícolas y sus protagonistas* (Chapingo: Colegio de Postgraduados, 1984), p. 159; Fernández y Fernández, "La Ley"; Jiménez Sánchez *El autoabastecimiento*, p. 18; LAS, *Boletín de Informaciones* (July–August 1939): 30, 43; FRAM, *Programa*, pp. 55–56.

47. SAF, *Memoria 1939–1940*, p. 263; SEP, *Instituto*, p. 61.

48. Friedrich E. Schuler, *Mexico between Hitler and Roosevelt: Mexican Foreign Relations in the Age of Lázaro Cárdenas* (Albuquerque: University

of New Mexico Press, 1998), pp. 142–43; Marco Antonio Durán, *Los sofismos de la Reforma Agraria* (México, D.F.: LAS, 1939), pp. 60, 84; Ignacio Malo Álvarez, *La burgesia y la reforma agraria* (México, D.F.: Unión Cooperativa de Artes Gráficos del D.F, 1940), p. 42; Vicente Lombardo Toledano, "Discurso en el Consejo Extraordinario de la C.T.M." (n.d.), ARFF, Box 142; CNC, *Bases*, p. 9; FRAM, *Programa*, p. 45; *Proposiciones del Partido Comunista Mexicano para el Plan Sexenal de 1940–1946* (México, D.F.: 1939), p. 3; LAS, "Una revisión"; Nettie Duskis to Gonzalo Robles, January 13, 1939; Sociedad de Amigos de la URSS to organizaciones de obreros, de campesinos, de intellectuals, de profesionistas, etc. (n.d.), AGN, Robles papers, Correspondence Section, Box 39, Exps. 5 and 31.

49. Stephen R. Niblo, *War, Diplomacy, and Development: The United States and Mexico, 1938–1954* (Wilmington, Del.: Scholarly Resources, 1995), pp. 65–70; Schuler, *Mexico between Hitler and Roosevelt*, pp. 66–69; "Informe general y presupuesto de la Oficina de Abonos Nacionales," 1938, 1939, and 1940, AGN, Robles papers, Agriculture Section, Box 1, Folder 8; Cárdenas to C. Cosme Hinojosa, March 8, 1937, AGN, Cárdenas papers, Exp. 502.1/14; Hector Calzado, "Productos del coco," in CNA, *Memoria . . . Oleaginosas*, pp. 320–21; Edmundo Flores, *Historias de Edmundo Flores, autobiografía 1919–1950* (México, D.F.: Martín Casillas Editores, 1983), p. 139; LAS, "Noticiero de la Liga," *Boletín de Informaciones* 2, 9–10 (August 1940): 1; SAF, *Memoria 1939–1940*, p. 258.

50. Leandro Martínez, "Descripción de la agricultura en Tabasco," *Agricultura* 2, 5 (September–October 1939): 36–42; Zenon Lemański, "Las características del servicio de la organización de la defensa de las plantas en los Estados Unidos de Norte America," *Agricultura* 1, 5 (March–April 1938): 48–49; José Lázcano Romero, *Analysis de la situación algodonera de México* (México, D.F.: Instituto Mexicano de Estudios Agrícolas, 1938), p. 192; Eduardo Turcott B., "El arroz: Estudio de su situación actual y posibilidades de exportación," October 24, 1938, p. 1, ARFF, Box 17; "Technológicos azucareros," *El Universal*, November 3, 1938; J.W. Torres to Pan American Union, July 14, 1938, AHJ, Box AG-27-I, Exp. Asuntos Vários, Revista *Jalisco Ganadero*; "Summary of Visits to Three Colleges of Agriculture in States Contiguous to Mexico (Arizona, New Mexico, and Texas); to the Louisiana State University, and to the School of Agriculture at Juárez, Mexico," May 1941, RFA, RG 1.1, Series 323, Box 1, Folder 2, RAC; R.E. McDonald, "Biología y metodo de control del 'Gusano Rosado del algodón,' " *Agricultura* 1, 8 (September–October 1938): 61–62; Enrique J. Herás to Cárdenas, April 5, 1938, AGN, Cárdenas papers, Exp. 506.1/8; "Se va a permitir el cultivo de zoca de algodón," *El Siglo de Torreón*, June 12, 1938; "No habrá zoca—se sembrará algodón," *El Siglo de Torreón*, October 27, 1939; "Weekly Report, Pink Bollworm and Thurberia Weevil Control," November 14 and December 18, 1939, December 9, 1940, NA,

RG 7, Box 1399; Josephus Daniels, 'The Dedication of Mexico's Leland Ossian Howard Entomological Laboratory," *Journal of Economic Entomology* 31, 6 (1938): 773; Interdepartmental Committee on Cooperation with the American Republics, "Minutes," August 2, 1939, NA, RG 353, Box 29, Folder Mexico, 1939; "Report of Department of Agriculture to Interdepartmental Committee on Cooperation with the American Republics," August 3, 1938, NA, RG 353, Box 3, Folder Budget Justifications Agriculture and Civil Aeronautics; H.L. Wilson to Robert G. Caldwell, November 6, 1940, NA, RG 229, Box 95.

51. Earl N. Bressman, "Statement before the Defense Investigating Committee of the Senate" (n.d.), pp. 4–5, Rockefeller Family Archives, Nelson Rockefeller papers, RG 4, Series 0, Box 10, Folder 78; "Section IV Economic Penetration" (n.d.), p. 19, 27, RG 4, Series 0, Box 11, Folder 85, RAC; Medina, *Del cardenismo*, p. 43; Edgar B. Brossard, "United States Program of Scientific and Cultural Cooperation with Other American Republics," *Bulletin of the Pan American Union* 80, 10 (October 1946): 547; Kelchner to Laurence Duggan, June 22, 1938, NA, RG 353, Box 2; "Cultural Relations and Inter-American Cooperation," Volume 1; Committee on International Cooperation in Agricultural Education, "Minutes," November 7, 1940, NA, RG 59, Decimal File 1940–1944, Box 196.

52. "The United States Plans Cooperation with the Other American Republics," *Bulletin of the Pan American Union* 73, 1 (January 1939): 56; Duggan to Sumner Welles, April 13, 1938; Ellis O. Briggs, "Proposed Expansion of Activities of the Department of Agriculture in the American Republics," July 12, 1938, NA, RG 353, Box 2, V. 1; Wallace, "Inter-American Agricultural Cooperation," *Bulletin of the Pan American Union* 74, 4 (April 1940): 275–76; Committee on Cooperation with the American Republics, "Minutes of Meeting," October 28, 1938, NA, RG.353, Box 29, Folder Minutes, May 25 to December 29, 1938.

53. Bressman, "Statement," pp. 8–9; "Proposed Paragraph for Substitution on Page 2 of Statement re Institute of Tropical Agriculture" (n.d.), NA, RG 353, Box 2; "Cultural Relations and Inter-American Cooperation," Volume 2; General Advisory Committee of the Division of Cultural Relations of the Department of State, "Minutes," September 17–18, 1941, NA, RG 353, Box 29, Folder General Advisory Committee, Minutes.

54. Rebeca De Gortari Rabiela, "Educación y conciencia nacional: Los ingenieros despues de la revolución Mexicana," *Revista Mexicana de Sociología* 49, 3 (1989): 130; Martin H. Greenberg, *Bureaucracy and Development: A Mexican Case Study* (Lexington, Mass.: D.C. Heath, 1970), p. 16; Wilson to Caldwell, December 6, 1940; Elvin C. Stakman, W.L. Popham, and R.C. Casell, "Observaciones del chahuixtle en México durante el mes de febrero de 1938," *Agricultura* 2, 1 (January–February 1939): 3–5; "Es-

tudios sobre la genética del maiz en EE UU," *El Nacional*, April 13, 1938; Nathaniel Weyl and Sylvia Weyl, *The Reconquest of Mexico: The Years of Lázaro Cárdenas* (New York: Oxford University Press, 1939), p. 223; Jesús Patiño N. to Aaron Saenz, September 21, 1939, AGN, Robles papers, Correspondence Section, Box 39, Exp. 5; "La forma para clasificación en el algodón," *El Nacional*, July 24, 1939; "Informe . . . Oficina de Abonos," p. 5; James G. Byington, "Progress Made to Date in the Construction of the Dam at El Palmito, Durango, Mexico," April 23, 1939, NA, RG 166, Box 149; Dario L. Arrieta M., "Comisión Internacional de Plagas" (n.d.), p. 5, RFA, RG 6.13, Series 1.1, Box 10, Folder 121, RAC; "Acuerdo con el Presidente de la República," May 27, 1940, AGN, Cárdenas papers, Exp. 506.17/22; Division of Inter-American Educational Relations, "Alumni of United States Colleges of Agriculture Who Are Citizens of the Other American Republics," April 1944, NA, RG 59, Decimal File 1940–1944, Box 196.

55. Niblo, *War*, pp. 39–40; Briggs to Duggan, July 27, 1938; Duggan to Welles, July 28, 1938, NA, RG 353, Box 2, V. 1.

56. Cordell Hull to Harold L. Ickes, March 6, 1939, NA, RG 59, Microfilm Roll m1370; Briggs to Bursley, May 2, 1939; Briggs, "Fisheries Cooperation with Mexico," December 18, 1939; "Assignment of a Fisheries Expert to Mexico," January 20, 1940, NA, RG 353, Box 2, V. 1 and V. 2; Milton J. Linder, "Memorandum on the Mexican Fisheries Problem," January 25, 1940, NA, RG 353, Box 51, Folder Fisheries; Interdepartmental Committee on Cooperation with the American Republics, "Minutes of Meeting," April 28, 1939, NA, RG 353, Box 29, Folder Minutes, 1939; Niblo, *War*, p. 51; LAS, "Noticiero," p. 1; Arrieta M., "Comisión," p. 2; "Proposed Projects under the Cultural Relations Program," November 7, 1940, Nelson Rockefeller papers, RG 4, Series 0, Box 5, Folder 35, RAC; "Convention on the Inter-American Bank," *Bulletin of the Pan American Union* 74, 7 (July 1940): 532; "Relación de los productos," SAF, *Memoria 1939–1940*, p. 263; Ramón Beteta to Josephus Daniels, August 28, 1940, NA, RG 166, Box 151; E.W. Brandes, "The Outlook for Plantation Rubber in Tropical America," 1942, ARFF, Box 105.

57. "Directive Council," December 21, 1937, and others, NA, RG 353, Box 2, V. 1; S.M. to Raymond B. Fosdick, May 25, 1938; RBA to Fosdick, "South American Program," August 8, 1940; "Latin American Program," December 3–4, 1940; "Interviews: DHS," October 24, 1938, RFA, RG 1.2, Series 300, Box 1, Folder 7, RAC.

58. Lester D. Mallory, "The Possible Development of a Quinine Industry in Mexico," May 6, 1940, NA, RG, 166, Box 146; Briggs, "Decision of the Rockefeller Foundation to Appoint a Permanent Representative in

Buenos Aires," August 8, 1940, NA, RG 353, Box 2, V. 2; Interdepartmental Committee on Cooperation with the American Republics, "Minutes," December 21, 1939, NA, RG 353, Box 29, Folder Minutes 1939.

59. Deborah Fitzgerald, "Exporting American Agriculture: The Rockefeller Foundation in Mexico, 1943–1953," in Marcos Cueto, ed., *Missionaries of Science: The Rockefeller Foundation & Latin America* (Indianapolis: Indiana University Press, 1994), pp. 72–96; JD, FBH with Dr. Sanchez, February 21, 1941, RFA, RG 1.1, Series 323, Box 11, Folder 70; Carl O. Sauer to Joseph Willits, March 12, 1941, RFA, RG 1.2, Series 323, Box 10, Folder 63, RAC.

60. Warren Weaver to J. George Harrar, January 27, 1949, RFA, RG 1.1, Series 323, Box 5, Folder 34; John A. Ferrell to George C. Payne, January 21, 1941; Ferrell to Fosdick, "Aid to Mexico," January 27, 1941, RFA, RG 6.13, Series 1.1, Box 33, Folder 366; "Staff Conference," February 18, 1941, RFA, RG 1.1, Series 323, Box 1, Folder 2, RAC.

61. Jorge L. Tamayo, "Nuestra agricultura," *Excelsior*, July 29, 1941; Mauricio Magdaleno, "La agricultura nacional: Dirección técnica," *El Universal*, September 2, 1941; "Report on the Agricultural Survey of Mexico for Complimentary Crops, submitted by the Special Mission of Agricultural Experts from the USDA," September 1941, AHSRE, Exp. III-438-9; "Wallace in Mexico," *Wallace's Farmer and Iowa Homestead*, February 22, 1941; General Advisory Committee of the Division of Cultural Relations of the Department of State, "Minutes," February 27, 1941, NA, RG 353, Box 29, Folder General Advisory Committee, Minutes; Dayle C. McDonough to Hull, April 18, 1941, NA, RG 166, Box 148; Nathan Whetten, *Rural Mexico* (Chicago: University of Chicago Press, 1948), p. vii; "Report of the Division of Cultural Relations," February 1942, p. 2, NA, RG 353, Box 23; "Training Professionals in the Agricultural Sciences" (n.d.), p. 5, NA, RG 166, Box 308.

62. Harry M. Miller, "Mexican Log," March 2–28, 1941, p. 4, RFA, RG 6.13, Series 1.1, Box 33, Folder 366, RAC.

63. Ferrell, "Memorandum of Conference with Wallace and Fosdick Regarding Mexico," RFA, RG 1.1, Series 323, Box 1, Folder 2; Wallace to Fosdick, November 24, 1941, RFA, RG 1.1, Series 323, Box 11, Folder 72; "Agricultural Conditions and Problems in Mexico, Report of the Survey Commission of the Rockefeller Foundation," 1941, pp. 1, 3–4, RFA, RG 1.1, Series 323, Box 5, Folder 37; Wallace to Fosdick, May 13, 1941, RFA, RG 1.1, Series 323, Box 12, Folder 79; "The Rockefeller Foundation's Survey of Agriculture in Mexico," 1941, p. 7, RFA, RG 1.1, Series 323, Box 11, Folder 70; Pauline Madow, "The Reminiscences of Elvin C. Stakman," June 2, 1970, p. 1061; Cobb, "Interview with Edwin J. Wellhausen," June 28, 1966, p. 165, RFA, RG 13, RAC.

64. "The Rockefeller Foundation's Survey," p. 4.

65. A.R. Mann to Miller, February 26, 1941, p. 3, RFA, RG 1.2, Series 323, Box 10, Folder 63, RAC.

66. "The Rockefeller Foundation's Survey of Agriculture in Mexico," 1941, p. 3, RFA, RG 1.1, Series 323, Box 11, Folder 70, RAC.

67. Cobb, "Interview with Paul Mangelsdorf," pp. 52; Cobb, "Interview with Wellhausen," pp. 7, 48, RFA, RG 13; Erlif A. Miller, John B. Pittner, Ricardo Villa J., and Carlos Romo G., "Population Density of Unirrigated Maize and Its Influence upon Fertilizer Efficiency in Central Mexico," 1948, p. 7, RFA, RG 6.13, Series 1.1, Box 46, Folder 531, RAC.

68. Stakman et al., "Agricultural Conditions and Problems of Mexico," pp. 43, 46.

69. Mann to Miller, February 26, 1941, p. 3.

70. Carlos Ayala Origel to Avila Camacho, October 31, 1941, AGN, Avila Camacho papers, Exp. 506.1/7; Cobb, "Interview with Bradfield," p. 21; Eduardo Morilla Zafa, "Un Congreso Panamericano de Agrónomos," April 1941, in SAM Segundo; "Asuntos gremiales," April 1941, in SAM Segundo; "Puntos ideológicos de la Sociedad Agronómica B.A.R.," August 22, 1941, AGN, Avila Camacho papers, Exp. 404.1/23.

71. "Con ayuda de tres sabios americanos resucitará la agricultura," *Carta Seminal*, August 2, 1941; "A ser estudiada la forma de aumentar la producción agrícola de nuestro país," *Excelsior*, July 28, 1941; George C. Payne to Frank Blair Hanson, August 20, 1941, RFA, RG 1.1, Series 323, Box 11, Folder 72, RAC; Mallory to Hull, October 31, 1941, NA, RG 166, Box 151; "Este año, por primera vez, México alimentará a todos sus habitantes," *Excelsior*, November 13, 1941; Alfonso González Gallardo to Secretario de Relaciones Exteriores, March 27, 1941, AHSRE, Exp. III-438-9; "El cultivo del plátano," *El Universal*, March 29, 1941; "Se combate con exito la plaga de chamusco en Tabasco," *El Universal*, May 4, 1941; Thomas H. Lockett, "Periodic Economic Review—Mexico," August 4, 1941, NA, RG 166, Box 147.

72. Bressman, "Statement," pp. 29–31; Rafael Zubarán Capmany, "El cultivo del caucho," *El Universal*, August 12, 1942; E. Suárez del Real, "El problema del caucho sintético," *El Universal*, October 4, 1942; "Possible Conference of Ministers of Foreign Affairs of the American Republics, Preliminary Draft Outlines of U.S. Position with Supporting Considerations on Certain Economic Topics Which Latin American Nations Might Wish to Discuss," December 30, 1944, Nelson Rockefeller papers, RG 4, Series 0, Box 5, Folder 39, RAC.

73. Interdepartmental Committee on Cooperation with the American Republics, "Minutes," May 28, 1941, NA, RG 353, Box 29, Folder Minutes—1941; Luis L. León, "Memorandum sobre hule: Conveniencia de establecer plantaciones de arboles de hule en México," February 1942, pp. 1–24, in Hule, BCM Bound Volume #43–3; "México debe producir el

hule necesario," *El Universal*, January 17, 1944; E. Padilla to Messersmith, March 5, 1942, NA, RG 166, Box 302; Charles S. Dolley to Elmer W. Brandes, December 3, 1942, AGN, Robles papers, Agriculture Section, Box 13, Exp. 110; "El guayule ofrece una gran fuente de riqueza para el país," *El Universal Gráfico*, February 27, 1942; Claude R. Wickard to E.C. Auchter, October 1, 1942; "Plantation Rubber Production in the Western Hemisphere," December 1942, NA, RG 310, Box 26, Folder Rubber; "Objectives of Hemispheric Policy," June 24, 1942, Nelson Rockefeller papers, RG 4, Series 0, Box 8, Folder 61, RAC.

74. Henry T. Dwyer to Hull, May 19, 1943; "Site Offered by Mexico for the Proposed Inter-American Institute for Tropical Agriculture," February 14, 1942; Mallory to Hull, December 2, 1942; L.B. Clark, "Monthly Economic Review—Mexico," January 1943; Messersmith to Hull, May 25, 1943; Mallory to Hull, November 25, 1942; Lester R. Geyer and Thomas D. Mallory to Mallory, July 10, 1943, NA, RG 166, Boxes 39, 302, 308, 312, and 319; "Informe rendido por el Comité Central Ejecutivo de la SAM a la Tercera Asamblea del Consejo Nacional Directivo" (n.d.), ARFF, Box 180; "Saludo a los Congresistas Interamericanos de Agricultura," July 6, 1942; Jesse May Harper to P.N. Annand, November 30, 1942, NA, RG 7, Box 1398; Adalberto Polo Celis, "El cultivo de la higuerilla," *Siembra* 1, 3 (June 1943): 29–30; "Mayor escala en la producción de café en Chiapas," *El Nacional*, May 25, 1942; "Avispas para combatir la plaga del pulgón," *El Universal*, July 6, 1943; Arrieta M., "Comisión," p. 5; F. Castillo Najera to Hull, August 6, 1943, NA, RG 353, Box 51; Kempton to G. Laguardia, December 18, 1941; Caldwell to Wilson, February 25, 1941; Messersmith to Nelson Rockefeller, June 23, 1943, NA, RG 229, Boxes 95 and 1687.

75. Manuel Gamio to Miller, May 10, 1941; "Mexican Log," March 2–21, 1941; Hanson to Harrar, November 17, 1942; Mann, "Observations in Mexico," April 9–25, 1943; Stakman to Mangelsdorf and Bradfield, March 27, 1943; Stakman to Hanson, March 17, 1943; Harrar to Hanson, May 11, 1943; Stakman to Hanson, August 7, 1943, RFA, RG 1.1, Box 1, Folders 2, 3, 4, 5, and 6; Stakman, "Report on Agricultural Activities in Mexico," February 3–May 20, 1943, pp. 23–24; Miller to Hanson, February 10, 1943, RFA, RG 1.2, Series 323, Box 10, Folders 60 and 63, RAC; SAF, *Informe 1941–1942*, pp. 42–43.

76. Harrar to Hanson, March 4, 1943; Manglesdorf to Hanson, March 9, 1943, RFA, RG 1.1, Series 323, Box 1, Folder 4; Harrar, "Diary," August 15 to November 30, 1943, RFA, RG 1.1, Series 323, Box 1, Folder 6, RAC; Mann, "Oberservations," p. 5; Jesús Gracia Fadrique, "El desarrollo de la industria química y la industria de los fertilizantes en México, 1759–1948," in Jesús Gracia Fadrique et al., *Estado y Fertilizantes* (México, D.F.: Fondo de Cultura Económica, 1988), p. 161; Gracia Fadrique, "Enseñanza de la química en México," in Fadrique et al., *Estado*, p. 194.

77. Mangelsdorf to González Gallardo, December 10, 1943, p. 7, RFA, RG 1.1, Series 323, Box 1, Folder 6, RAC.

78. Fosdick to Wallace, November 27, 1941, RFA, RG 1.1, Series 323, Box 11, Folder 72; Bressman to Miller, May 8, 1941; "Second Meeting of the Committee on Inter-American Cooperation in Agricultural Education," May 1–2, 1941; Payne to Ferrell, March 24, 1943; Payne to Hanson, April 1, 1943; R.E. Moore to Miller, May 26, 1943; T.H. Rogers to Harrar, June 21, 1943, RFA, RG 1.1, Series 323, Box 1, Folders 2, 4, and 5; Harrar, "Diary," February 9, 1943, RFA, RG 6.13, Series 1.1, Box 25, Folder 280, RAC; W.C. Brister to James D. LeCron, June 5, 1943, NA, RG 229, Box 1687, Folder June 1943; KM to Pattee, November 15, 1941, NA, RG 59, Box 196, Decimal Files 1940–1944.

79. Miller to Ingeniero Murillo [*sic*], April 21, 1941, RFA, RG 1.2, Series 323, Box 10, Folder 63; "Meeting of the Mexican Agricultural Commission," June 5, 1941, RFA, RG 1.1, Series 323, Box 11, Folder 71; Fosdick to Stakman, April 8, 1942; Stakman to Fosdick, July 3, 1942; Harrar to Hanson, May 11, 1943; Stakman to Hanson, March 17, 1943; Ferrell, "Notes on Mexico," April 21–May 7, 1943; Harrar to Hanson, "Publicity in Mexico and Request by the Military Intervening Officer in El Paso for Periodic Reports," June 22, 1943; Stakman to Mangelsdorf, December 2, 1943, RFA, RG 1.1, Series 323, Box 1, Folders 3, 4, 5, and 6; Mann, "Observations," p. 3; Harrar, "Diary," August 15–November 30, 1943; Eduardo Limón G. to Harrar, October 25, 1943, RFA, RG 6.13, Series 1.1, Box 69, Folder 784, RAC; Miller, "Mexican Log," March 2–28, 1941, p. 43; Stakman, "Report on Agricultural Activities," p. 10.

80. Edward G. Trueblood to Hull, May 16, 1942, NA, RG 166, Box 308; Hector Lazos, "Creación de Becas Interamericanas destinadas a especialización de professionales en ciencias agrícolas," June 1942; E.J. Kyle, "Desirable Extension of the Relationships between the Agricultural Educational Institutions in the Americas," 1942, ARFF, Box 180.

81. Albert R. Driesbach to Emilio G. Collado, November 13, 1943, NA, RG 229, Country Files, Mexico, Box 95; Flores, *Historias*, p. 239; Bursley to Drier, January 26, 1942, NA, RG 59, Box 3688, Decimal File 1940–1944; Mann, "Observations"; Bradfield to Miller, November 22, 1943, RFA, RG 1.1, Series 323, Box 9, Folder 56; "Mexicans Who Studied in the U.S.," December 1944; Harrar to Hanson, October 13, 1943, RFA, RG 1.1, Series 323, Box 1, Folders 8 and 6, RAC.

82. Flores, *Historias*, p. 236; Division of Inter-American Educational Relations, "Alumni of United States Colleges of Agriculture"; "Training Professionals in the Agricultural Sciences" (n.d.), p. 4, NA, RG 166, Box 308.

83. Bressman, "Pan Americanism for the People," speech at Hamburg, Iowa, May 31, 1941, RFA, RG 1.1, Series 323, Box 1, Folder 2, RAC; General Advisory Committee of the Division of Cultural Relations,

"Minutes," September 17–18, 1941; "Report of Division of Cultural Relations," March 1942, NA, RG 353, Box 23, Folder Reports on Activities of Division of Cultural Relations; Rockefeller to Gerald G. Smith, May 20, 1943; Rockefeller to Messersmith, June 17, 1943; George C. Dunham to Rockefeller, July 7, 1944; Dunham to Messersmith, October 22, 1943, NA, RG 229, General Records 1.1, Box 95, Folder Food Supply; Rockefeller to Hull, December 7, 1943, NA, RG 229, Records of Basic Economy Food Supply Division, Box 1687, Folder June 1943–Mexico II; Advisory Committee on Inter-American Cooperation in Agricultural Education, "Minutes," May 25–26, 1942, NA, RG 59, Box 196, Decimal Files 1940–1944; William C. Brister, "Increasing the Food Supply of the Western Hemisphere," *Bulletin of the Pan American Union* 79, 7 (July 1945): 388; "Text of Report of Mexican-American Commission for Economic Cooperation," July 17, 1943, NA, RG 353, Committee on Inter-American Development 1944–1945, Box 1; "Cultural Relations and Propaganda Activities of other Governments in the American Republics" (n.d.), NA, RG 353, Box 9, Folder Operations, Special Studies, Competition.

84. Manuel J. Gándara and Armando Esquivel to Avila Camacho, February 18, 1941; Gonzalo Peralta A. and others to Avila Camacho, July 11, 1941, and other documents, AGN, Avila Camacho papers, Exps. 437.3/52 and 133.2/67; Peter H. Smith, *Labyrinths of Power: Political Recruitment in Twentieth Century Mexico* (Princeton, N.J.: Princeton University Press, 1979), p. 167; "Nuevos altos funcionarios del gobierno" (n.d.), in Agrónomos; "Asuntos gremiales," in SAM Segundo; Cesar Martino to Avila Camacho, December 23, 1942, AGN Avila Camacho papers, Exp. 437.3/127; Salvador Camarena M., "Para incrementar la producción agrícola en México," April 1942, ARFF, Box 97; Crawford and Mallory, "Mexico—Wartime Food Supply Problems, a Summary of Legislation and Government Activities," August 22, 1944, p. 4, NA, RG 166, Box 302.

85. SAM, *Resoluciones y memoria de la Segunda Reunión del Consejo Nacional Directivo* (México, D.F.: 1941), pp. 20, 36; Taboada and Correu, "La producción," pp. 5–6; Durán, "Experimentación agrícola," April 1941, p. 7, in SAM Segundo; Ricardo A. Gamiz, "Nexos que debe existir entre las fracciones o delegaciones de la SAM con los Gobiernos locales, con miras al major desenvolvimiento de los programas agrícolas estatales" (n.d.), ARFF, Box 88.

86. "Puntos ideológicos."

87. Santos Amaro, "Los fertilizantes en México," January 23, 1941, AGN, Robles papers, Agriculture Section, Box 1, Exp. 3.

88. José del Riego and others to Avila Camacho, July 9, 1941, AGN, Avila Camacho papers, Exp. 133.2/67.

89. Durán, "Experimentación," pp. 3, 6; Proyecto de Puntos Resolutivos, "Enseñanza agrícola," April 1941, in SAM Segundo.

90. Adalberto Polo Celis, "Plan para la organización agrícola de México," pp. 2–3, in Política Agrícola, BCM Bound Volume #179–53; Mallory and Crawford, "The Annual Agricultural and Livestock Exposition of Mexico," 1941, p. 12, NA, RG 166, Box 146; Miguel E. Schultz to Avila Camacho, April 8, 1942, ARFF, Box 87.

91. José Lázcano Romero to Mauro Ruiz Ayala, March 11, 1942, ARFF, Box 135, "Puntos ideológicos"; Rodrigo Olvera, "La organización ejidal y el fomento agrícola," April 1941, p. 2, in SAM Segundo.

92. SAF, *Informe 1940–1941*, pp. 23, 36–37, 42; SAF, *Informe 1941–1942*, p. 79; SAF, *Informe de labores 1942–1943* (México, D.F.: Editorial Cultura, 1943), p. 99; Manuel Mesa A., *Agricultural Resources of Mexico* (Washington, D.C.: Interamerican Development Commission, 1946), p. 89; Fernández y Fernández, *Chapingo*, p. 136; Oscar Brauer H., "La tecnología y la ciencia en el desarrollo de la agricultura," p. 13, in Brauer H., ed., *Diagnóstico de las ciencias agrícolas en México* (México, D.F.: Ediciones Productividad, 1968); Rodrigo Acosta Proudinat, *La productividad agricola en Mexico* (México, D.F.: 1970); José Gutiérrez to Avila Camacho, October 24, 1942; Morilla Safa to Gajón Sánchez, October 31, 1941, and others, AGN, Avila Camacho papers, Exps. 485.1–10 and 506.1/8; Alton P. Willis and Mills H. Byron, "Production of Vegetable Fibers in the Yucatan" (n.d.), NA, RG 59, Box 4032, Decimal Files 1940–1944; Gómez, "Discurso pronunciado en la apertura de la IV Convención de la SAM," July 26, 1943, ARFF, Box 88; Roberto Osoyo Acala, "Ensayos de maíz norteamericanos en León, Gto.," unpublished thesis, ENA, 1946, p. 18; Miller, "Mexican Log," March 2–28, 1941, p. 33; Gómez, *Biografías*, pp. 229–30; Taboada and Damién Correu, "La producción agrícola," April 1941, p. 4; "Un aniversario," 1942, in ENA; Edmundo Sterling, "Necesidad de diversificar la agricultura," April 1941, p. 5; Carlos Anzures, "Los principales problemas agrícolas," April 1941, p. 1, in SAM Segundo; "Acta de la Sesión Inaugural," SAM, Tercera Reunión del Consejo Nacional Directivo, April 1942. ARFF, Box 180.

93. SAF, *Informe 1940–1941*, pp. 29, 36, 72; SAF, *Informe 1941–1942*, pp. 37–38; Jane Swift Powell, *Agriculture in Mexico* (Washington, D.C.: Pan American Union, 1942), p. 18; "La selección de la semilla de maiz," *Campeche* 1, 1 (November 1942): 5; Eduardo Limón, "El aumento de los rendimientos del maiz," April 1941, in SAM Segundo; "Campeche en marcha ascendente al progreso," *Excelsior*, September 1, 1943; Rivas Tagle, "Como aumentar la producción de los trigales," *Siembra* 1, 7 (October 1943): 22–23; Actividades que desarrollan en Chiapas las dependencias de la SAF," January 27, 1943, and others; Ramón Jiménez Delgado to Avila Camacho, October 21, 1942, AGN, Avila Camacho papers, Exps. 506.1/22 and 404.1/23; SAF, *Plan de Movilización Agrícola* (México, D.F.: 1942), p. 13; "La implantación de cultivos de frijol de soya," *El Nacional*,

April 27, 1942; Gabriel Bonilla M., "Observaciones al programa agrícola nacional," June 12, 1942; "Program del Consejo Mixto de Económia Regional," October 21, 1942; Arrieta M. to Julián Rodríguez Adame, February 12, 1942, AGN, Avila Camacho papers, Exps. 433/84, 404.1/23 and 506.1/7; Baltazar Reyes Garcia, "Cultivo racional del camote," *Siembra* 1, 6 (September 1943): 26; Gustavo Lesser Jones, "El girasol o maiz de Texas," *Siembra* 1, 6 (September 1943): 4; Gómez to Jaime Torres Bodet, July 3, 1941, AHSRE, Exp. III-601-4; José Octavio Vargas, "La irrigación ejidal," April 1941; Fernández y Fernández, "El problema agrario de México," April 1941, p. 5, in SAM Segundo; "Arados y alambre a Nayarit," *Siembra* 1, 9 (December 1943): 63; Hewitt de Alcantara, *Modernizing*, p. 65; Niblo, *War*, p. 171; Rafael López Ocampo, "Campesino: Por que debes preferir el sistema colectivo de trabajo," *Siembra* 1, 3 (June 1943): 27; García Cruz, "Opinión reveladora," *Boletín de Informaciones* 2, 12 (May 1941): 33–36.

94. David S. Ibarra, "Un problema nacional: La erosión del suelo y la perdida de su fertilidad," April 1941, in SAM Segundo; Lane Simonian, *Defending the Land of the Jaguar: A History of Conservation in Mexico* (Austin: University of Texas Press, 1995); SAF, *Plan de Movilización*, p. 6; "The Americas and the War," *Bulletin of the Pan American Union* 77, 9 (September 1943): 511–525; SAF, *Informe 1940–1941*, p. 30; "Totalmente fue vendido el algodón de La Laguna," *El Nacional*, January 5, 1942; CNC, "Exposición doctrinaria del Programa de la CNC" (n.d.), ARFF, Box 143; "La fabricación nacional de fertilizantes, como grande y pequeña industria," *El Nacional*, October 1, 1941; Manuel Noriega, "Application for Authority to Begin Construction Pursuant to Order L-41 on Behalf of Guanos y Fertilizantes de Mexico, S.A.," October 15, 1941, NA, RG 229, Box 592, Folder Guanos y Fertilizantes de Mexico, S.A.

95. Cobb, "Interview with Wellhausen," p. 113; "El problema de la langosta fue resuelto en Campeche," *Campeche* 1, 2 (December 1942): 10–11; "Agricultural Conditions and Problems of Mexico," 1941, n.p.; Crawford, "Bananas–Mexico," May 20, 1943; Mallory to Hull, November 29, 1943, NA, RG 166, Boxes 312 and 302; Veracruz Branch, SAM, "Minutes," January 31, 1942; Guillermo Brondo Whitt, "Las invasiones de langosta," 1942, ARFF, Boxes 180 and 139; SAF, *Informe 1940–1941*, pp. 51–53; Relaciones Exteriores to Legamex, San José, Costa Rica, and others, AHSRE, Exp. III-2413-9; Roberto Anthony Crespi to Gómez, February 28, 1941; José Ch. Ramírez to Avila Camacho, April 21, 1941; Bonifacio Gonzalez and others to Avila Camacho, March 17, 1943, AGN, Avila Camacho papers, Exps. 506.17/2 and 506.17/11; "Es dificil combatir la plaga de chamusco," *Novedades*, August 21, 1941; Luis Vera Pérez, "El Lacco Cupro '50,' una promesa para controlar el Chamusco, v breve información sobre su experimentación en Tabasco," *Tierra* 2, 5 (May 1947): 269–70;

Crawford, "Bananas—Mexico, Production and Export Situation," August 25, 1942, and others, NA, RG 166, Box 312; "Valiosa cosecha perdida por falta de transportes," *El Universal*, February 25, 1944.

96. Mallory and Crawford, "Mexico—Vegetable Oil and Oilseed Situation," March 16, 1942, NA, RG 166, Box 308; "Escasez de maiz en Coatepec, Veracruz," *El Universal*, September 24, 1942; "Siguen dificultades con el azúcar," *Carta Seminal*, April 19, 1941; "Exportación del azúcar," *El Universal*, September 21, 1943; "Gran escasez de maiz en Yucatán," *El Universal*, September 22, 1943; "Viene maiz procediente de Guatemala," *El Universal Gráfico*, September 8, 1943; "Terrible escasez de maiz en Celaya," *El Universal*, October 7, 1943; Borlaug, "The Impact," p. 279; "Por el ojo de la llave—el que siembra su maiz," *El Universal*, September 11, 1943; Formula de salvación: Pan en vez de tortilla," *El Universal*, September 27, 1943; "Recomendase el uso de harina de trigo," *El Universal*, October 8, 1943; Niblo, *War*, pp. 130–33; Hewitt de Alcantara, *Modernizing*, p. 11; "El maiz, cosecha de 1943, sus efectos en el abastecimiento de México," January 22, 1944, ARFF, Box 122.

97. Medina, *Del cardenismo*, pp. 147–48, 213–20, 239; Flores, *Historias*, pp. 179–81; "Perifonemas," *Últimas Noticias*, June 18, 1942; "Actividades de la C.N.C.," *Siembra* 1, 5 (August 1943): 20–21.

98. "El viaje de los 5 agrónomos es descabellado y muy caro," *Últimas Noticias*, October 22, 1941; Alfonso García Rodríguez to Francisco Xavier Gaxiola, April 21, 1942; Ricardo G. de la Llave to Jesús González Gallardo, July 3, 1941, AGN, Avila Camacho papers, Exps. 133.2/67 and 506.1/7; José Ch. Ramírez, "Problemas nacionales: La caña de azúcar," *Hoy*, November 22, 1941.

99. Benjamín Hinojosa, "Voz del agora," *Últimas Noticias*, November 2, 1943; "The Failure of Agriculture," *Hoy*, August 21, 1943, translation from NA, RG 166, Box 302.

100. Jesús Díaz Montiel and José Pavia Crespo to Waldo Castro, July 23, 1941; José Francisco Villegas to Avila Camacho, January 7, 1942; Manuel Gudiño to Avila Camacho, December 1943; José Dolores Marín and others to Avila Camacho, February 27, 1942; Ignacio Gallegos and Bernardo Higareda L. to Avila Camacho, January 28, 1941, and others, AGN, Avila Camacho papers, Exps. 133.2/67, 506.1/7, 151.3/59, 506.1/27, and 133.2/6.

101. Jesús Díaz Moret to Avila Camacho, July 10, 1941, AGN, Avila Camacho papers, Exp. 506.1/7; Roberto Gavaldón, "Algunos datos sobre los ingenieros agrónomos," *Últimas Noticias*, October 14, 1941.

102. "Quejanse los productores de plátano de Chiapas," *Carta Seminal*, May 17, 1941; Enclosure to Dispatch No. 15,599, American Embassy, Mexico City, January 21, 1944, NA, RG 59, Box 4032, Document File 1940–1944; "Obligación ineludable es el cultivo del maiz," *El Universal*,

January 31, 1944; José Fernández Villegas to Avila Camacho, July 15, 1941, AGN, Avila Camacho papers, Exp. 133.2/67; Whetten, *Rural Mexico*, p. 223.

103. José Zálvidar Flores, "Un plan de explotación para la Cienega de Chapala," unpublished agronomy thesis, ENA, 1942, pp. 20–27; Herminio Corral, "Consideraciones sobre el sistema de trabajo colectivo de la Cienega de Chapala," April 1941, in SAM Segundo; Whetten, *Rural Mexico*, p. 280; Cobb, "Interview with Wellhausen," pp. 164–65; Cobb, "Interview with Norman Borlaug," p. 155, RFA, RG 13, RAC.

104. Crawford and Mallory, "Mexico—Wartime Food Supply Problems," p. 3.

105. Ricardo G. de la Llave to Jesús González Gallo, July 3, 1941, AGN, Avila Camacho papers, Exp. 506.1/7; Fermín Luna, "La organización ejidal como base para el aumento de la producción agrícola nacional," April 1941, in SAM Segundo; Anzures, "Los principales," p. 4; Jesús Valenzuela Rivera, "Los agrónomos y el problema agrario de México," April 1941, p. 3, in SAM Segundo; Zálvidar Flores, "Un plan," p. 59; Miller, "Mexican Log," pp. 35, 46; Flores, *Historias*, pp. 222, 227–28.

106. Mary Kay Vaughan, *Cultural Politics in Revolution: Teachers, Peasants, and Schools in Mexico, 1930–1940* (Tucson: University of Arizona Press, 1997), pp. 103, 198.

107. Miller, "Mexican Log," pp. 47–49; "Rural Reconstruction News Camohmila Y.M.C.A. Center, Mexico," December 1942, p. 2, NA, RG 229, Box 95.

108. Harry M. Cleaver, "Origins of the Green Revolution," unpublished Ph.D. dissertation, Stanford University, 1974, pp. 247–48, 304; Flores, *Historias*, pp. 220–21; Comité Ejecutivo to LAS, October 28, 1940, ARFF, Box 180; Steve Weismann, "Why the Population Bomb Is a Rockefeller Baby," *Ramparts*, May 1970, p. 43; Andrew Pearse, *Seeds of Plenty, Seeds of Want* (New York: Oxford University Press, 1980), p. 79; Stakman et al., "Agricultural Conditions and Problems of Mexico," p. 43; Bruce H. Jennings, *Foundations of International Agricultural Research: Science and Politics in Mexican Agriculture* (Boulder, Colo.: Westview Press, 1988), p. 47; "Memorandum on the Origin of the Cultural Relations Program, Its Geographical Growth, Broad Accomplishments, Major Projects, and Personnel," October 22, 1943, NA, RG 353, Box 1, Folder History and Development; Claude Curtis Erb, "Nelson Rockefeller and U.S.–Latin American Relations, 1940–1945," unpublished Ph.D. dissertation, Clark University, 1982, p. 495.

Chapter 5

Agrónomos Técnicos: The Rockefeller Foundation, the United States, and the Transformation of Mexican Agronomy, 1944–1950

U.S. World War II agricultural programs influenced the professionalization of Mexican agronomy profoundly. Contacts between scientists made U.S. agricultural science the dominant model for Mexican agronomy. After the Allies triumphed, the U.S. government reduced its projects in Mexico to concentrate on the struggle with the Soviet Union for influence in Europe and Asia, which became the primary theaters of conflict during the early Cold War. The end of wartime programs created some problems and ill will, but the Rockefeller Foundation stayed the course and expanded the MAP, which had improved corn and wheat ready for distribution to farmers in 1947 and 1948. The foundation exerted substantial influence on the evolution of Mexican agronomy and learned that the MAP was an excellent model for agricultural programs in the developing nations. By 1950 Mexico's Green Revolution was under way, and the foundation was laid to extend it to other regions to diffuse the population bomb and contain Communism. In the process, the foundation abandoned wartime concerns about poverty and the distribution of benefits to advance a program of capitalist modernization and urban industrialization.

To establish credibility as scientific professionals, *agrónomos* wanted to learn specialized knowledge from U.S. scientists. The MAP fulfilled this objective and improved their public image because the foundation let the government take credit for the new seeds. Miguel Alemán's administration (1946–1952) favored urban interests, industrialization, and commercial agriculture and did little to address the problems of the *ejidos*. The MAP supported his industrialization program by helping Mexico produce more food that could be subsidized to keep wages low. A few *agrónomos* who already bred plants resented the MAP and its U.S. scientists, mostly because they could not compete with them professionally, but most supported the program because it established their credibility as *técnicos*, sustaining their alliance with the State after *agrarismo* fell out of vogue.

ADDRESSING THE DEMANDS OF WAR, INDUSTRIALIZATION, URBANIZATION, AND PROFESSIONALIZATION: GÓMEZ'S SAF, THE MAP, AND THE *AGRÓNOMOS*, 1944–1946

Mexico still experienced food shortages from 1944 to 1946, and the war made it harder for the United States to help, causing more political problems for *agrónomos* and the SAF. In February 1944 the agency said that programs to increase corn production would bring the largest harvest in history, but shortages led to complaints in several States. In San Luís Potosí, women stood in line to buy corn, and prices skyrocketed in Sinaloa. Sugarcane, tomato, sesame, cotton, and wheat growers had to plant 10% of their land in corn. The BNCA withheld the credit and water of farmers who did not comply, and the BNCE imported U.S. seed to distribute to *ejidatarios*, but maize deficits continued in 1945 and 1946. *Chahuixtle* struck Mexico's wheat fields, and a 1945 shipment of Australian grain contained flag smut, bringing a new plague and criticism in Mexico City newspapers. In 1946 Messersmith thought that public reaction to the wheat deficit, which included demonstrations by Mexico City's small-bakery owners, was a serious economic and political threat. Low price subsidies and transportation problems caused shortages of rice, sugar, beans, oilseeds, and animal fats. *Chamusco* still infested plantations in Veracruz, and the U.S. government set a lower price for the State's bananas.[1]

Eduardo Villaseñor and Alejandro Noye, heads of Mexico's Chambers of Commerce, claimed that land reform caused the deficits. *Agrónomo* Ignacio L. Figueroa, an ex-*agrarista* turned friend of Jalisco's *pequeñas propietarios*, joined him, criticizing the *ejido* in

El Economista. The *agrónomos'* enemies denigrated their scientific abilities, devotion to politics, and neglect of cultivators' technical problems. Private farmers detested their *agrarismo*. At Tapachula vandals tore the labels off cacao trees used in crossbreeding experiments, and farmers living near the El Progreso Station in Morelos refused to stop growing rice, which demanded lots of irrigation, or to rent lands to the SAF for corn experiments. In December 1945 *El Universal* forecast a catastrophic shortage of corn if immediate steps were not taken to prevent it. The State and the *agrónomos* felt considerable pressure to solve Mexico's agricultural problems. Avila Camacho told the SAF to end food imports that drained exchange reserves, combat wartime inflation, produce strategic farm goods for the war effort, and diversify agriculture to provide export revenue to purchase industrial equipment and new products for city dwellers. In 1944 the *Comisión de Coordinación Económica*, which participated in economic planning, called "Mexican agricultural research . . . indispensable."[2]

The SAF tried to increase production of many farm goods. In 1945 the National Institute of Nutrition pronounced the corn tortilla nutritionally sound, but some *agrónomos* still called the *campesinos'* diet inadequate and told them to grow vegetables. The SAF distributed fruit trees and planted vegetable gardens in Chiapas, and soybeans in three states, but most *agrónomos* and the press supported Marte Gómez's effort to make Mexico corn self-sufficient by creating high-yield varieties, which continued at Chapingo, León, Pabellón, Briseñas, and Cortazar, Guanajuato. To feed the growing urban middle and working classes, the SAF crossbred wheat and ran extension campaigns to convince *ejidatarios* to grow it rather than corn. The SAF bred and tested beans at Tlalnepantla and Briseñas, evaluated imported rice in the Yaqui Valley, and purified native strains in Michoacán's Nueva Italia and Lombardia collective *ejidos*. To reduce imports and satisfy the Union of Sugar Producers' demand for State-supported research, the SAF hired Don Mc H. Forbes to develop new strains and tested them on land rented from *ejidatarios* near Zacatepec, Morelos. To be self-sufficient in cacao, Gómez founded a station in Tapachula, Chiapas, to study cultivation practices and develop new varieties. To help ranchers feed their stock and provide raw material for the beer industry, the SAF tested 60 imported sorghums. To end the oilseed shortage and supply the U.S. war effort, Miguel E. Schultz, SAF *Agente General* in Oaxaca, tested castor beans in 13 *ejidos,* and Llera crossbred sesame and studied cultivation practices to find a substitute for corn in irrigation districts. To produce hard fibers for the war effort, the SAF tested Roselle hemp at Llera and searched for wild sources in the south and the northern deserts. To reduce imports, the agency helped certified cotton seed growers in Chihuahua and Nuevo León and seed potato producers in the Bajío. In 1946, to make Mexican fruit competitive in international markets, the SAF began to work on pineapples.[3] The MAP's scientists assisted several of these projects.

In 1944 U.S. soil scientist William Vogt studied soil erosion and wildlife in Mexico. Citing scientists' failure to avert the declining yield of Ohio's exhausted soils, he advised an emphasis on conservation while modernizing Mexican agriculture. In 1945 MAP soil scientist William E. Colwell tested green manure crops, but *agrónomos* in the fertilizer business and others like Nicolas Moreno of the El Mante hacienda joined *Guanos y Fertilizantes de México* to call chemical fertilizers a panacea for low farm productivity. The firm developed plans for factories and extension campaigns, mined guano islands in the Sea of Cortez, obtained rights to sell Allied Chemical and Dye Company nitrates while waiting for permission to import machinery, and saw irrigation districts as an excellent market. The ENA sent a student to the United States to study the chemical fertilizer industry, and the SAF tested fertilizers on potatoes at León and sugarcane in Morelos. NS director Warren Weaver gave soil research a high priority, and the MAP's scientists assisted experiments, sometimes with agribusinesses' help, on wheat in seven states. The potential of new herbicides and pesticides like DDT also intrigued *agrónomos* and the SAF. In 1945 Brondo Whitt urged the State to produce insecticides and fungicides and hire *agrónomos* to teach farmers how to use them. To protect Mexican manufacturers, Avila Camacho raised tariffs on arsenical insecticides. From 1944 to 1946 the SAF tested many insecticides and fungicides, sometimes with manufacturers' representatives. For example, at the ENA the MAP's scientists helped a Shell Oil employee demonstrate the soil disinfectant "D-D."[4] The SAF still had many problems. Poor record keeping ruined tests of foreign sugarcane. Wartime inflation led to budget cutbacks that slowed the experimentation stations' work and forced Avila Camacho to reject Veracruz banana growers' pleas for help to fight *chamusco*. Rumors that SAM politics would lead to a purge in the SAF undermined morale. Corruption afflicted the agrobureaucracy. In 1945 *agrónomo* Ramiro Sanabria accused several BNCE employees, particularly *agrónomo* Rafael López Ocampo, of improprieties while buying farm machinery in the United States. Marte Gómez was also implicated, and *El Universal* said that he forced SAF employees to "donate" funds to build a fancy house on his farm in Llera. Reinforced by their contacts with U.S. scientists, *agrónomos* denounced *campesinos'* "primitive practices" and blamed them when their projects failed. Sánchez Colín said farmers' unwillingness to cooperate ruined many SAF programs. ENA student Roberto Osoyo Alcala decried "*campesinos'* apathy toward experts' advice."[5] The *Banco de México* claimed that growers had no faith in *agrónomo* Augusto Pérez Toro, director of the *Instituto Técnico Henequeñero*. Sugar mill spokesman claimed that *ejidatarios* used poor cultivation practices and failed to select their seed properly. Reflecting the CNC's growing conservatism, the organization proposed that "each ejidal zone [should] be directed by honest and competent technical experts," and Nabor Cuervo Martínez wanted *agrónomos* to "govern" *ejidos* and distribute land in new irrigation districts only to *campesinos* with "knowledge, experience as farmers, and moral virtue" while excluding "bad elements, rascals, and charlatans."[6] *Guanos y Fertilizantes* told banks to give farmers chemical fertilizers as loans in kind. Some *agrónomos* admitted that they did not know how to work with poor, uneducated farmers. To address these problems, manage many projects, and respond to politicians' requests for assistance, Gómez asked the MAP to work on cacao, sesame, sugar beets and sugarcane, entomology, and opening lands for corn in San Luís Potosí. In 1946 the director of Antonio Narro asked the foundation to begin a training program at his school like the MAP's role at the ENA. The officers decided that these requests detracted from primary goals and declined proposals to expand the program, but the MAP increased the SAF's ability to address demands from the agricultural sector. Gómez recognized this, and he admitted that only "one or two" *agrónomos* knew enough plant genetics to breed new corn varieties.[7]

The United States ran many technical assistance programs in Mexico to produce war materials such as rubber, insecticides, medicines, fibers, castor oil, and other seed oils. The USDA helped the SAF's projects to increase sugar and wheat production, and Mexico assisted the war effort by sending *bracero* farmworkers to the United States. The 1944 founding of the Mexican–U.S. Agricultural Commission epitomized the wartime spirit of cooperation and led to a USDA–SAF program to study Mexico's cotton industry. The Mexican-American Commission for Economic Cooperation also planned to help Mexico increase farm production, and the United States accepted *Guanos y Fertilizantes'* request for machinery factories. Pointing to the Nazis' use of technicians to make friends in Latin America during the 1930s, many members of the U.S. foreign policy community wanted agricultural assistance programs to continue after the war and extended to other underdeveloped regions to win a postwar struggle for influence with Great Britain and the U.S.S.R., claiming that they fostered democracy, political and economic influence, and rising standards of living. Supported by resolutions at the May 1944 Conference of Commissions of Inter-American Development, Nelson Rockefeller tried to begin a Food Supply Program in Mexico, claiming that shortages, high prices, and *campesino* poverty might lead to revolution, but Messersmith told him that it was unfeasible because the Rockefeller Foundation already worked with the SAF to solve these problems. Discussing agricultural assistance programs, R.E. Moore of the USDA, W.C. Brister and B.R. Bell of the CIAA, R.S. Atwood of the State Department, and C.N. Gibboney of the Foreign Economic Administration (FEA) acknowledged that in "subsistence agricultural communities . . . changes in the pattern of production" led to "changes in social institutions and economic and political relationships" that cause "conflicts between interest groups and political tensions" and suggested including "social, anthropological, and economic studies" to "supply the insights by means of which research in the natural science fields becomes fruitful."[8] At the February 1945 signing of the UN Economic Charter in Mexico City, Secretary of State E.J. Stettinius promised that the United States would help modernize Latin American agriculture.

In Mexico some USDA projects had little impact, such as the effort to establish cinchona in Chiapas, but others were significant successes. Baker perfected a way to sterilize oranges, mangos, and grapefruits, the United States lifted the fruit fly quarantine, and the SAF took credit for solving the overproduction problem of Nuevo León's orange growers. In 1945 USDA bollworm inspectors in Tamualipas reported that insects had caused no significant damage to the cotton crop, but other USDA programs caused problems. The rubber project led to turmoil in several *ejidos* in Veracruz and Oaxaca when some peasants earned cash growing *Hevea*, making corn cultivators resent their neighbors' prosperity. Other *ejidatarios* protested when the SAF made them plant rubber trees instead of corn on their best land. In Chiapas landowners with wild *cryptostegia* said that U.S. buyers refused to pay fair prices and indiscriminately destroyed their trees. Farmers who grew castor beans in several states, including *ejidatarios* forced by the SAF to participate, protested when U.S. payments (360 pesos/ ton) did not cover production costs because yields were much lower than the SAF promised, partly due to unknown pests and diseases. Many growers abandoned the crop, and Gómez complained when they defaulted on BNCE loans. U.S. economic policies also caused discontent. Some officials began to rethink their position on new wartime industries in Latin America, claiming that they should get no tariff protection or guaranteed market access through bilateral trade agreements. In 1944, ignoring several warnings that the decision would generate ill will, the United States rejected such measures, as well as a resolution to restrict synthetic rubber production proposed by several Latin American nations at a meeting of the Inter-American Development Commission. U.S. limits on Mexico's purchases of Cuban sugar to ensure European and domestic supplies led to bad feelings, and growers claimed that the Office of Price Administration practiced "racial discrimination" by setting a ceiling price for Mexican bananas from regions still affected by *chamusco* at $1.75 per hundredweight lower than Central American fruit grown on the plantations of U.S. capitalists.[9]

More problems emerged when the United States terminated the programs as victory approached. With the success of research on insecticides like DDT, Bajío farmers who planted pyrethrum complained to reporters when the FEA refused to pay the contract price for their harvest. In 1944, concluding that a postwar rubber surplus would make Latin American *Hevea* uncompetitive, the United States cut funding for research. To avoid political problems, the USDA continued the project but changed its emphasis to small-scale production to supplement subsistence farmers' income. Research on synthetic rubber produced an economically viable substitute by 1945, and the United States shipped it to Mexican manufacturers to accustom them to using it. In 1944 the United States decided not to buy *cryptostegia*; in 1945 the USDA canceled the project and closed the station in Ciudad Victoria. Contrary to U.S. officials' optimistic 1944 predictions, in 1945 the wartime guayule industry also faced imminent collapse. In 1947 Mexicans who started *Hevea* plantations with USDA assistance begged the government for tariff protection.[10]

Despite the problems, trade unions, *agraristas*, and *pequeños propietarios* supported Avila Camacho's programs to increase farm production for the war, and farmers wanted U.S. technical assistance. In 1944 the CNC urged returning *braceros* to teach *ejidatarios* U.S. agriculture. In 1945 New York's General Products Corporation agreed to study uses of henequen by-products for the *Henequeñeros de Yucatán* free of charge, sheep ranchers were so pleased with Madsen that they hired him again, UFCO gave advice to Alberto Álvarez Murphy when he started a banana plantation in San Luís Potosí, and farmers in Michoacán and Chiapas asked the USDA for seeds. In 1946 *La Hacienda* praised the MAP's role in the SAF, especially hybrid corn research. To learn more about it, the *Confederación Nacional de la Pequeña Propiedad Agrícola* invited MAP director J. George Harrar to a meeting. Trade unions in Veracruz asked Avila Camacho to send experts with "the marvelous insecticide DDT."[11]

Reflecting the growing influence of U.S. agronomy, *agrónomos* lauded the MAP and the USDA's wartime programs. Like their conservative critics, to modernize agriculture, the SAM wanted to bring more foreign scientists and farmers to Mexico. The SAF claimed that "the most important achievement in plant genetics is the hybrid corn made by U.S. scientists," at the 1944 National Agricultural and Livestock Exposition veterinarian Guillermo Quesada Bravo said that they "greatly aided many of our agricultural programs," and Salvador Lira López, Fernández y Fernández, and Quintín Ochoa thought that they could help Mexico's poorest farmers. In 1944 the USDA sent Walter G. Cadmus to the ENA to build a new laboratory, and the CNI employed an engineer and a geologist from the Bureau of Reclamation. In 1945 the Federal District hired L.E. Evans of Colorado to test potatoes and oats, the *Banco de México* asked the Armour Research Foundation to study the henequen industry, the USDA sent Frederick P. Frutchey to discuss extension programs, and the SAF had the USDA analyze corn seed. To improve the *campesinos'* diet, *agrónomo* Jorge Manrique urged U.S. investors to build a rice-processing plant. Mexico's delegates to the first meeting of the United Nations' Food and Agriculture Organization (hereafter FAO) wanted international research to solve "the Americas' most important agricultural problems," Messersmith said that "the attitude of the Mexican government is highly favorable to American capital and technicians," and geologist L.C. Graton, who studied the new volcano Paricutín, reported that "a very genuine interest in American science exists among the scientists of Mexico" and that "friendly cooperation has deep and vigorous roots."[12] The Rockefeller Foundation gave the ENA many USDA publications. Durán thought that the MAP, El Palmar, and Limón's corn experiments were the SAF's only effective research programs, and Fernández y Fernández said, "English is indispensable for a good professional *agrónomo*."[13]

The State Department appreciated the contributions of private firms and organizations like the Rockefeller Foundation to the Cultural Relations Program. For example, U.S. investors built an orange dehydrating plant in Nuevo León. Foundation officers attended meetings of the General Advisory Committee on Cultural Relations, and Bradfield and Miller served on the Advisory Committee on Inter-American Cooperation in Agricultural Education. The MAP's scientists assisted UNAM's search for medicinal plants in Guerrero, collaborated with the USDA's and the University of Minnesota's *chahuixtle* research, and performed political favors like analyzing seed corn that General Eulogio Ortiz sent to the SAF. In 1944 the State Department decided that postwar assistance to Latin American industry and agriculture should foster "private participation," but Harrar felt that "too close relationships with existing U.S. agencies might give a false impression and be misinterpreted."[14] Thinking about postwar conditions, the officers wanted to support programs like the MAP, but with some autonomy from the U.S. government. Stakman claimed that the MAP assisted the war effort by alleviating food shortages, freeing land for strategic farm products, and improving U.S.–Mexican relations. The officers considered beginning programs like the MAP in other regions, taking advantage of the USDA's and CIAA's experience in Latin America. In 1944 the NS proposed offering agricultural assistance programs to other Latin American nations, starting in Colombia, and training their scientists at the ENA. Fosdick thought that a project like the MAP could rebuild European and Chinese agriculture after the war. In February 1945, after receiving State Department approval, Harrar and Miller visited Guatemala, Honduras, Panama, Costa Rica, Colombia, and Venezuela. Stakman knew that the SAF would be proud to export science to other Latin American nations.[15]

The MAP developed its research agenda after consulting SAF researchers like Taboada. Like the *agrónomos*, the MAP's scientists knew that Limón's corn increased yields significantly in the Bajío and that *campesinos* would not use hybrids because they did not buy seed each year. Mexican and U.S. scientists looked for maize for *de temporal* agriculture and developed open-pollinated strains that could be used for seed, but they agreed that corn created by scientists was superior to the *campesinos'* traditional varieties. With their success in the United States and Gómez's support, the MAP's long-term objective was to develop hybrid corn and wheat. Mangelsdorf knew that many *campesinos* intercropped corn and beans, and the MAP's scientists took this "ancient custom" into account when they began working on beans, believing that "it was likely to persist for generations to come," but they also said that it lowered yields of both crops.[16] Bean research targeted commercial growers who monocropped and used insecticides. The MAP did not create seeds to solve the problems of peasant farmers.

In 1943 Harrar visited a YMCA agricultural project in Morelos run by Spencer Hatch that Nathan Whetten admired. Harrar acknowledged Hatch's "local support" and called the "Indians . . . good imitators" who "accept ideas and methods shown to be of practical value," but he called the project "different" from the MAP because it was "restricted and very expensive in relation to the population affected" and "concerned with practical agriculture and social problems of the simplest type rather than agricultural research."[17] In 1944 he admitted that "Mexico is still in the course of an agricultural revolution occasioned by the change from the hacienda to the ejido system," but some foundation officers and scientists disliked the agrarian reform.[18] In 1970 Stakman, who admired paternalistic *hacendados* who cared for their peons, said that "from a national standpoint, it would have been better if the food production had been in the hands of the more intelligent people and the larger owners who could operate on a bigger scale. . . . But the Mexicans preferred the poverty and the freedom to operate a piece of land themselves."[19] In 1945, finding "questions for social science," Sauer told Willits that the MAP's research reflected the government's move "away from subsistence or village agriculture to the needs of the city and the factory with the attendant emphasis on standardization of product and on yield . . . , and on the commodities which the privileged fraction of the population can absorb."[20] In 1946 Weaver, Mann, and trustee William I. Myers admitted that "limitations on the ejidal assignments of land" might "be so uneconomic as to defeat efforts to raise the level of agriculture."[21]

In 1944 the foundation's Inter-Divisional Committee on Latin America added public health and agricultural economics as fields deserving grants, and officer R.M. Taylor proposed a *cardenista* approach to rural development that included research on nutrition, public health, education, and community economics to improve the "physical well-being" and "happiness of the country-folk."[22] The IHD and Robert S. Harris of the Massachusetts Institute of Technology (MIT) urged the MAP to breed high-protein corn and work on general nutrition. Harrar sent samples of Mexican maize to MIT for analysis, gave the SAF the results, and studied the diet of the impoverished Otomi Indians, finding that a wild green that they ate was an excellent source of vitamin C. In 1945, claiming that Avila Camacho's agricultural policy "served, defended, and profited privileged farmers," *agrónomo* Aurelio Castillo Robles told *Discussión* that Mexican corn was better nutritionally, easier to cook, and tastier than foreign hybrids, and the foundation reacted by considering a cooperative MAP–IHD program, but by 1946 nutrition research "had largely collapsed."[23] Fighting malnutrition quickly became a secondary objective of the MAP.

By 1944 the foundation believed that the MAP had considerable influence in the SAF. When José Rodríguez returned from the University of Minnesota, Stakman was proud of his scientific zeal and thought that he would be a positive role model for *agrónomos*. Mexican fellows attended the best U.S. universities and visited the USDA's research stations. The foundation and the U.S. government cultivated the goodwill of influential *agrónomos*. In 1944 Gómez asked Harrar to represent Mexico on the Mexican-American Agricultural Commission, and SAF Undersecretary Alfonso González Gallardo dined with Fosdick in New York. In 1945 the foundation invited him to a meeting of the Advisory Committee on Mexican Agriculture. Stakman and Harrar attended SAM events, and in 1946 Harrar became a member. That year Henry Wallace visited the ENA with Gómez, ex-president Cárdenas, and Gustavo P. Serrano, secretary of the national economy; Stakman took Mexican students to a meeting of the American Phytopathological Society; and Rodolfo Moreno Dahne of the *Banco de México* attended the American Agronomy Conference in Omaha. The Mexico City press praised the MAP, and Avila Camacho asked the foundation and the USDA to continue their agricultural assistance programs. A few *agrónomos* at the SAF's experimentation stations, most notably Limón, saw the MAP's scientists as professional rivals and tried to undermine their research by forgetting to fertilize, weed, or water test plots, but most recognized the benefits of the program, like the 42 jobs that it created by 1946, and offered their support. For example, *agrónomo* Dario L. Arrieta M. told the relatives of a *campesino* laborer at the MAP's Morelos station who died suddenly after spraying pesticides that the SAF accepted all responsibility for the incident because he thought that it was essential to keep the foundation's reputation in Mexico "uninjured" to ensure the project's success.[24]

Contact with U.S. scientists reinforced *agrónomos'* interest in research. Many ENA students wanted to take Harrar's field plot technique courses, and faculty asked the MAP's scientists to help plan curriculum. *Agrónomos* such as Jesus Alarcón and Fernández y Fernández praised the school's scientific orientation. Thanks to the foundation's and the USDA's help, the ENA's directors proudly called it "one of the best . . . for scientific research in Latin America."[25] In 1946 Cárdenas praised the ENA's improvements. *Agrónomos* sought opportunities for advanced training at U.S. universities. Antonio Marino A., who bred corn with Limón, asked the foundation for a scholarship to study plant genetics. The *Instituto Politécnico*'s School of Biological Sciences and Antonio Narro asked the foundation to help their research programs. The *Banco de México* wanted to send students to Hawaii, Cuba, and Puerto Rico to study the pineapple industry. The SAM knew that poorly run chemical fertilizer tests led to costly failures and proposed sending *agrónomos* to the United States to study soil science. In 1944 four took courses at U.S. universities on State Department scholarships, and in 1945 five more joined them. By 1946, 44 Mexicans completed advanced agricultural studies in the United States, and 10 others were enrolled.[26]

The MAP taught the *agrónomos* plant breeding, giving them specialized knowledge that they said only experts could understand. *Agrónomos* wanted to create better sugarcane, rice, henequen, and cotton. In 1944, to avoid "confusion," Gómez said that only specialists should discuss plant genetics because "the formation of hybrids or synthetics requires expensive and delicate operations performed by people with advanced technical training and experience; *ejidatarios* or private citizens without knowledge, dedication, experience, and material means cannot make this corn."[27] In 1945 he told UNAM professors that the *Campaña en Pro del Maiz* was doomed to fail and told the CNC that the SAF would soon have many marvels, including "new crops of great benefit like the olive" as well as "improved seeds of the most important plants."[28] In 1946 Gontran Noble confessed to the CNC that *agrónomos* "committed the sin of utopians in asking each *campesino* to select the seeds he sows" and promised that in the future they would produce them with "hybridization and genetic research."[29]

Expanding the bureaucracy that Cárdenas created, Avila Camacho's support for research gave *agrónomos* more jobs, and the profession reached its greatest popularity with young Mexicans entering engineering fields. In 1945 the SAM claimed that 90% of its members held official or semi-official jobs. With the new employment opportunities, from 1944 to 1950 the ENA granted 360 degrees and Antonio Narro 184, but most students still came from urban families. The journal *Tierra* urged farmers to send their children to the ENA. Responding to the growth of parastatal and agribusiness firms, in 1945 the SAM slew a professional sacred cow by deeming private sector work acceptable, and *agrónomos* planned to produce and sell certified seeds. *Cardenista* scientists accepted these positions, such as Alfonso Romero of the *Instituto Biotécnico*, who worked for *Guanos y Fertilizantes*. Contact with U.S. scientists did not make the *agrónomos* renounce their *agrarista* heritage. They still helped peasants get land and government services and sometimes used the cultural campaign's language to claim that research would save the *ejido* and transform the *campesinos*. For example, to Gómez "the material, intellectual, and moral redemption of the rural laborer was and is the primary goal of agricultural policy," and *agrónomo* Francisco Cabanillas López said that "applying cheap and easy-to-use fertilizers to tilled soil" would raise "the economic, social, and moral conditions of the *ejidatarios*."[30] To improve *campesinos'* diet and make them healthier workers, the SAM urged the SAF to teach *campesinos* to grow fruits and vegetables. Many leftist *agrónomos* thought that small *parcelas* would never be profitable and advocated collectivization. Praising Stalin's agricultural programs and Lysenko's view of evolution, Hilario Mesa Cienfuegos wanted to send students to the Soviet Union for advanced training, and Mesa A. visited the U.S.S.R. in 1945, but other *agrónomos*, including leftists like Durán, knew that political trends favored the *pequeña propiedad* and individual *ejido* and urged the State to support research to solve their problems, even though they thought that only minimal gains were possible. Reflecting the Revolutionary Family's new position, Rubén López tried to reconcile *agrarismo* and Avila Camacho's interest in production. Claiming that *ejidatarios* needed technical assistance, not rhetoric on collectivization, he said, "In the end how land is distributed is not important, it is important to make the land produce . . . the SAM has always been and will continue to be *agrarista*."[31] The conservative León and the socialist Durán agreed that land reform did not bring prosperity to the *campesinos*, who needed an agricultural revolution (*revolución agrícola*) to vindicate the agrarian revolution (*revolución agraria*). DGA director Arrieta M. told the National Agricultural and Livestock Exposition that a "scientific revolution" would finish "the Social Agricultural Revolution" begun by "millions of disinherited farmers."[32]

Many *agrónomos* worried about their public image. Durán, Fernández y Fernández, and Fábila lamented that "people think agronomy does not require extensive scientific training," and Brondo Whitt told the SAM to study pest control practices carefully before "advising procedures that often fail, diminishing the SAF's and agronomists' prestige."[33] *Agrónomos* admitted that their attempt to convince the farmers failed because it was based on theory, advising practices that were ineffective or inferior to the *campesinos'* practical farming knowledge. Like the MAP's approach to research, Durán wanted experimentation stations to concentrate on a few projects. The SAF and *Tierra* agreed that fruit tree campaigns suffered many setbacks because they were not run by trained specialists. To define the SAF as a "scientific" agency, Gómez said, "In a matter always seen as political . . . we have abstained from political debates and putting . . . financial resources or influence to the service . . . of any political interest. Good or bad, we do agricultural work, exclusively agricultural. We seek technical solutions, apolitically" and ENA director Hector Lazos claimed that critics' assertions that *agrónomos* were "politicians" rather than scientists were true only if politics meant a devoted effort to make farmers prosperous and productive.[34]

Gómez used scientific successes to refute critics and enhance *agrónomos'* credibility. He told the CNC that "the first successes will bring faith and optimism to the peasants and affirm their old bonds of fraternity with *agrónomos*," and he praised successful fertilizer tests at Pabellón that included *ejidos*. He used the progress of cotton research in La Laguna to urge Mexicans to "have confidence in the technical works of the *agrónomos*, in the same way they have faith in the studies of the doctor, the civil engineer, or the lawyer," and not to "consider the money they give the SAF for research wasted."[35] Limón's corn and the new chiles created at Pabellón were the SAF's first plant-breeding successes. The foundation wanted the SAF to show Mexican farmers "a better concept of what big-time research . . . requires and contributes" and told the MAP to minimize publicity while the SAF ran articles in the Mexico City press and published research results to improve its public image by demonstrating scientific competence.[36] Gómez hoped that the seeds would end farmers' "inveterate mistrust" of new crop varieties. To produce them for distribution, he planned to work with cultivators who followed *agrónomos'* advice, like the Yaqui Valley's wheat growers. *Agrónomos* praised the SAF's research, such as DGA director Salvador Sánchez Colín, who called it "purely Mexican."[37] Reflecting the Revolutionary Family's agricultural policies, *agrónomos* tried to transform their professional identity, stressing their expertise as *técnicos* and downplaying *agrarismo*.

ALEMÁN'S SAF: *AGRÓNOMOS* AND FARMERS REACT TO THE MAP AND ITS NEW TECHNOLOGIES

While campaigning, Alemán promised to spend more money on agricultural research and education, irrigation projects, tropical agriculture, soil conservation, and promoting chemical fertilizers. *Agraristas* and *pequeños propietarios* supported his pledge to modernize agriculture, and the CTM agreed that Mexico needed foreign technical assistance to attain prosperity. Influenced by Avila Camacho, Alemán wanted to include "scientists and technicians in the government," and in 1946 Gómez reinforced his ideas by stressing the need to raise farmers' standard of living, especially that of *ejidatarios*, "to avoid suffering an economic collapse in the future with social consequences that would produce ideological disenchantment with our regime."[38] Whatever they thought of his statements about land reform, as Revolutionary Family members, *agrónomos* such as BAR leader Jesús Merino Fernández and FRAM head Adan Ramírez López supported Alemán and his agricultural policies. Alemán wanted to increase the production of corn and other crops, established the *Instituto de Investigaciones Agrícolas* (Agricultural Research Institute, hereafter IIA) to manage the SAF's research, and told the agency to work on wheat, beans, corn, sugarcane, potatoes, rubber, chiles, rice, cacao, sorghum, soybeans, and sesame. To reap the political benefits of distributing better corn seed, he founded the *Comisión Nacional del Maiz* (National Corn Commission) and made his ally Senator Gabriel Ramos Millán, who promised to bring high-yield varieties to "farmers, private concerns, and ejidatarios."[39]

Although historian Blanca Torres says that Alemán decided to restrict agricultural programs to the most promising regions during his campaign, this policy developed gradually. Green Revolution technologies had their greatest impact in the northwestern irrigation districts, but by 1950 the Corn Commission had promoted hybrids in 9 states, and the MAP had conducted experiments in 19 and distributed new seed in 22 and over 100 kilograms of it in 10. By 1949 the MAP conducted corn research at Chapingo, Celaya, Guadalajara, and Morelos; worked on hybrids for the tropics; and tested wheat in Chapingo, Sonora, and La Laguna. Responding to commercial farmers and other interests, the MAP studied seed potatoes, safflower, an African oilseed, insect pests of tomatoes, potato diseases, soybeans, and sorghum. In 1948 the MAP tested new wheats at 68 "ranchos, haciendas, or farms," 13 "*ejidos*," and 22 unspecified rural properties, but, as Torres notes, Alemán favored private farmers and discriminated against *ejidatarios* because he had little faith in them. He put the

SEP and the states in charge of several cultural campaign projects, such as vegetable cultivation and small-scale irrigation. Reflecting rural teachers' role as purveyors of agricultural knowledge to the *campesinos*, while *agrónomos* and U.S. scientists worked with more educated farmers, in 1949 the MAP gave the teachers agricultural courses. The CNI reduced expenditures on small-scale projects, and the SAF distributed most of the MAP's literature to *agrónomos* in irrigation districts. In Veracruz agricultural banks lent *pequeños propietarios* money to fight *chamusco*, but not *ejidatarios*, even though this made it harder to eradicate the spore. The CNA often ignored *ejidatarios* who tried to protect their lands from the encroachment of growing cities.[40]

The CNC advocated food self-sufficiency, and the MAP's scientists bred hybrid corn and wheat to attain it. Some of their projects were similar to the cultural campaign. They admitted that Mexican corn was high in protein but said that this was irrelevant because people obtained it from beans. In 1946, to address the ENA's and the foundation's goal of improving *campesinos'* diet, Harrar added vegetable cultivation and breeding to the MAP's project list. In 1949 Wellhausen tried to get an ENA professor into the University of California, Davis, to improve his courses on the subject. In 1950 the MAP encouraged ENA field hands to grow vegetable gardens. To put more protein on the *campesinos'* dinner table, Harrar wanted to promote home fish production, and in 1949 Herbert S. Jackson built several demonstration ponds for the MAP. The foundation again considered adding nutritional concerns to the MAP's plant-breeding program and gave samples of new corns and wheats to the National Institute of Nutrition, but in 1948 Harrar was angry when data indicating that some hybrids were low in protein were leaked to the press. Weaver told him to "keep in touch with developments" because "better nutrition for the Mexican people" was an "important ultimate objective," but by 1950 no program existed, even though the National Institute of Nutrition wanted to study beans.[41]

In 1948 the foundation's social scientists tried to collaborate with Harrar in IHD-NS-Social Sciences Division programs in the Third World. They urged him to give "very careful thought" to "the distribution of your results to the millions of small landholders in Mexico" to make research bring "the largest amount of good to the largest number of people" and warned that if only "the larger land-holders" benefited from the new technologies, they could "defeat much of the good which came about from the early revolutions."[42] To Whetten, increasing *ejidatarios'* productivity was "one of Mexico's most serious and urgent problems," and he advocated "programs that will fit the needs of the local inhabitants . . . gradually lead them to accept more efficient techniques and practices . . . [and] avoid introducing changes so suddenly that the social equilibrium will be thrown out of balance and result in the creation of new problems."[43] Edgar Anderson, director of Washington University's Shaw School of Botany, urged the MAP's scientists to study traditional farming and crop varieties, but the foundation rejected the idea, claiming that it reflected the self-serving professional interests of social scientists who wanted to use *campesino* culture as a laboratory to study "picturesque folk ways."[44]

The MAP had already decided to exclude many *campesinos* from the corn program. In 1946 the foundation planned to distribute seed to *ejidatarios* but worried that "passive and active resistance of less intelligent farmers because of apathy, suspicion, and a lack of facilities" would inhibit their adoption. In 1947 Wellhausen said that "areas in which a surplus can be produced should be attacked first; the pure subsistence areas may be left to a long time in the future," and *campesinos'* reaction to the hoof-and-mouth disease campaign convinced Harrar that "unlettered Indians and Mestizos" were "quite superstitious and very suspicious of outsiders."[45] In 1948 Weaver admitted that "dealing with uneducated and poor farmers, and localities where production and distribution of 'fancy' seed is difficult, a lower yielding seed, which for one or more reasons is more readily available and more easily utilized, may well result in a greater total increase in corn production," and in 1950 Harrar told the officers that "rapid acceptance and utilization of hybrid varieties by the ejidos and some of the less progressive farmers" were impossible, partly because the *campesinos* would not buy new seed every year, but the MAP still favored them.[46] In 1946 and 1948 Weaver wanted to add an agricultural economist to the MAP, but Harrar resisted because he thought that "the economic aspects of agriculture" posed "dangers to the program," and the officers agreed to restrict its involvement in other problems and regions.[47] In Alemán's Mexico the MAP's scientists adopted Mann's definition of the project's goals and forgot the equity concerns that Wallace, Fosdick, and others raised. In 1947 Mangelsdorf admitted that the MAP might not attain the goals of the Mexican Revolution. He had

> little doubt that Mexican agriculture will be improved as the result of the research already accomplished and now in progress. I am less certain that the individual Mexican will benefit . . . , if the population increases as rapidly as the food supply the net gain [in] individual prosperity will be negligible. The only hope . . . is to improve agriculture so rapidly . . . there will be a period of prosperity sufficiently prolonged and pronounced to result in a reduction in the birth rate. I am not certain that this objective can be obtained but . . . it is worth striving for.[48]

Wellhausen wanted to "reduce the number of people required for food production leaving a higher percentage to make things which create a higher standard of living" and said that Mexico needed "high hectare production," which meant "less man hours per unit of grain produced" to "release some of the farm population to work in the industrial plants."[49]

The foundation began to discuss extension work in 1946. Alemán's support for research and personal interest in the MAP pleased the officers and scientists, but the founding of the Corn Commission upset them. He did not ask their advice on distributing new corn seeds, and Ramos Millán's appointment suggested that politics would influence the process profoundly. The officers wanted to work with the SAF to give the MAP more control over the distribution of the new seeds. In December 1947 they hired M.F. Barrus to develop the SAF's extension service, but the project was a disappointment. Agents did not complete their tasks because they had to attend to others like pest control, enforcing quarantines, and gathering statistics or lacked funds to visit demonstration plots regularly. Barrus thought that the *agrónomos* assigned to him resented performing new tasks without a salary increase, claimed they felt no responsibility to tell the MAP about their activities, and suspected that some disliked "farm life." In 1948 the foundation renounced all responsibility for harvesting, shelling, grading, and preparing hybrid corn seed for distribution, but for the first time in over 10 years, Mexico was briefly self-sufficient in maize, and the officers were more satisfied. In 1949, to avoid a messy political conflict, the trustees refused to use their influence to change the situation, and Barrus quit the MAP, claiming that progress was impossible without major changes in the SAF. Stakman complained that "ineptitude on the part of some officials of the Mexican Department" ruined some demonstration plots, and Bradfield said that "a shortage of technically trained men" would be "a bottleneck in the development of Mexican agriculture for many years" because many public agencies and private businesses wanted to hire U.S.-educated *agrónomos*.[50]

In 1947 the MAP's new corn seeds were ready to deliver to farmers, and in 1948 the SAF distributed *chahuixtle*-resistant wheat. *Agrónomo* José Luís de la Loma urged farmers to use the *Campaña en Pro del Maiz* until hybrid corn was available, but some did not want to wait. The Corn Commission received so many requests to multiply seed that some were rejected. *Ejidos* in Tamualipas, Zacatecas, Veracruz, and Mexico State asked Alemán, the MAP, or the ENA to help them get hybrid maize. In 1948 Wellhausen reported that farmers' demands for his new open-pollinated and hybrid corns surpassed supply. The governors of Chiapas and Nuevo León, politicians in Tamualipas, rural teachers in Mexico state, farmers in La Laguna, and *Ejidal* Commissions in Chiapas and Veracruz asked Alemán to found experimentation stations or send speakers to popularize the new seeds. Organizations like the *Sociedad Local de Crédito Agrícola* of Juventino Rosas, Guanajuato, inquired about new wheats. In 1947 *Tierra* said that the MAP's scientists would produce "*supertrigos*" (super wheats). By 1948 Harrar claimed that farmers were so "excited" about new wheat that there was a black market for seed. In 1949, 20 to 40 people a day visited him, including the governors of five states. To avoid negative publicity and criticism, the foundation rejected several requests to assist the hoof-and-mouth campaign, and the Mexico City press gave the MAP favorable reviews.[51]

Torres claims that the MAP was committed to agrochemicals, and in the 1940s mainstream scientists criticized the "muck and mystery" school of farming, "the mystical belief that nothing must be used in agriculture that was not supplied by nature," but scientists still tested green manure crops at the ENA, in the Yaqui Valley, La Laguna, and the Bajío, and one of their first publications explained how to grow Hubam clover.[52] Bradfield believed in chemical fertilizers, but he also stressed the importance of organic methods. The foundation urged the SAF to devote more effort to this program, claiming that central Mexico's climate and soils often made chemical fertilizers inefficient, especially on *de temporal* corn, and also because they were too expensive for most Mexican farmers. In 1949 the foundation gave the green manure crop program a high priority. *Agrónomos* such as Fourtón also advocated organic ways to improve soil fertility, but with the State's investment in the fertilizer industry, Alemán wanted to popularize their use, and he reduced funding for Avila Camacho's soil conservation program. The ENA sent Rodolfo Peregrina Robles to study soil science and conservation at Cornell. His M.A. thesis discussed chemical fertilizers. The directors of *Guanos y Fertilizantes* organized the *Campaña Nacional de Fertilizantes*, and MAP, SAF, and SEP personnel and U.S.-trained Mexican soil scientists and chemical engineers with the *Banco de México* assisted the firm's employees. Bradfield, Robert Chandler, and John B. Pittner urged agricultural banks to distribute fertilizers as loans in-kind and include salesmen in extension programs. The MAP's seeds helped secure an ample market for new fertilizer factories, its scientists told the SAF that hybrid maize and wheat needed them to reach their potential, and the Corn Commission promoted them while demonstrating new varieties. In 1948 the MAP tested chemical fertilizers on *de temporal* corn in 11 states—and many *ejidatarios* and sharecroppers participated—and in 1949 on corn and wheat in six states. Some *agrónomos* thought that chemical fertilizers were not properly tested, inferior to organic methods, or a complement to them, but many in and outside the industry urged farmers to use them. In 1949 U.S. organic farming enthusiast J.I. Rodale was dismayed when SAF officials showed no interest in his ideas. In the Bajío Hubam clover gained some acceptance, but the SAF's ambivalence toward this program dismayed the foundation.[53]

The SAF and the MAP also tested and promoted pesticides and herbicides. In 1947 the MAP used DDT and Benzine-hexachloride to control "thrips," a corn pest, but admitted that "these insecticides are too expensive for most Mexican farmers."[54] Nevertheless, in 1948 the MAP published a pamphlet on DDT, showed the farmers of the Bajío and the Valley of Mexico how to use 2,4-D to control weeds, and in 1949 applied Parathion, Chlordane, and other pesticides to corn, beans, and wheat. In 1950 the MAP's Mexican employees tested DDT and chemical fertilizers on beans in the Bajío. Reflecting continued interest in biological pest control, in 1949 the *Banco de México* granted a scholarship to send an *agrónomo* to Hawaii to study natural predators of insects that attacked sugarcane.[55]

Many farmers wanted to try agrochemicals. In 1948 *ranchero* Alexander Holste used Dow Chemical's Esteron 44 to kill weeds. After failing with DDT, B-H-C, and Toxaphene, in 1949 the El Mante sugar mill used Chlordane to control cane borers. Claiming that organic materials were inadequate, in 1946 the Aguascalientes branch of the CNC urged Alemán to build agrochemical and application equipment factories. The Bajío's farmers sent many inquiries to the SAF, which passed them on to vendors. In Tacámbaro, Michoacán, *agraristas* asked Alemán for a loan to buy chemical fertilizers for their cane fields. Agrochemicals helped the *campesinos* of Hueyapan, Morelos, replenish their sterile soils, making small-scale farming more profitable than wage labor on sugar haciendas. In 1949 La Laguna's farmers planned to conduct their own experiments. Testing chemical fertilizers on *de temporal* corn, the MAP found that "the farmer's interest in experimentation and desire to cooperate were great . . . almost without exception," but Bradfield believed that *ejidatarios'* lack of education, capital, and access to credit would make regular use difficult.[56]

Farmers from the United States such as the Arizona-born manager of the president of Monsanto's Hacienda El Carrillo, quickly adopted the new seeds. *Hacendados* who protected their estates from land reform, like José Gómez, who divided La Venta near San Juan del Rio, Querétaro among his brothers, also planted hybrids. Some less affluent farmers such as *ranchero* Bernardo González Cárdenas of San José de Gracia learned to use scientific agriculture successfully on their own. Demonstrations of new technologies attracted *ejidatarios* as well as private farmers, and the MAP responded to *campesinos'* inquiries about them. By 1950 Wellhausen created "Celita" hybrid corn for *de temporal* farming, but Alemán's policies, the lessons of variety and chemical fertilizer tests, the foundation's influence, and the *agrónomos'* backgrounds guaranteed that prominent and educated farmers rather than *campesinos* had the best technical support and easiest access to the new technologies. In the belief that influential farmers would demand more government support, Wellhausen urged the foundation to stimulate their interest in research. In 1948, reflecting Alemán's objectives, the MAP's scientists encouraged farmers to form crop improvement associations and cultivated friendships with private organizations like the *Pequeña Propiedad Agrícola de la Comarca Lagunera* and the *Asociación Nacional de Cosecheros* and important politicians like the governor of Hidalgo and Rodolfo Elias Calles, who gave wheat seeds to prominent growers in Sonora. Reflecting rising faith in science, the Union of Citrus Growers of Apatzingán, Michoacán, built a laboratory and asked the SAF to staff it, El Mante hired five *agrónomos* to conduct research, and Tapachula's director praised two growers who supported his work on cacao. In 1949 private sugar producers formed the *Union de Fabricantes de Conservas S.A.* (Union of Canned Goods Manufacturers, UNFASA) and hired González Gallardo to manage its experimentation stations. In 1949 Stakman reported that "many of the more intelligent and influential growers" appreciated "the value and need of agricultural science in crop production."[57]

Problems still affected the SAF and other branches of the agrobureaucracy. The financial crisis caused by the peso devaluation and the hoof-and-mouth campaign led to budget cutbacks in the SAF. In 1947 underfunding slowed work at Tapachula, other stations could not hire enough laborers, and the government withheld the MAP's subsidy. In 1949 the SAF paid the MAP's Mexican employees three months late, vehicles sat unrepaired, and laboratories lacked essential equipment. The SAF's directors and *agrónomos* resented the Corn Commission's high profile and salaries, and political intrigues eroded morale. Some agencies did not cooperate, such as the CNI, which refused to supply water to an experimental farm in Guanajuato. The Corn Commission did not supervise the seed increase harvest, and farmers added other kernels to profit from the high price of hybrids, making varieties impure. Transferring *agrónomos* from post to post disrupted SAF projects. In 1948 Alanis Patiño said that the agency suffered from "inconsistency of programs, absence of coordination, bureaucratic vices, the lack of personal initiative, and inadequate funding."[58]

The IIA and the MAP communicated regularly and planned some activities jointly, and the MAP's scientists wanted *agrónomos* to use the new seeds to enhance their credibility, but the IIA's plant breeders resented their superior financial resources. The foundation thought that the IIA hired political favorites rather than the most qualified researchers. Tensions flared when the MAP's new seeds showed them up in meetings and correspondence with the SAF's directors. They became rivals, struggling over land use at experimentation stations, and in 1947 the IIA tried to seize administrative control of the MAP. Foundation fellows who worked for the MAP demanded salary increases commensurate with their education and a more demanding workload. Stakman noticed some SAF employees' "apathy, envy or jealousy," but the IIA's plant breeders and minister of agriculture Ortiz Garza knew that the MAP's seeds enhanced their credibility, and he insisted that his agency deliver them to the Corn Commission. Some *agrónomos* resented Alemán's support for the MAP, but most cooperated with the program. The agrobureaucracy employed more *agrónomos*, like the 47 researchers who worked for the IIA in 1947. The MAP helped them get pay raises and hired 96 more *técnicos* from 1947 to 1950. In a 1950 report on 118 MAP employees, 16 still worked there; 31 had SAF, Corn Commission, or agricultural bank jobs; 12 taught at agricultural colleges; 18 became state employees; 25 were self-employed or hired by private or parastatal firms, 8 pursued advanced degrees, and one was a *Presidente Municipal*. Reflecting prevailing gender roles, of seven women in the group, five abandoned the profession after marriage. Antonio Narro students asked Harrar for funds to visit the MAP's experimentation stations, and the director thanked him for hiring his graduates. In a 1950 report to the FAO, Mesa A. praised the MAP's fellowship program, contributions to the ENA, and new seeds. To foster collegiality with Mexican scientists, the MAP forged bonds with the *Sociedad Mexicana de Geografía y Estadística*. *Tierra* called Bradfield "a great friend of Mexico."[59]

Mexican culture and institutional behavior sometimes annoyed the MAP's scientists. Mangelsdorf disliked its work ethic, claiming that trainees quickly became "careless and slipshod in their habits" when not supervised by MAP personnel, and Wellhausen thought that many *agrónomos* "think in terms of their personal gain" rather than a sincere desire to improve agriculture.[60] *Agrónomos'* "tendency to promise anything in the enthusiasm of the moment" (*proyectismo*) bothered Stakman, and he could not accept the patron–client system, complaining that returning fellows "mix unduly their professional and personal undertakings and their unreasonable expectations with regards to financial help must tax the patience and diplomatic skill of even the most altruistically minded individual."[61] Not understanding why foundation-trained *agrónomos* expected more assistance for their professional advancement, in 1948 he said that "the expectations of some of the returned fellows are preposterous. Some of them seem to think that when the Foundation gave them a fellowship it also assumed the obligation of an indulgent foster parent . . . they still think too much in terms of special favors granted by friends rather than in terms of sound and stable institutional organizations."[62]

Some *agrónomos* still believed that better technologies alone could not solve Mexico's rural problems, but training at U.S. universities and contact with U.S. scientists reinforced their faith in research. The new seeds and agrochemicals helped *agrónomos* attain credibility as scientific professionals, even with some farmers who detested *agrarismo*. SAF *Agente General* Federico Sánchez Navarette showed tomato growers in Morelos how to use DDT to "show by example that *agrónomos* really work to benefit agriculture."[63] Cecilio Cuanalo Guevara of the *Delegación de Promoción Ejidal* in Tula, Hidalgo, gladly accepted the MAP's offer to give a lecture on wheat, corn, sorghum, and soybean cultivation with chemical fertilizers and green manure crops. Roberto Palacios was pleased when 800 *campesinos* from Toluca's *Juventud Católica Mexicana* (Mexican Catholic Youth) attended his talk on hybrid corn. After growing "Supremo" wheat successfully, Holste, a friend of ex-*jefe máximo* Calles, praised Antonio Marino Ambrosio's technical advice. Marino began to breed his own wheat. Felix Agramont Cota, a foundation fellow, said that with improvements at the ENA and the MAP's research, his colleagues understood "the most advanced aspects of genetic science in the area of maize."[64] "As a result of the corn campaign," Norberto Aguirre told Alemán, the State earned "the prestige of a doer; which gives people confidence," and he urged him to take advantage of the situation by immediately distributing new hybrid wheats.[65]

But the *agrónomos* still fretted about their public image. "Excessive" publicity of hybrid corn worried Fabila because "the same thing will occur as happened with so many technologies: it excited the farmers then provoked a reaction, eroding confidence in and cooperation with technical experts."[66] With their credibility as scientific professionals enhanced, many *agrónomos* overlooked the new seeds' shortcomings. De la Loma admitted that hybrid corn needed inputs and irrigation to which few farmers had access, and Fábila recognized that it might drive those who could not use it off the land, but neither criticized the MAP's research. In 1948 *Tierra* praised "Rocamex R-7" hybrid corn, even though the editors said that *de temporal* farmers could not use it. Leftists like Narciso Bassols demanded more land reform, but the wonders of science enchanted many *agrónomos*. A few, such as Carlos J. Aguilar, became Green Revolution farmers.[67]

Some *agrónomos* deemed *campesino* farming "condemned to technical backwardness," had many negative stereotypes about peasants, and claimed that they refused to follow their advice. Luís Vera Pérez called the *Unión Productores de Plátano de Tabasco* "apathetic" and "negligent" because they did not treat their plantations for *chamusco* as he recommended, Ch. Ramírez "struggled" to get El Mante's *campesinos* to select their seed and denounced their "irresponsible attitude," Mesa A. claimed that they habitually resisted new practices because of "routine," ignorance, laziness, and misery," and Arnulfo Landaverde decried their "indifference" to improvement.[68] *Agrónomos* wanted to force farmers to use new technologies, some of whom distrusted and even detested *agrónomos* and the SAF. The use of the *rifle sanitario* (shooting diseased cattle) in the hoof-and-mouth campaign enraged *campesinos* and led to the murder of six employees near Zitacuaro, Michoacán, and one in the Bajío. In Jalostoc, Morelos, *campesinos* killed *agrónomo* Francisco Reyes Domínguez after he diverted their water to Corn Commission fields. Rafael Aguilar, a prominent farmer who cooperated with the MAP, told Borlaug that since the Revolution *agrónomos* had been "social parasites" because they participated in the land reform. Other foundation employees said that many cultivators distrusted the agrobureaucracy, especially large farmers who feared expropriation.[69] The new technologies did not render the *agrónomos'* professional history irrelevant.

As several authors note, in 1949 trustee John S. Dickey accurately predicted the economic, social, and political changes that the MAP's technologies produced:

> [T]his program before long might begin to have a considerable impact upon the whole land-use policies of Mexico, and I am perfectly sure that within three to five years [it] will raise some very acute problems with respect to the political control of these benefits. . . . if these benefits hold the promise which they seem to hold and if the present land-use policies

of the Mexican government are continued, these very benefits may introduce fresh economic disparities . . . particularly [in] the agricultural economy, which will present political problems not now even dimly perceived by many Mexicans.[70]

To avoid soiling the foundation's image in a way that might "limit the prospects for extending the [MAP] to other Third World nations," he wanted to end the project before his predictions were realized but also urged the officers to send a social scientist to study "emerging economic and social problems" and to use "as much wisdom as possible . . . [in] any new projects of this sort in foreign countries" in the "critical phase" when new seeds are distributed.[71] In 1949, again reflecting a revised approach to modernizing Mexican agriculture, to address his concerns, the foundation sent agricultural economist Herrel de Graff, who believed that increased production would alleviate Third World hunger, to work with the SAF.[72]

THE UNITED STATES, THE ROCKEFELLER FOUNDATION, AND POSTWAR AGRICULTURAL ASSISTANCE FOR MEXICO

After the defeat of the Axis the State Department evaluated the Food Supply Program and concluded that it made valuable contributions to the war effort and political stability in Latin America. In October 1945 Secretary of Agriculture Clinton H. Anderson told the FAO that the United States would help modernize agriculture in other nations. In 1946 the USDA sent three scientists to seven Latin American countries, including Mexico, to evaluate agricultural assistance, and they found that these projects were received favorably. The Commerce Department studied Mexico's insecticide market and concluded that the government's programs provided an excellent opportunity for U.S. manufacturers. Identifying "long term interests," Chester C. Davis of the president's Famine Emergency Committee pleaded for international cooperation to increase food production by transferring scientific agriculture to China, India, and South America, and the State Department decided that scientific aid, including to produce foodstuffs that the United States exported, was an important tool to "build a friendly working relationship between two governments which will be useful when political and economic agreements are . . . discussed . . . , gain the appreciation of foreign peoples . . . , demonstrate that the U.S. leads the world in the agricultural sciences . . . , [create] familiarity and reliance on American equipment, books, materials, and methods," and foster political stability.[73]

When tensions between the Allies escalated, pundits claimed that hunger was a weapon for revolutionaries linked to the U.S.S.R. Leaders like FAO director Lord John Boyd Orr and scientists like Sir John Russell, Sir Henry Tizzard, M.N. Grady of Canada's St. Francis Xavier University, and biologist Fairfield Osborne, president of the New York Zoological Society, discussed an impending "food crisis" and urged free-world nations and scientists to concentrate on increasing production to avert "chaos and misery." Scientists lauded new agrochemicals like the herbicide 2,4-D. Fritz Went of Cal Tech said that scientific farming in the Third World would feed its growing population. Paul Becquerel, nephew of discoverer of radiation Henri, claimed that irradiating seeds would dramatically increase farmers' yields. Harvard chemist Eugene G. Rochow predicted that science would avert starvation, "even though man [*sic*] may have to eat treated coal and wood," and the *New York Times* opined that "it is worth gambling on science."[74] Despite occasional concerns over incidents like a 1946 Soviet offer to buy cordage from the *Henequeñeros de Yucatán*, the State Department saw Latin America as a safe region and worried more about Communism in Europe and then Asia. Despite fears that European famines would breed radical politics, in 1946 the United States sent food to Mexico to alleviate shortages, but in February the State Department decided that Mexico had to solve its supply problems on its own. In 1947 the agency said that "no project should duplicate those of other organizations or agencies," and George C. Marshall declared that the United States would devote most foreign aid to "countries where conditions are so unstable that proper safeguards against ideological coercion have weakened."[75] The Interdepartmental Committee on Scientific and Technical Cooperation began training Greek and Filipino scientists. The USDA did not withdraw completely from Mexico but focused on prewar agendas like protecting U.S. farmers and promoting complementary crops. The agency helped González Gallardo obtain funds for research on sugarcane. The pink bollworm, fruit fly, citrus black fly, and *chahuixtle* control programs continued, and USDA employees began to study sugarcane diseases and pests that attacked tomatoes grown for export in Tamualipas. To protect U.S. ranchers, the USDA sent many experts to Mexico to fight hoof-and-mouth disease, but the *rifle sanitario* generated resentment of U.S. technical experts. Dismayed by the appearance of South American Leaf Blight, to maintain goodwill the USDA assigned three scientists to breed resistant *Hevea* at El Palmar and promote it as an income supplement for subsistence farmers. The USDA also sent an agronomist to assist Mexico's soil conservation program. Recognizing that training Latin Americans in the United States fostered demand for U.S. goods and sympathy for the anti-Communist cause, from 1946 to 1949 the USDA gave fellowships to seven Mexican soil conservation, two extension, and four agricultural statistics students. Harrar evaluated their applications. In 1948 Congress passed Public Law 402 authorizing the use of government funds for "a world-wide program of scientific and technical exchange." In January 1949, claiming that the programs advanced "foreign economic and social policy objectives" and influenced "attitudes toward the United States," President Harry S Truman said that "technical assistance for economic development of backward areas" was one of four major U.S. foreign policies.[76] He wanted private concerns to manage scientific and technological exchanges, with the government in an advisory role, but to build support for the Point IV Program, secretary of agriculture Charles F. Brannan told Congress that agricultural assistance programs in Latin America raised standards of living, which created buyers for U.S. manufactured goods.

Bolstered by Boyd Orr, who told *Scientific American* that transferring developed nations' farm technologies to the Third World averted revolution by fostering economic growth, business prosperity, and full employment, with the fall of China and the outbreak of the Korean conflict Cold Warriors turned their attention to Asia's farmers and left the modernization of Mexican agriculture to private organizations like the Rockefeller Foundation, agribusinesses, and engineering firms and universities, assisted by EXIM Bank loans and U.S.-trained *técnicos*. Firms like Du Pont taught Mexican farmers to select the best insecticide. Ferguson Tractors got Harrar's help to begin ENA courses on their use. Associated Engineers of Fort Wayne, Indiana, studied Mexico's pineapple industry, and the *Nacional Distribuidora y Reguladora* hired the Jones-Hettelsater Construction Company to build grain elevators to store bumper harvests of the MAP's new seeds. The University of Texas held meetings of *agrónomos* and U.S. scientists. The Hawaiian Sugar Growers Association collaborated with the SAF, sending scientist R.H. Van Zwaluwenberg to speak to González Gallardo and sugar industry workers. The EXIM Bank gave a $5 million loan to the *Financiera Nacional* for tractors and $6 million to *Guanos y Fertilizantes* to build an ammonium sulfate factory. The *Banco de México* hired Ford, Bacon, and Davis to study new irrigation projects in the northwest and gave scholarships to nine Mexican students who learned about sugarcane research at Louisiana State University. The Armour Research Foundation proposed a program for Yucatán similar to the MAP, including research on food crops and U.S. training for Mexican *técnicos*, but the *Henequeñeros* rejected it because they had to pay half the cost. In 1948 the Rockefeller Foundation exchanged information with OFAR, and Nelson Rockefeller visited Mexico to study the MAP. The officers and trustees rejected formal cooperation, but they wanted to assist U.S. foreign policy in Latin America. In 1949 OFAR officials praised the MAP, Stakman evaluated OFAR projects in Guatemala and Ecuador, and the foundation accepted a request to bring two Guatemalan students to Mexico but rejected Point IV funds. The USDA's lukewarm interest in using the MAP as a model for agricultural assistance programs in the Third World disappointed Weaver.[77]

After the atomic bomb ended World War II, to avoid nuclear holocaust the foundation pleaded for international harmony and cooperation between the victors, including the U.S.S.R., but the officers wanted the United States to retain hegemony in Latin America. In 1945 the foundation awarded a three-year grant to the National Agricultural School in Medellín, Colombia, to send a student to study with the MAP and brought a Chilean *agrónomo* there for the same purpose. Agreeing with Bradfield's opinion that the foundation's private status made it a more effective adviser than government experts sworn to promote U.S. interests, in 1946 the trustees accepted the MAP as a model for agricultural programs in other regions. In 1947 the foundation brought three Colombian agronomists to Mexico; trained Honduran, Salvadorean, and Guatemalan students at the ENA; considered a proposal for an agricultural assistance program in India; and gave formal approval to begin projects like the MAP in other nations. Stakman described the modernization of Mexican agriculture in the language of the Cold War, calling U.S. assistance to develop the fertilizer industry "an effective Voice of America."[78]

In 1948 Borlaug and Wellhausen visited Colombia, and the officers discussed the MAP with a representative of the China Colleges. Fosdick worried about critics who said that the foundation's public health programs dropped a "population bomb" on the developing nations, and his annual report claimed that projects like the MAP would diffuse it. In 1949 Stakman joined other scientists who proclaimed that science could solve the food supply problem, telling the *New York Times* that "we are in the midst of an agricultural revolution as a result of which the surface of the earth is being stretched to make possible a more abundant life for an increased population. Science is providing an answer to the gloomy Malthusian prophets of doom who predict that the world's population is growing too large for the earth to support it."[79] Reflecting Weaver's faith in science, to address needs of the distant future, he proposed grants for research on climate modification, solar energy, and farming the oceans. To find more immediate solutions, several of the trustees visited Mexico to study the MAP and left very impressed. The MAP held a plant-breeding symposium, inviting scientists from 11 Latin American nations, and trained students from Colombia, Guatemala, Honduras, and Bolivia. Harrar found "no evidence of national barriers in the conference" and noted that "it is easier for men from different countries to cooperate on an objective-scientific program than it is on one dealing with social, economic, or political problems."[80] In 1950 the foundation decided to tell the world about the MAP by publishing a review of its activities during the first six years. The *New York Times* called the MAP's work with *campesino* corn growers in Juchitepec, Oaxaca, a "Blueprint for Hungry Nations." After visiting Colombia, Stakman said that an agricultural program could restrain "extreme political activity" in that strife-torn nation.[81] By 1950 the foundation knew that the MAP was a valuable weapon in the struggle to contain Communist expansion.

CONCLUSIONS

By 1945 the MAP was using its substantial influence on Mexican agricultural research to "manipulat[e] professional communities toward the production of specific kinds of knowledge" chosen by

U.S. scientists, as several scholars note.[82] The foundation knew that training *agrónomos* in the United States served U.S. economic and political interests. For example, Flores said that his classes at the University of Wisconsin diluted his Marxism and changed his view of Mexico's agrarian problem, and Stakman claimed that Mexican apprentices had "higher" "social" as well as "scientific ideals,"[83] but the foundation's goals met the *agrónomos'* professional objectives and reflected their definition of Mexico's problems. By the end of the Alemán administration *agrónomo* Manuel J. Gándara, head of the SAM during the 1930s and again in 1947, was CNC secretary-general. He said that *agrónomos* supported Alemán's agricultural policy, which "maintained the principles of the Revolution," and PRI leader General Rodolfo Sánchez Taboada affirmed the SAM's "together[ness] with its "trajectory."[84] *Agrónomos* recognized the connection between *agrarismo* and their new role as *técnicos*. In 1945 the SAM said, "The erroneous concept that the guild was dedicated only to politics should be clear to all members; this was no more than a necessary way to put the profession where it rightfully belongs and be able to develop its scientific and technical functions as Mexican agricultural policy demands."[85] More support for research and other services gave the *agrónomos* jobs and training that improved their position in the Revolutionary Family. As historian Stephen R. Niblo observes, "[I]t is very difficult to avoid the conclusion that the agricultural bureaucracy was, in effect, an employment-generating scheme for the urban middle class."[86] Hewitt de Alcantara's claim that the IIA's *agrónomos* rejected hybrid corn to defend the *campesinos'* interests is questionable. Cleaver's assertion that the foundation put "project members and their proteges in a position removed from the Mexican peasantry" is anachronistic.[87] The *agrónomos'* distance from the *campesinos* predates the founding of the MAP, and its technologies enhanced their prestige, as *agrónomos* and the foundation wanted. Plant breeding and soil science became specialized knowledge to validate their expertise, at least in the eyes of important groups in rural Mexico. Contrary to some *agrónomos* of the 1980s, during the 1940s the profession made a significant contribution to the State's shift from *agrarismo* to technical solutions to rural problems.[88]

After World War II the United States concentrated on rebuilding Europe to stop the advance of Communism and reduced aid for

Latin America, which generated ill-will in the region, but the foundation decided to continue its work in Mexico.[89] By initiating the MAP and then extending it to other countries, the foundation supported U.S. foreign policy and economic objectives. The MAP served the Revolutionary Family's new definition of the Revolution during Avila Camacho's and Alemán's administrations: to industrialize and win the support of progressive cultivators; the *agrónomos*, who sought professional credibility and upward socioeconomic and political mobility; many of Mexico's private farmers, who profited from using new technologies; and even the *campesinos*, who at least kept a legal right to petition for land. Without the Green Revolution, Mexico might have ended the agrarian reform long before the administration of Carlos Salinas de Gortari.

NOTES

1. "Ningun maiz de los E.E.U.U," *El Universal*, May 25, 1944; "Epoca de oro para la agricultura Nacional," *El Universal*, February 18, 1944; "En Orizaba ha faltado maiz," *El Universal*, February 16, 1944; "Se carece de maiz para el pueblo, en San Luis Potosí," *El Universal*, January 1, 1944; "Habrá un déficit de 100,000 toneladas de maiz en Jalisco," *El Universal*, January 27, 1944; "Absoluta falta de maiz en Sinaloa," *El Universal*, March 11, 1944; Stephen R. Niblo, *War, Diplomacy, and Development: The United States and Mexico, 1938–1954* (Wilmington, Del.: Scholarly Resources, 1995), pp. 133–38; "Especulación con el maiz en San Luis," *El Universal*, March 28, 1945; "Guanajuato teme posible escasez del maiz," *El Universal*, July 28, 1945; "No les costea sembrar maiz," *El Universal*, February 1, 1945; "Trigo australiano que es causa de epidemias," *El Universal*, May 11, 1945; Norman E. Borlaug, J. George Harrar, and Elvin C. Stakman, "Flag Smut of Wheat in Mexico," *Phytopathology* 36, 6 (June 1946): 479; "Piden mayor precio para el arroz los productores Sonorenses," *El Universal*, November 20, 1944; "Escasez de arroz en varias regiones," *El Universal*, January 27, 1945; "Gran escasez de arroz en Colima," *El Universal*, April 10, 1945; "Inundaciones y sequia: Por ello menos azúcar," *El Universal*, October 3, 1944; "Hay en Tamualipas escasez de azúcar," *El Universal*, August 9, 1944; "Agriculture" (n.d.), RFA, RG 1.1, Series 323, Box 2, Folder 14, RAC; "Una aclaración sobre el frijol exportado," *El Universal*, September 4, 1945; "Problemas fitosanitarios de importancia en Veracruz," *Siembra* 3, 38 (May 1946): 24; Messersmith to Secretary of State, May 11, 1946, NA, RG 59, Decimal File 1945–1949, Box

5113; Mervin G. Smith, "The Vegetable and Animal Fats and Oils Situation in Mexico," September 19, 1945, NA, RG 166, Box 308; Dana Markiewicz, *The Mexican Revolution and the Limits of Agrarian Reform 1915–1946* (Boulder Colo.: Lynne Rienner, 1993), p. 133.

2. Blanca Torres, *Hacia la utopía industrial* (México, D.F.: El Colegio de México, 1984), p. 41; Luís Medina, *Del cardenismo al avilacamachismo* (México, D.F.: El Colegio de México, 1978), p. 266; Frederick R. Margold to Cordell Hull, December 20, 1944, NA, RG 166, Box 305; "La SAM refuta al Director del Banco de México," *Excelsior*, April 27, 1945; Concha de Villareal, "La SAM solo ha hecho política en sus 25 años de existencia," *Excelsior*, May 28, 1945; William C. Cobb, "Interview with Edwin J. Wellahuasen," June 28, 1966, p. 41, RFA, RG 13; "Estación Experimental 'El Progreso' en el Estado de Morelos" (n.d.), p. 2, RFA, RG 6.13, Series 1.1, Box 68, Folder 767, RAC; "Grave escasez del maiz para el año entrante," *El Universal*, December 2, 1945; "Mayor producción con cultivos científicos," *El Universal*, April 10, 1944; Marte R. Gómez, *Biografías de agrónomos* (Chapingo, México: Colegio de Postgraduados, Escuela Nacional de Agricultura, 1976), p. 194; SAF, *Informe de labores 1945–1946* (México, D.F.: Editorial Cultura, 1946), p. 361.

3. "Las clasicas tortillas de maiz son insustitutibles," *El Universal*, October 2, 1945; "El mejoramiento del campo," *El Universal*, February 21, 1945; Alfonso Tico, "México debe autoabastecerse de maiz," *El Universal*, June 22, 1946; SAF, *Informe 1945–1946*, pp. 14, 59, 94–96, 205–10, 255–64, 293–300, 397–98, 404–16; SAF, "Capoc o Kapoc" (n.d.), AGN, Robles papers, Agriculture Section, Box 3, Exp. 54; "Si hay estimulo habrá mas extensos trigales," *El Universal*, November 19, 1945; Mervin G. Smith, "Mexico—Sugar Situation," August 10, 1945, NA, RG 166, Box 320; "Un acuerdo presidencial para resolver el problema del cacao," *El Universal*, April 15, 1945; Salvador Lira López, Ramón Fernández y Fernández, and Quintín Ochoa, *La pobreza rural en México* (México, D.F.: mimeograph, 1945), p. 71; Celestino E. Escalante, "El cultivo de la soya," *Tierra* 1, 5 (February 1946): 143; Dario L. Arrieta M. to Tomas Leal Moreno, June 26, 1944, RFA, RG 6.13, Series 1.1, Box 3, Folder 36, RAC; "Perspectivas agrícolas de Veracruz," *Siembra* 3, 38 (March 1946): 12.

4. William Vogt, "Confidential Memorandum Submitted to the Comisión Impulsora y Coordinadora de la Investigación Científica de México," November 1944, NA, RG 59, Decimal File 1940–1944, Box 4032; "HHS Diary Notes," October 13, 1945, RFA, RG 1.1, Series 323, Box 2, Folder 10; Arrieta M. to Harrar, March 23, 1946; Joaquin Laredo Goytortua to SAF, January 3, 1945, RFA, RG 6.13, Series 1.1, Box 3, Folder 36; Weaver, "Notes on Mexico Trip," September 12–October 7, 1946, RFA, RG 1.1, Series 323, Box 2, Folder 13; Miguel Brambila to Gonzalo Robles, "Memorandum referente a la naturaleza de los suelos de México

a su necesidad de fertilizantes," January 6, 1946, RFA, RG 6.13, Series 1.1, Box 46, Folder 531; "Progress Report of the Oficina de Estudios Especiales," July 1, 1944–December 1, 1945, RFA, RG 1.1, Series 323, Box 6, Folder 40, RAC; "Facilidades para que el labriego estime el abono de los terrenos," *El Universal*, July 19, 1944; Carlos H. Benítez, "Memorandum formulado por Guanos y Fertilizantes de México" (n.d.), pp. 5–11; Rodolfo Moreno Dahme, "Memorandum sobre estudios para la fabricación de abonos químicos," July 10, 1945, AGN, Robles papers, Agriculture Section, Box 1, Exp. 10 and 11; "Fertilización barata para lograr aumentar la producción del azúcar," *El Universal*, August 5, 1945; Jesús Merino Fernández, "La agricultura nacional, base de una economía propia y solida," May 1945, p. 9, in SAM V Consejo Nacional Directivo, BCM Bound Volume #186–10 (hereafter SAM V Consejo); W.E. Colwell, "Studies on the Effect of Nitrogen, Phosphorus, and Potash on the Yield of Corn and Wheat in Mexico," *Proceedings of the Soil Science Society of America* 11 (1946): 332–40; Guillermo Brondo Whitt, "Servicios fitosanitarios coordinados," May 1945, p. 3, ARFF, Box 145; William A. Conkright, Henry B. Allen, and Frederick R. Margold, "Comments on the Motives and Effects of the Mexican Import Control Measure," December 8, 1945, NA, RG 166, Box 313; Salvador Sánchez Colín, "El DDT, insecticida de moda," *Tierra* 4 (January 1946): 185–86; "En los terrenos de la ENA se efectuaran pruebas del nuevo fumigante 'D-D,' " *Excelsior*, October 26, 1946.

5. "Esperan renuncias en la SAF en estos dias," November 29, 1945, RFA, RG 1.1, Series 323, Box 2, Folder 10; Mervin G. Smith to Secretary of State, "Commission Appointed to Investigate the Purchase of Agricultural Machinery by the BNCE," July 6, 1945, NA, RG 166, Box 318; Smith to Secretary of State, "Reporting Statement by Marte R. Gómez, Minister of Agriculture and Development, Answering Charges Brought against Him," July 5, 1945, NA, RG 166, Box 312; Felix Ireta V., "Último informe ante la legislatura del Gobernador Constitucional de Michoacán," *El Universal*, September 22, 1944; Enrique G. Najera, "Plan General de Siembras de Maiz y Metodo de Ejecución: Sector Ejidal," January 25, 1945, ARFF, Box 123; Sánchez Colín, "La cooperación de los agricultores," *El Universal*, November 21, 1946; Roberto Osoyo Alcala, "Ensayos de maiz norteamericano en León, Gto.," unpublished thesis, ENA, 1946, p. 2.

6. Roberto Villaseñor, "Junta con los productores de henequen," March 1945; "Informe de la visita al Instituto Tecnológico Industrial de Henequeñeros de Yucatán" (n.d.), AGN, Robles papers, Agriculture Section, Box 2, Exp. 46; "Es el alcohol una causa de la gran escasez de azúcar," *El Universal*, July 21, 1945; "Resoluciones y Memorias del V Consejo," pp. 23, 36–39; "El campesinado nacional en las Conferencias de Mesas Redonda, en Guanajuato," *Siembra* 3, 41 (August 1946): 21.

7. SAF, *Informe 1945–1946*, pp. 93, 332–33; SAF, *Informe de labores 1943–1944* (México, D.F.: Editorial Cultura, 1944), pp. 50–52; Juan M. Vázquez and Cosme Sastre P. to Manuel Avila Camacho, April 16, 1944; Acosta to Javier Rojo Gómez, June 27, 1946, AGN, Avila Camacho papers, Exps. 506.1/48 and 506.1/30; "Mexican Agriculture—Annual Meeting," October 17, 1946, RFA, RG 6.13, Series 1.1, Box 32, Folder 359; "Corn Breeding Program," 1948, RFA, RG 6.13, Series 1.1, Box 11, Folder 136; Harrar to Frank Blair Hanson, January 13, 1944, RFA, RG 1.1, Series 323, Box 1, Folder 7; Gómez to Harrar, July 17, 1945, RFA, RG 1.1, Series 323, Box 2, Folder 9; Harrar, "Preliminary Report Concerning Sugar Beets Requested by Alfonso González Gallardo" (n.d.), RFA, RG 6.13, Series 1.1, Box 23, Folder 258; Lorenzo Martínez M. to Harrar, January 9, 1946, RFA, RG 6.13, Series 1.1, Box 70, Folder 786; A.R. Mann, "Telephone Conversation with President John Towers, of Ford, Bacon, and Davis Engineers," October 31, 1946; Warren Weaver, "Notes on Mexico Trip," September 12–October 7, 1946, RFA, RG 1.1, Series 323, Box 2, Folder 13, RAC; Benítez, "Memorandum," p. 9; Lira López et al., *La pobreza*, p. 81.

8. "Progress of the Mexican–United States Agricultural Commission," *Bulletin of the Pan American Union* 79, 3 (March 1945): 165–67; "Conferencia tenida con González Gallardo," December 13, 1944, AGN, Robles papers, Agriculture Section, Box 1, Exp. 9; Committee on Inter-American Economic Development, "Minutes," March 24, 1944; Mexican-American Commission for Economic Cooperation, "Minimum 1944 Program," March 10, 1944, NA, RG 353, Box 1; John E. Lockwood to A.R. Dreisbach, May 8, 1944, NA, RG 229, Box 1687, Folder June 1943–Mexico II; George S. Messersmith to Cordell Hull, September 7, 1944, NA, RG 166, Box 309; Donald R. Heath to Charles A. Thompson, November 18, 1943, RFA, RG 1.2, Series 300, Box 1, Folder 7; Fred Gardner to Nelson Rockefeller, October 31, 1944; *Food Supply Division Monthly Report*, September 15, 1945; Alfredo M. Saco, "Production of Corn in Mexico," July 1944, Nelson Rockefeller papers, RG 4, Series 0, Box 3, Folders 19, 20, and 22; "Section IV: Economic Penetration" (n.d.), Nelson Rockefeller papers, RG 4, Series 0, Box 11, Folder 85; Frank A. Waring to Rockefeller, August 26, 1944, and others, Nelson Rockefeller papers, RG 4, Series 0, Box 5, Folder 39, RAC; "Policies and Instrumentalities for Inter-American Economic Development, Report of Subcommittee II—Field of Agriculture," December 16, 1944, p. 9, NA, RG 353, Box 2, Folder Miscellaneous Draft Papers.

9. E.R. Stettinius Jr., "Memorandum for the President: Suggested U.S. Policy toward Latin America," January 3, 1945, Nelson Rockefeller papers, RG 4, Series 0, Box 11, Folder 88, RAC; Abraham Vigil, "Commodity Report on Cotton, Tamualipas, Mexico," January 1–June 30, 1945; Yvonne-José Marchant, "Questionnaire on Foreign Sources for Crude

Drugs and Essential Oils," December 10, 1945; "Translation of Memorandum Addressed by Banana Producers to the President of the Mexican Republic," November 28, 1944; Messersmith to Secretary of State, November 22, 1944, NA, RG 166, Boxes 303, 304, 312, and 320; Sección de Investigación y Divulgación, "Las Moscas de la Fruta ya no sea un problema para la exportación de frutos de naranjo, pomelo y mango a los Estados Unidos," *Fitofilo* 4, 4 (July–August 1945): 240–42; "El Ing. Gómez estimula a los naranjeros," *El Universal*, November 11, 1945; Presidente del Comisariado Ejidal de San Agustín, Ver., Cosolapa, Oax., and Est. Tezonapa, Ver. to Avila Camacho, July 1944; Loreto Lara S. and others to Avila Camacho, February 28, 1945, AGN, Avila Camacho papers, Exps. 506.1/5 and 506.1/15; Medina, *Del cardenismo*, p. 255; "Resulataron de datos inexactos," *El Universal*, May 15, 1944; "No cultivaron ya higuerilla," *El Universal*, March 6, 1944; Lester D. Mallory to Hull, February 23, 1944, NA, RG 59, Decimal File 1940–1944. In the 1930s the SAF promised castor bean growers 2,800 kgs/ha and in 1943 800–1000 kgs/ha. Farmers obtained less than 800. Smith, "The Vegetable and Animal Fats," p. 4; "Cultivadores que salen defraudados," *El Universal*, December 18, 1945; "Interview with Efrain Hernández Xolocotzi," in Leobardo Jiménez Sánchez, *Las ciencias agrícolas y sus protagonistas* (Chapingo, México: Colegio de Postgraduados, 1984, p. 209; "Minutes," November 29, 1944, and others; "Importance of the Mexican-American Commission for Economic Cooperation," January 27, 1944, NA, RG 353, Boxes 1 and 2; Lawrence W. Witt, "Effects of the War on Latin American Agriculture," *Bulletin of the Pan American Union* 77, 6 (June 1943): 343–46; Advisory Committee on Inter-American Cooperation in Agricultural Education, "Minutes of Meeting," May 3–4, 1944, NA, RG 59, Box 196.

10. Stephen G. Rabe, *Eisenhower and Latin America: The Foreign Policy of Anticommunism* (Chapel Hill: University of North Carolina Press, 1988), p. 16; "No será hule natural sino sintético el que llegará," *El Universal*, November 14, 1945; "Terminaron el estudio de un substituto del hule Heveas," *El Universal*, May 4, 1945; "The Latin American Cooperative Plantation Rubber Program of the USDA," September 1 and 18, 1944, NA, RG 353, Box 28, Folder Projects, Scientific and Technical, Rubber; "Resuelve el guayule el problema del hule," *El Universal*, March 8, 1944; "Por fin de la guerra estan en grave situación los guayuleros," *El Universal*, October 7, 1945; Luís Cabrera to Miguel Alemán, June 26, 1947, AGN, Alemán papers, Exp. 506.1/24.

11. "En busca de un mayor rendimiento agrícola," *El Universal*, December 7, 1944; Gonzalo Robles, "Estudios sobre henequen," March 1, 1947; "Memorandum: Cera de henequen," October 29, 1949, AGN, Robles papers, Agriculture Section, Box 2, Exps. 27 and 50; Carlos Jiménez to Avila Camacho, March 17, 1944; Sindicato de Trabajadores

Petroleras, Sección 31 to Avila Camacho, November 26, 1945, AGN, Avila Camacho papers, Exps. 506.1/50 and 506.17/11; Medina, *Del cardenismo*, p. 297; Edward L. May to Hull, February 27, 1945, NA, RG 166, Box 312; Wickard to Hull, June 11, 1945, NA, RG 353, Box 51; H. Claremont Moses to Hull, July 12, 1944, NA, RG 166, Box 320; Manuel M. Cárdenas to Tesero General del Estado, August 8, 1944, and others, Archivo Estatal de Michoacán, Section Agricultura y Ganadería, Depto. De Gobernación, Exp. 6–8; Alfonso Castillo de A. to Harrar, May 24, 1946, RFA, RG 6.13, Series 1.1, Box 20, Folder 216, RAC; Patricia Fent Ross, "México planea su futuro agrícola," *La Hacienda* 41, 4 (April 1946): 30–31.

12. SAM, "Quinta Convención Nacional," May 16–18, 1945, pp. 31–32, AGN, Avila Camacho papers, Exp. 433/84; Frederick R. Margold to Secretary of State, "Remarks of President of Mexican Chambers of Commerce on Mexico's Economic Future," December 20, 1944; Mervin G. Smith to Secretary of State, February 21, 1945; Guillermo Quesada Bravo, "Closing Address, San Jacinto Agricultural Exposition," trans. by Smith, November 30, 1944, p. 3, NA, RG 166, Boxes 305, 308, and 302; Lira López et al., *La pobreza*, pp. 109–10; SAF, *Informe 1943–1944*, p. 49; SAM, *Informe 1945–1946*, p. 60; L.E. Evans to Miguel M. Acosta, June 28, 1946, and others, AGN, Avila Camacho papers, Exp. 506.1/30; Mexican Ambassador to Secretary of State, May 25, 1944; Ricardo Coronado Padilla to Delegado Administrativo, April 17, 1945, RFA, RG 6.13, Series 1.1, Box 3, Folder 36; "Ponencia sobre investigaciones agrícolas," 1945, RFA, RG 6.13, Series 1.1, Box 23, Folder 258, RAC; Armour Research Foundation, International Division, "Proyecto para el estableciemiento de una Corporación Mexicana del Sureste para el desarrollo técnico y económico de la región compuesta por Yucatán, Campeche, Quintana Roo," August 31, 1948, AGN, Robles papers, Agriculture Section, Box 2, Exp. 35; Jorge Manrique, "Arroz: Un nuevo metodo de molienda que permite consumir granos con abundantes vitaminas," May 1945, ARFF, Box 16; L.C. Graton to Raymond Zwemer, September 14, 1945; G. Howard Shaw to Claude R. Wickard, March 27, 1944, NA, RG 353, Boxes 28 and 51, Folders Mexico–Laboratory Construction and Projects, Scientific and Technical, Paricutín.

13. Marco Antonio Durán, *Del agrarismo a la revolución agrícola* (México, D.F.: Talleres Gráficos de la Nación, 1947), pp. 180–81; Fernández y Fernández to Consejo Directivo, ENA, August 6, 1945, in ENA, Bound Volume #168–2; (hereafter ENA); Fernández y Fernández, "Nota bibliografica: *Informe de labores de la SAF 1944–1945* (n.d.), ARFF, Box 153; Arrieta M. to Weaver, November 6, 1946, RFA, RG 1.1, Series 323, Box 2, Folder 13, RAC.

14. "Operation of a Governmental Program of Cultural Relations," February 19, 1944; General Advisory Committee on Cultural Relations,

"Dinner Session," June 29, 1944, NA, RG 353, Box 1, Folders General Advisory Committee, February and June 1944; "Empacadora en Montemorelos," *El Universal*, July 6, 1944; Miller to Harry H. Person, January 8, 1945, NA, RG 59, Decimal File 1945–1949, Box 469; "Plantas Mexicanas de gran valor medicinal," *El Universal*, April 10, 1944; Borlaug, "Annual Survey of Wheat Growing Areas of Mexico for Determination of Severity of Damage Caused by Rust Fungi and to Collect Samples for Physiological Race Determinations," March 16, 1945, RFA, RG 6.13, Series 1.1, Box 48, Folder 544; Arrieta M. to Harrar, January 8, 1946, RFA, RG 6.13, Series 1.1, Box 3, Folder 36; Harrar, "Diary," February 26–August 31, 1944, RFA, RG 6.13, Series 1.1, Box 33, Folder 365; Arthur Paul to Executive Committee on Economic Foreign Policy, September 14, 1944, Nelson Rockefeller papers, RG 4, Series 0, Box 5, Folder 39, RAC.

15. WB to DHS, HMM, RFE, and ME, December 2, 1943, RFA, RG 1.2, Series 300, Box 1, Folder 7; Stakman to Hanson, May 27, 1944, RFA, RG 1.1, Series 323, Box 1, Folder 7; Hanson to Ruth B. Shipley, January 22, 1945; Stakman to Hanson, February 10, 1945, RFA, RG 1.1, Series 323, Box 2, Folder 9; "Progress Report, Rockefeller Foundation Agricultural Program in Mexico," November 1, 1944, p. 12, RFA, RG 1.1, Series 323, Box 6, Folder 40; "Exhibit of the NS to Inter-Divisional Committee on Latin America" (n.d.); To ME, JHW, and RFE, Social Sciences and Latin America, January 7, 1944, RFA, RG 1.2, Series 300, Folder 12, Box 2, RAC; Raymond B. Fosdick, *The Rockefeller Foundation, A Review for 1944* (New York: The Rockefeller Foundation, 1944), p. 23; Advisory Committee on Inter-American Cooperation in Agricultural Education, "Minutes," January 25–27, 1945, NA, RG 59, Decimal File 1945–1949, Box 6469.

16. Paul C. Mangelsdorf to González Gallardo, December 10, 1943, pp. 3–4; Edwin J. Wellhausen to Miller, May 30, 1944; Fosdick to Hanson, September 12, 1944, RFA, RG 1.1, Series 323, Box 1, Folders 6, 7, and 8; "Resultados del programa de cruzas de maiz llevado a cabo por la Fundación Rockefeller en cooperación con la SAF," 1945, RFA, RG 6.13, Series 1.1, Box 12, Folder 142; "Plan de producción de semilla certificada de maiz híbrido (variedad sintética) para substituir maices de bajo rendimiento existentes en la altiplanicie mexicana," August 1946, p. 5, RFA, RG 6.13, Series 1.1, Box 55, Folder 635, RAC; Adolfo Vazquez Humasque, "La técnica en la agricultura—maiz híbrido," *Hoy* 507 (November 9, 1946): 50–51.

17. Harrar, "Diary," February 12, 1943, p. 2, RFA, RG 6.13, Series 1.1, Box 25, Folder 280, RAC; Nathan Whetten, *Rural Mexico* (Chicago: University of Chicago Press, 1948), p. 453.

18. "Annual Report, Rockefeller Agricultural Program in Mexico," February 1–December 31, 1943, p. 29, RFA, RG 1.1, Series 323, Box 6, Folder 38, RAC.

19. Pauline Madow, "Reminiscences of Elvin C. Stakman," May 29, 1970, pp. 900, 901, 906–7, RFA, RG 13, RAC.

20. John H. Perkins, "The Rockefeller Foundation and the Green Revolution, 1941–1956," *Agriculture and Human Values* 26 (Summer–Fall 1990): 8; Carl O. Sauer to Joseph Willitts, February 12, 1945, RFA.

21. "Topical Diary of Visit to MAP," 1946, p. 20, RFA, RG 1.1, Series 323, Box 3, Folder 12, RAC.

22. R.M. Taylor to George K. Strode, June 7, 1944; "Report of the Inter-Divisional Committee on Latin America to RBF," January 31, 1944, RFA, RG 1.2, Series 300, Box 1, Folder 7, RAC.

23. Payne to Hugh H. Smith, December 8, 1944, RFA, RG 1.2, Series 300, Box 1, Folder 6; Harrar to González Gallardo, August 7, 1944; Harrar to Gómez, March 9, 1945, RFA, RG 6.13, Series 1.1, Box 23, Folder 258; Weaver, "Interviews with Strode and Smith," November 18, 1946; Dr. Downs to Smith, December 11, 1946, RFA, RG 1.1, Series 323, Box 12, Folder 82, RAC; Mallory to Messersmith, August 16, 1944, NA, RG 166, Box 319; Aurelio Robles Castillo, "¿La Revolución Mexicana ha cumplido con el campesino de los temporales?" *Discussión*, May 5, 1945, p. 7.

24. Stakman to Hanson, May 27, 1944, RFA, RG 1.1, Series 323, Box 1, Folder 7; "Mexican Agriculture, Annual Meeting," October 15, 1945, RFA, RG 6.13, Series 1.1, Box 32, Folder 359; Harrar to SAM, February 12, 1946, RFA, RG 6.13, Series 1.1, Box 3, Folder 36; Stakman to Hanson, August 2, 1944, RFA, RG 1.1, Series 323, Box 1, Folder 8; Cobb, "Interview with Wellhausen," pp. 38–43; Arrieta M. to Harrar, August 29, 1946, RFA, RG 6.13, Series 1.1, Box 69, Folder 785; "Personnel Técnico," 1944–1946, RGA, RG 6.13, Series 1.1, Box 42, Folder 535; Mangelsdorf, "Report on a Trip to Mexico for the Rockefeller Foundation," February–March 1946, pp. 1–4, RFA, RG 1.2, Series 323, Box 10, Folder 61, RAC; "Corn Breeding," 1948, pp. 7–9; "Los agrónomos en la vida social y profesional," *Tierra* 1, 15 (December 1946): 854; Brillante homenaje de la SAM *El Universal*, March 20, 1944; ". . . Wallace habla," pp. 660–63; "La Fundación Rockefeller da importante ayuda a nuestra agricultura," *La Prensa*, March 15, 1946; SAF, *Informe 1945–1946*, p. 58; Rodolfo Moreno Dahne, "Algunas temas que se trataron en el Congreso Agronomico Americano," November 19–22, 1946, AGN, Robles papers, Agriculture Section, Box 3, Exp. 57.

25. "HHS Diary Notes," October 13, 1945, RFA, RG 1.1, Series 323, Box 2, Folder 10; Hector Lazos to Wellhausen, Harrar, and Colwell, October 11, 1944, RFA, RG 6.13, Series 1.1, Box 18, Folder 206, RAC; Jesús Alarcón, "Prologo," October 12, 1945, in Agrónomos, BCM Bound Volume #161–40; "Observaciones al ante-proyecto de Ley sobre Enseñanza Agrícola," July 6, 1945, in Enseñanza; "Inauguración de cursos en la ENA," *Tierra* 1, 6 (March 1946): 345.

26. Antonio Marino A. to Harrar, August 2, 1945, RFA, RG 6.13, Series

1.1, Box 69, Folder 784; Miller to Harrar, April 5, 1945, RFA, RG 1.1, Series 3232, Box 2, Folder 9; Eulogio Flores Aguirre to Harrar, May 6, 1946, RFA, RG 6.13, Series 1.1, Box 70, Folder 786, RAC; "Puntos para redactar memorandum relativo becas para estudio del cultivo e industrialización de la piña" (n.d.), AGN, Robles papers, Agriculture Section, Box 7, Exp. 79; "A los agrónomos de la República," *El Universal*, September 22, 1945; "Resoluciones del V Consejo Nacional Directivo de la SAM," May 1945, p. 20, in SAM V Consejo; Advisory Committee . . . , "Minutes," January 25–27, 1945; "Henry A. Wallace habla para 'Tierra,' " *Tierra* 12 (September 1946): 662; Sección Mexicana, Comisión Agrícola México-Americana, "Preparación de técnicos e intercambio de información científica," July 1946, RFA, RG 2-1946, Series 323, Box 339, Folder 2300, RAC.

27. Jacobo Aragón Aguillón to Julián Rodríguez Adame, "Estudio de la región mas apropriada de Chiapas para establecer un ingenio azucarero," May 28, 1945, in Azúcar, BCM Bound Volume #10–13; Roberto Galván López, "El cultivo del arroz en Morelos," in SAM V Consejo, pp. 31–32; "La conferencia de mesa redonda en Yucatán," *Siembra* 3, 39–40 (June–July 1946): 34; "Habla Agricultura sobre la polinización del maiz," *El Universal*, November 12, 1944.

28. Marte R. Gómez, *Tendencias, medios, y fines de la política agrícola de México* (México, D.F.: 1945), p. 36; Gómez, "Es un mito lo de nuestro fracaso agrícola," *Siembra* 3, 42 (September 1946): 12.

29. "El problema agrícola nacional III," *Siembra* 3, 38 (May 1946): 6.

30. Sección Agronómica de Jalisco, "Acción gremial por defunción," May 1945, p. 1, in SAM V Consejo; SAF, *Informe 1945–1946*, pp. 56–57; Gilberto Mendoza Vargas, "La enseñanza y la extension agrícola," February 1954, p. 5, in Enseñanza Agrícola Superior, Bound Volume #145–43; "Editorial: Chapingo," *Tierra* 14 (November 1946): 749; "Resoluciones del V Consejo," pp. 74–77; Moreno Dahne, "Los abonos potasicos," February 16, 1946, AGN, Robles papers, Agriculture Section, Box 1, Exp. 15; Sección Agronómica de Chihuahua, "Establecimiento de campos productoras de semilla seleccionada," May 1945, ARFF, Box 88; José Rodríguez Elias, "Necesidad de formación de un nuevo centro de población rural en Pozo de Gamboa, Municipio de Riogrande, Zacatecas," May 1945, in SAM V Consejo; Gómez, *Tendencias*, p. 43; Carl O. Sauer to Joseph Willits, February 12, 1945, RFA, RG 1.1, Series 323, Box 2, Folders 9 and 15; Cobb, "Interview with Norman Borlaug," June 1967, pp. 155, 173, RFA, RG 13, RAC; David E. Lorey, *The University System and Economic Development in Mexico since 1929* (Stanford, Calif.: Stanford University Press, 1993), pp. 31, 66; Francisco Cabanillas López, "Resultados de la aplicación de algunos abonos en la zona ejidal de San Martín Texmelucan, Puebla," February 23, 1945, ARFF, Box 1.

31. SAM, "Investigación Agrícola: Proyecto de Resoluciones," May 12,

1945, ARFF, Box 83; "La hortiliza familiar," *Siembra* 3, 41 (August 1946): 28; Merino Fernández, "La agricultura"; Julián Rodríguez Adame, "El problema agrario mexicano y la maquinización agrícola," in Centro de Estudios Sociales, *Cuestiones agrarias de México* (México, D.F.: El Colegio de México, 1945), p. 72; Durán, "El problema agrario y nuestra economía agrícola," in Centro de Estudios Sociales, *Cuestiones agrarias de México*, pp. 24–31; Hilario Mesa Cienfuegos, "Designación de agregados agrónomos a la embajada Mexicana en la U.R.S.S.," in SAM V Consejo; Gómez, *Biografías*, p. 385; Durán, *Del agrarismo*, pp. 186–87; "Quinta," pp. 57–58, 71; Sección Agronómica de Jalisco, "El agrónomo como factor de colonización," May 1945, in SAM V Consejo.

32. Jesús Silva Herzog, *El agrarismo mexicano y la reforma agraria: Exposición y critica* (México, D.F.: Fondo de Cultura Económica, 1959), p. 469; Durán, *Del agrarismo*, pp. 11, 181–82; Arrieta M., "Opening Address at Agricultural Exposition," trans. Mervin G. Smith, November 30, 1944, p. 5, NA, RG 166, Box 302.

33. Durán, Fernández y Fernández, and Gilberto Fábila to Consejo Directivo, ENA, August 25, 1945, in ENA; Brondo Whitt, "Servicios"; Edmundo Flores, *Historias de Edmundo Flores, autobiografía 1919–1950* (México, D.F.: Martín Casillas Editores, 1983), p. 396.

34. Fernández y Fernández, "Nota bibliográfica"; SAF, *Informe 1945–1946*, pp. 12–13, 360–66; "Editorial," *Tierra* 1, 5 (February 1946): 233; "Los agrónomos no hacen política," *Así*, May 26, 1945.

35. Gómez, "Es un mito," p. 12; Gómez to Harrar, Alfonso González Gallardo, and others, September 24, 1945, RFA, RG 6.13, Series 1.1, Box 23, Folder 257, RAC; SAF, *Informe 1945–1946*, pp. 237, 354.

36. SAF, *Informe 1945–1946*, pp. 104, 182–84; SAF, *Resultados obtenidos con la applicación de abonos en el cultivo de chile* (México, D.F.: 1944); "Hacia la solución científica del grave problema del maiz," *El Universal*, October 13, 1944; "Metodos científicos para la agricultura," *El Universal*, December 9, 1945; Durán, *Del agrarismo*, p. 182; "Importantes estudios sobre el maiz realizados por Agricultura," *El Universal*, March 21, 1946; "La cosecha triguera y el problema del maiz," *El Universal*, March 15, 1946; Harrar to Hanson, July 22, 1944, RFA, RG 1.1, Series 323, Box 1, Folder 7; Stakman to Hanson, February 10, 1945, RFA, RG 1.1, Series 323, Box 9, Folder 56, RAC.

37. Salvador Sánchez Colín, "El maiz y la experimentación agrícola," *El Universal*, July 5, 1946.

38. Torres, *Hacia*, p. 58; Medina, *Del cardenismo*, pp. 329–31; "Como entiende Miguel Alemán el problema agrario," *Siembra* 2, 28 (July 1945): 2–6; Guillermo H. Rodríguez to Avila Camacho, December 30, 1940, AGN, Avila Camacho papers, Exp. 133.2/6; "Plan de producción," p. 2.

39. "La SAM comenta el discurso que en materia agrícola y agraria

pronuncio Miguel Alemán en el Consejo de la CNC," *El Universal*, July 22, 1945; "Organizar nuestra agricultura: Grande e immediato objectivo," *Siembra* 2, 33 (December 1945): 18–20; "El repudiado Merino Fernández rompe la tradición democrática de la agronómica," *El Universal*, October 29, 1945; "Nuevo Comité Ejecutivo de la SAM," *Tierra* 2, 8 (August 1947): 493; "A los agrónomos de la República," *El Universal*, September 22, 1945; "Plan para aumentar el cultivo del maiz," *El Universal*, December 23, 1946; Nazario S. Ortiz Garza to Director de IIA, January 15, 1948; "Respuestas al cuestionario sobre agricultura formulado por Jack C. Corbett," October 20, 1947, AGN, Alemán papers, Exps. 506.1/15 and 161.1/9; Manuel Mesa A., "La agricultura en Mexico," March 1950, p. 70, AGN, Robles papers, Agriculture Section, Box 11, Exp. 96; "Empezo a funcionar la Comisión del Maiz," *El Universal*, January 17, 1947; Wellhausen, "Corn Breeding Program" (n.d.), pp. 10–15, RFA, RG 1.1, Series 323, Box 12, Folder 141, RAC.

40. Torres, *Hacia*, pp. 44, 58–59, 73–76; J. George Harrar, *Mexican Agricultural Program: A Review of the First Six Years of Activity under the Joint Auspices of the Mexican Government and the Rockefeller Foundation* (New York: The Rockefeller Foundation, 1950), p. 13; "El cultivo del olivo en la Baja California," *El Universal*, February 9, 1947; "Importancia sin precedente, el informe del Gobernador de Jalisco," *El Universal*, February 2, 1947; Roberto Palacios Morales, "Sobre el incremento de la producción agrícola a través de la enseñanza rural," May 3, 1948; Alberto Salinas Ramos to Alemán, July 25, 1949; Guillermo Vera and others to Alemán, November 17, 1947, AGN, Alemán papers, Exps. 506.1/38, 565.4/97, and 545.22/143; Harrar, "Report to the NS and the Advisory Committee on Agriculture," September 1949–October 1950, p. 6; "Corn Improvement Program," 1949, RFA, RG 1.2, Series 323, Box 10, Folder 61; "Report of the Oficina de Estudios Especiales," September 1, 1949–August 31, 1950, RFA, RG 1.1, Series 323, Box 6, Folder 38; "Distribución de la semilla de híbridos en la República Mexicana," 1948, RFA, RG 6.13, Series 1.1, Box 11, Folder 136, RAC; Salvador Gómez Gómez to Harrar, November 3, 1947, RFA, RG 6.13, Series 1.1, Box 54, Folder 621; Harrar to Guillermo Bonilla y S., January 10, 1949, RFA, RG 6.13, Series 1.1, Box 55, Folder 633; "Wheat Demonstrations," 1948, RFA, RG 6.13, Series 1.1, Box 43, Folder 486; "Relación de las muestras de TRIGO SUPREMO Y TESTIGO que cosecha Manuel Lezama Mayorga," December 17, 1948; Palacios to Sánchez Colín, September 7, 1948, RFA, RG 6.13, Series 1.1, Box 14, Folder 167, RAC; Gómez, *Biografías*, pp. 416–17.

41. "Suggestions as to the Terms of a New Agreement between the Mexican Government and the Rockefeller Foundation," December 15, 1948, RFA, RG 1.2, Series 323, Box 11, Folder 64; Harrar to Weaver, December 23, 1948, RFA, RG 1.1, Series 323, Box 11, Folder 64; Weaver,

"Interview with Barnard," February 17, 1949; Harrar to Weaver, July 27, 1949, "General Information Concerning Mexico Program," May 26, 1949; Weaver to Harrar, August 9, 1949, RFA, RG 1.1, Series 323, Box 3, Folders 17 and 18; Rene O. Cravioto, "Proyecto de un programa de investigaciones sobre el frijol por desarrollar en el Instituto Nacional de Nutriología, en colaboración con la Oficina de Estudios Especiales, de la SAF," November 16, 1950, RFA, RG 6.13, Series 1.1, Box 44, Folder 498, RAC.

42. Allbaugh to Harrar, March 8, 1948, RFA, RG 1.1, Series 323, Box 3, Folders 16, RAC.

43. Whetten, *Rural Mexico*, pp. 281, 566.

44. Mangelsdorf, "Memo," July 26, 1949, RFA, RG 1.1, Series 323, Box 3, Folder 18, RAC.

45. "El problema agrícola nacional I," *Siembra* 4, 36 (March 1946): 3; Advisory Committee, MAP, "Memorandum supplementing Minutes," October 17–18, 1946, pp. 2–3, RFA, RG 1.1, Series 323, Box 10, Folder 67; Advisory Committee, MAP, "Annual Meeting," October 30–31, 1947; "Extension Activities," 1946, RFA, RG 6.13, Series 1.1, Box 32, Folders 359 and 360; Wellhausen to K.A. Ryerson, September 13, 1949, RFA, RG 6.13, Series 1.1, Box 18, Folder 206; Harrar, "Report to the NS; Harrar to Miller, February 10, 1948, RFA, RG 1.1, Series 323, Box 5, Folder 32A; George K. Strode, "Diary," January 10 and 20, 1947; Harrar to Weaver, February 13 and September 20, 1948, RFA, RG 1.1, Series 323, Box 12, Folder 83; Francisco de P. Miranda to Wilbur G. Downs and Harrar, July 29, 1947, RFA, RG 6.13, Series 1.1, Box 44, Folder 498; "Protein Content of Corn Grains" (n.d.), RFA, RG 6.13, Series 1.1, Box 12, Folder 143, RAC; Harrar, "Diary," October 23–24; Wellhausen, "Organization of Corn."

46. Weaver, "Corn Breeding: Notation and Comments," August 13, 1948, p. 15, RFA, RG 1.2, Series 323, Box 10, Folder 59; Harrar, "Report to the Natural Sciences Division and the Advisory Committee on Agriculture," September 30, 1949–October 1, 1950, p. 5, RFA, RG 1.2, Series 323, Box 10, Folder 61, RAC.

47. "Topical Diary of Visit to MAP," 1946, p. 21, RFA, RG 1.1, Series 323, Box 3, Folder 12; Weaver to Chester I. Barnard, "Joint Operations Committee and Mexican Agriculture," October 18, 1948; Allbaugh to Harrar, March 8, 1948, RFA, RG 1.1, Series 323, Box 3, Folder 17, RAC.

48. Mangelsdorf, "Report," February 14–27, 1947, pp. 9–10.

49. Wellhausen, "Corn Breeding," p. 10; "Corn Breeding Program," 1948, p. 23, RFA, RG 6.13, Series 1.1, Box 11, Folder 136, RAC.

50. Barrus to Palacios, November 29, 1948, RFA, RG 6.13, Series 1.1, Box 14, Folder 167; Bradfield to Stakman, September 19, 1949, RFA, RG 1.1, Series 323, Box 9, Folder 60; "Plans for Organizing Extension Work in Mexico," 1949, RFA, RG 6.13, Series 1.1, Box 43, Folder 487; Barrus,

"Organization of Extension Work in Mexico" (n.d.), p. 13, RFA, RG 6.13, Series 1.1, Box 16, Folder 185; Weaver, "Lunch with Barrus and Miller," April 21, 1949, RFA, RG 1.1, Series 323, Box 3, Folder 17; Weaver to Alemán, October 21, 1946; Weaver, "Notes on Mexico Trip," September 12–October 7, 1946; Harrar to Weaver, February 24, 1947; Alemán to Fosdick, March 26, 1947; Mann to Harrar, January 3, 1947; Stakman to Weaver, May 16, 1947, RFA, RG 1.1, Series 323, Box 2, Folders 13, 14, and 15; Parker to Harrar, June 27, 1947, RG 6.13, Series 1.1, Box 49, Folder 555; Weaver to John S. Dickey, W.I. Myers, and Thomas Parran, October 28, 1949, RFA, RG 1.1, Series 323, Box 3, Folder 19, RAC; Stakman, "Report," April 25–May 17, 1949.

51. Reed Hertford, *Sources of Change in Mexican Agricultural Production, 1940–1965* (Washington, D.C.: Economic Research Service, U.S. Department of Agriculture, 1971), p. 9; José Luis de la Loma, "Mejoremos el maiz por medio de la selección en masa," *Tierra* 2, 4 (April 1947): 206; Gabriel Ramos Millán to Alemán, April 15, 1947; Comisión Ejidal of Mamaleón, Tamualipas to Alemán, December 20, 1948; Juan M. Esponda to Alemán, September 12, 1947; Arturo B. de la Garza to Alemán, May 2, 1947; Magdaleno Aguilar to Ramos Millán, April 30, 1949; Ortiz Garza to Rogerio de la Selva, February 24, 1949; Central y Investigación y Propagación de Cacao, Tapachula, Chiapas, IIA, "Circular No. 1," December 13, 1948; Vera and others to Alemán, November 17, 1947, AGN, Alemán papers, Exps. 110.1/2, 506.1/50, 506.1/19, 506.1/18, 506.1/29, 433/23, 506.1/14, and 545.22/143; "Editorial: Trigo," *Tierra* 2, 9 (September 1947): 523; Cobb, "Interview with Wellhausen," p. 65; Harrar to David Guerrero V., October 19, 1948, RFA, RG 6.13, Series 1.1, Box 20, Folder 216; Vicente Nuñez and Agustín Piñeda to Harrar, August 18, 1949, RFA, RG 6.13, Series 1.1, Box 46, Folder 529; Limón to Harrar, September 30, 1948, RFA, RG 6.13, Series 1.1, Box 69, Folder 784; Harrar to Weaver, August 12, 1947, RFA, RG 2–1947, Series 323, Box 379, Folder 2559, RAC; "Corn Breeding," p. 3; Harrar, "Report to the NS"; "Editorial: Mexico's Rockefeller Foundation," *Nosotros* (English section), October 11, 1947; Fosdick, *The Rockefeller Foundation*, pp. 29–30.

52. "British Scientists See Food Increase," *New York Times*, September 7, 1949.

53. Torres, *Hacia*, p. 76; John B. Pittner, *Trebol Huban*, Foleto de Divulgacion #2, (México, D.F.: Oficina de Estudios Especiales, Secretaría de Agricultura y Ganadería, 1948), p. 5; Bradfield to Weaver, November 1, 1949; "A Resolution Adopted by Unanimous Vote of the Advisory Committee on Agricultural Activities of the Rockefeller Foundation at Its Annual Meeting," November 14, 1949, RFA, RG 1.1, Series 323, Box 3, Folder 19; Erlif V. Miller, Pittner, Ricardo Villa J., and Carlos Romo, "Population Density of Unirrigated Maize and Its Effects upon Fertilizer Efficiency in

Central Mexico," 1948, pp. 5–6, RFA, RG 6.13, Series 1.1, Box 46, Folder 531; "Confidential Monthly Report No. 96, for the Information of the Trustees," October 1, 1947, RFA, RG 1.2, Series 323, Box 23, Folder 155; "Lista general del personel que actualmente se encuentran trabajando dentro el programa agrícola de emergencia para siembras de maiz, frijol, y trigo, planeado por la SAF" (n.d.), RFA, RG 6.13, Series 1.1, Box 42, Folder 478; Harrar to Weaver, April 14, 1949, RFA, RG 1.1, Series 323, Box 3, Folder 17; J.I. Rodale to SAF, December 27, 1949, RFA, RG 6.13, Series 1.1, Box 15, Folder 170; Harrar to Ortiz Garza, November 14, 1949, RFA, RG 6.13, Series 1.1, Box 37, Folder 412; "Report to the NS and the Advisory Committee on Agriculture," September 30, 1949–October 1, 1950, RFA, RG 1.2, Series 323, Box 10, Folder 61; Pittner and Robert F. Chandler Jr., "A Proposed Soil Management Research Program for Mexico" (n.d.), p. 5; "Crops and Soils Program for 1949" (n.d.), RFA, RG 6.13, Series 1.1, Box 46, Folders 530 and 531, RAC; Enrique Peredo Reyes, "Continuando nuestra labor," *Tierra* 1, 3 (October 1946): 726–28; "Respuestas al cuestionario"; "Radical medidas aplicadas a la revindicación de la agricultura," *El Universal*, January 19, 1947; SEP, *La resurrección de la tierra*, Cuaderno de Cultura Popular #8 (México, D.F.: n.d.); Norman Borlaug and J.A. Rupert, *Metodos que pueden aumentar la producción de trigo en México* (México, D.F.: Secretaría de Agricultura, 1949), p. 8; "Notes on Conversation, Richard Bradfield, Adolfo Sisto and Leopoldo Gutierrez about the Fertilization of Mexican Soils," April 15, 1948; Roberto Moreno to Adolfo Sixto S., March 29, 1948; Rodolfo Peregrina Robles, "Informe durante su existencia en la Universidad de Cornell" (n.d.), AGN, Robles papers, Agriculture Section, Box 1, Exps. 1, 2 and 18; Mesa A., "La agricultura," pp. 14, 36; "Interview with Norberto Aguirre Palancares," in Leobardo Jiménez Sánchez, *Las ciencias agrícolas*, p. 54; Fernández y Fernández, "Situacion del mercado del superfosfato de calcio," *Tierra* 2, 3 (March 1947): 166; José Ch. Ramírez to Alemán, October 8, 1949, AGN, Alemán papers, Exp. 506.1/20; Lane Simonian, *Defending the Land of the Jaguar: A History of Conservation in Mexico* (Austin: University of Texas Press, 1995), p. 115.

54. "Confidential Monthly Report," October 1, 1947, RFA, RG 1.2, Series 323, Box 23, Folder 155, RAC.

55. M.F. Barrus to Manuel Ayala, July 19, 1948, RFA, RG 6.13, Series 1.1, Box 14, Folder 67; Mauricio Muñoz, "Experimentos con frijol en el Bajío," December 15, 1950, RFA, RG 6.13, Series 1.1, Box 4, Folder 46, RAC; J.J. McKelvey and Dorothy Parker, *Nuevas insecticidas* (México, D.F.: Oficina de Estudios Especiales, Secretaría de Agricultura y Ganadería, 1948); R.H. Van Zwaluwenberg to González Gallardo, September 15, 1949, AGN, Robles papers, Agriculture Section, Box 13, Folder 110; "Corn Improvement."

56. "El campesinado nacional en las Conferencias de Mesa Redonda, en Aguascalientes," *Siembra* 3, 4 (August 1946): 19; "El problema agrícola nacional III," *Siembra* 3, 38 (May 1946): 6–7, 40; Cobb, "Interview with Wellhausen," p. 168; Alexander Holste to SAF, Dependencia Propagación de Semilla Mejorada, Jefatura Zona Sur, December 23, 1948. RFA, RG 6.13, Series 1.1, Box 47, Folder 541; Palacios to Barrus, November 6, 1948; Alexander Holste to Joaquín Archundia Sánchez, November 13, 1948, RFA, RG 6.13, Series 1.1, Box 14, Folder 167, RAC; Miller et al., "Population Density," pp. 3–6; "Notes on Conversation, Richard Bradfield, Adolfo Sisto," pp. 3–4; Ortiz Garza to Rogerio de la Selva, February 24, 1949; Ch. Ramírez "Conferencia en la Universidad de Texas, E.U.A.," June 10, 1949; Comisión Agraria Ejidal de Chupio to Alemán, April 7, 1947, AGN, Alemán papers, Exps. 506.1/20 and 506.1/3; Judith Friedlander, *Being Indian in Hueyapan: A Study of Forced Identity in Contemporary Mexico* (New York: St. Martin's Press, 1975), p. 63.

57. Luis González y González, *San José de Gracia: Mexican Village in Transition*, John Upton, trans. (Austin: University of Texas Press, 1974), pp. 227–28; Palacios to Sánchez Colín, October 8, 1948, RFA, RG 6.13, Series 1.1, Box 14, Folder 167; Holste to SAF; Stakman, "Report on Rockefeller Foundation Mexican Program," April 25–May 17, 1949, RFA, RG 1.2, Series 323, Box 10, Folder 60; "Annual Meeting of the Advisory Committee on Agriculture of the Rockefeller Foundation," October 1948, RFA, RG 1.1, Series 323, Box 10, Folder 67; "Report," September 1, 1949–August 31, 1950, p. 10; "Organization of Corn Breeding Work" (n.d.), RFA, RG 6.13, Series 1.1, Box 12, Folder 141; Harrar to Weaver, April 22, 1949; Weaver, "Diary," August 31–September 1–2, 1949, RFA, RG 1.1, Series 323, Box 3, Folders 17 and 18; Jesús Uribe Ruiz to Sánchez Colín, September 22, 1949, RFA, RG 6.13, Series 1.1, Box 15, Folder 20, RAC; Cynthia Hewitt de Alcantara, *Modernizing Mexican Agriculture: Socioeconomic Implications of Technological Change, 1940–1970* (Geneva: United Nations Research Institute for Social Development, 1976), p. 148; "La producción de cacao en Soconusco futura base d'la economía de Chiapas," *Gráfico*, December 11, 1948; "Entrevista, que para tratar el problema de la fertilización de los suelos mexicanos, entre Bradfield, Harrar, and Pittner of the Rockefeller Foundation and Sixto Velasco and Leopoldo Gutiérrez de Zubiaurre of the Banco de México," April 15, 1948, AGN, Robles papers, Agriculture Section, Box 1, Exp. 2; Ch. Ramírez to Alemán, October 8, 1949; Stakman, "Report," April 25–May 17, 1949; Gómez, *Biografías*, p. 253.

58. Harrar to Weaver, September 7, 1948, and April 12, 1949, RFA, RG 1.1, Series 323, Box 3, Folders 16 and 17; Epifano Duarte to Ricardo León Manzo, February 19, 1948, RFA, RG 6.13, Series 1.1, Box 15, Folder 170; Enrique Torres Ortega to Sánchez Colín, July 7, 1950, RFA, RG 6.13,

Series 1.1, Box 68, Folder 770; Parker to Harrar, December 19, 1947, RFA, RG 6.13, Series 1.1, Box 49, Folder 555; Taboada to Harrar, October 21, 1949, RFA, RG 6.13, Series 1.1, Box 27, Folder 299; Advisory Committee on Agricultural Activities, "Annual Meeting," November 4–5, 1948; Stakman, "MAP," March 10–April 3, 1948, RFA, RG 6.13, Series 1.1, Box 32, Folder 360; Cobb, "Interview with Borlaug," p. 77; Cobb, "Interview with Wellhausen," p. 99; Bernardo Cartas Díaz to Alfredo Campos, July 16, 1948, RFA, RG 6.13, Series 1.1, Box 47, Folder 541; Mangelsdorf, "Report on a Trip to Mexico for the Rockefeller Foundation," February 14–27, 1947, p. 4, RFA, RG 1.2, Series 323, Box 10, Folder 61, RAC; Emilio Alanis Patiño, "La Corporación Mexicana del Sureste," October 5, 1948, AGN, Robles papers, Agriculture Section, Box 2, Exp. 35; Gonzalo H. Garrido to Alemán, March 18, 1947; Ortiz Garza to Alemán, December 23, 1946, AGN, Alemán papers, Exps. 506.1/4 and 110.1/2; "Importantes investigaciones científicos harán de Chiapas gran productor de cacao," *El Heraldo*, December 23, 1948.

59. Cobb, "Interview with Wellhausen," pp. 44, 69–70, 172–73; Cobb, "Interview with Borlaug," p. 345; Cobb, "Interview with Pittner," p. 69; Harrar to Weaver, April 17 and 30, 1947; Harrar to Ortiz Garza, May 20, 1947, RFA, RG 1.1, Series 323, Box 2, Folders 14 and 15; Ortiz Garza to Harrar, November 2, 1949; Harrar to Weaver, November 29, 1949, RFA, RG 1.1, Series 323, Box 3, Folder 19; Harrar to Guillermo Pérez P., October 29, 1946, RFA, RG 6.13, Series 1.1, Box 20, Folder 216; Harrar to Procopio Ortiz, October 15, 1948; Harrar to Weaver, December 20, 1948, RFA, RG 1.2, Series 323, Box 11, Folder 64; Felix Agramont Cota and 26 other *agrónomos* to Ortiz Garza, April 15, 1947, RFA, RG 6.13, Series 1.1, Box 37, Folder 411; Ortiz Garza to León Manzo, September 30, 1948, RFA, RG 6.13, Series 1.1, Box 47, Folder 541; Harrar to Alberto P. León, August 5, 1947, RFA, RG 6.13, Series 1.1, Box 57, Folder 649; Ortiz Garza to Harrar, September 3, 1947, and May 13, 1948, RFA, RG 6.13, Series 1.1, Box 37, Folder 411; Edmundo Taboada R. to Harrar, June 4 and 5, 1948, RFA, RG 6.13, Series 1.1, Box 27, Folder 299; Jose Pérez Razo to Taboada R., November 9, 1949, RFA, RG 6.13, Series 1.2, Box 68, Folder 767; Stakman, "Report," April 25–May 17, 1949; Rodolfo Agustini V. and others to Harrar, July 2, 1947; Lorenzo Martínez M. to Harrar, December 21, 1948, RFA, RG 6.13, Series 1.1, Box 78, Folder 786; Wellhausen, "Origen del material de los híbridos y sintéticos actuales" (n.d.), p. 1, RFA, RG 6.13, Series 1.1, Box 37, Folder 441; Commissioned Personnel of the Oficina de Estudios Especiales" (n.d.), RFA, RG 6.13, Series 1.1, Box 42, Folder 478, RAC; Harrar, "Report to NS Division"; Mangelsdorf, "Report," February 14–27, 1947, p. 3; "Personal Técnico," 1947–1950; "Los agrónomos en la vida social y profesional," *Tierra* 3, 3 (March 1948): 232–33; Mesa A., "La agricultura."

60. Cobb, "Interview with Wellhausen," pp. 39, 44, 100–101; Stakman, "MAP," March 10–April 3, 1948; "Corn Breeding," 1948, p. 29; Mangelsdorf, "Report," February–March 1946, p. 9.

61. Stakman, "MAP," March 10–April 3, 1948, p. 10; Stakman, "Latin American Agricultural Institutions," May 8–July 14, 1947, RFA, RG 6.13, Series 1.1, Box 32, Folder 360, RAC.

62. Stakman, "MAP," March 10–April 3, 1948, p. 8.

63. Author unknown, "El problema agrario de México," 1946, AGN, Robles papers, Agriculture Section, Box 3, Exp. 58; Ing. Bravo, "Notas para el memorandum al Ing. Robles sobre la Corporación del Sureste," September 1948, AGN, Robles papers, Agriculture Section, Box 2, Exp. 35; Flores, *Historias*, p. 354; "Entrevista de Tierra," *Tierra* 2, 1 (January 1947): 33.

64. Friedrich E. Schuler, *Mexico between Hitler and Roosevelt: Mexican Foreign Relations in the Age of Lázaro Cárdenas* (Albuquerque: University of New Mexico Press, 1998), p. 77; "Review of the Work of the Oficina de Estudios Especiales, S.A.G.," 1946–1947, RFA, RG 1.1, Series 323, Box 6, Folder 38; Holste to SAF; Cecilio Cuanalo Guevara to Barrus, February 19, 1949, RFA, RG 6.13, Series 1.1, Box 14, Folder 167; Roberto Palacios to Harrar, January 20, 1949, RFA, RG 6.13, Series 1.1, Box 46, Folder 529, RAC; José M. de la Puente, "México esta a la par con los Estados Unidos en genética del maiz," *Tierra* 3, 12 (December 1948): 823; Weaver, "Diary," August 31–September 1–2, 1949.

65. "Annual Meeting of the Advisory Committee on Agriculture of the Rockefeller Foundation," October 13–14, 1949, RFA, RG 1.1, Series 323, Box 10, Folder 67, RAC; Noberto Aguirre, "Memorandum sobre el problema del trigo," October 20, 1949, AGN, Alemán papers, Exp. 506.1/5.

66. Gilberto Fábila, "La técnica y nuestra agricultura," *Tierra* 2, 8 (August 1947): 489–90.

67. De la Loma, "¿Que es el maiz hibrido?" *Tierra* 1, 10 (July 1946): 527–30; Fábila, "La técnica," pp. 487–90; Fábila, "La técnica y nuestra agricultura," *Tierra* 3, 10 (October 1947): 615–18; Fábila, "La técnica y nuestra agricultura," *Tierra* 3, 11 (November 1947): 688–90; "Editorial: Maiz," *Tierra* 3, 9 (September 1948): 639; Narciso Bassols, "Estudiamos la cuestión agraria," *Revista de Economía* 12, 1 (January 1949): 6–7; Gómez, *Biografías*, p. 16.

68. Luís Vera Pérez, "El Lacco Cupro '50,' una promesa para controllar el Chamusco, breve información sobre su experimentación en Tabasco," *Tierra* 2, 5 (May 1947): 270–71; Ch. Ramirez to Alemán, October 9, 1949; Mesa A., "La agricultura," pp. 5, 21, 36; Arnulfo Landaverde, *El maiz y su cultivo en México* (México, D.F.: Editores José C. Torres y Cia., 1949), p. 7.

69. Aguirre, "Memorandum"; "Los agrónomos en la vida social y profesional," *Tierra* 3, 3 (March 1948): 232–33; Cobb, "Interview with Borlaug," p. 155; Cobb, "Interview with Kenneth Wernimont," May 8, 1967, p. 1; Cobb, "Interview with John B. Pittner," June 15, 1967, pp. 65–66, RFA, RG 13; Harrar, "Diary," October 23–24, 1947, RFA, RG 2 (1947), Series 323, Box 379, Folder 2559, RAC.

70. John S. Dickey, "Several Observations on Mexican Trip" (n.d.), p. 3, RFA, Record Group 1.1, Series 323, Box 3, Folder 19, RAC.

71. Dickey, "Several Observations," pp. 5–6.

72. Weaver to Harrar, January 27, 1949, RFA, RG 1.1, Series 323, Box 3, Folder 17, RAC; "Food Needs in U.S. May Halt Exports," *New York Times*, August 13, 1948.

73. "Food Supply in Peacetime," *Food Supply Division Monthly Report*, September 15, 1945, Nelson Rockefeller papers, RG 4, Series 0, Box 3, Folder 20, RAC; "Procedure in the Field of Agriculture: Selected Exhibits Relating to the Work of the Food Supply Division," August 1945, NA, RG 353, Box 2; "Crop Curbs Decried by New Zealander," *New York Times*, October 19, 1945; U.S. Department of Commerce, "Insecticide Market in Mexico," *Industrial Reference Service, Part II, Chemicals, Drugs, and Pharmaceuticals* 4, 11 (January 1946): 3; " 'Know-How' Urged to Help Feed All," *New York Times*, August 27, 1946; Edgar B. Brossard, "U.S. Program of Scientific and Cultural Cooperation with the Other American Republics," *Bulletin of the Pan American Union* 80, 11 (November 1946): 633–42; William Benton to Haldore Hanson, June 24, 1946, NA, RG 353, Box 14, Folder General File 1938–1946; James G. Maddox, "Point IV in Mexico," *American Universities Field Staff Reports, Mexico and Caribbean Area*, 4, 3 (1957): 8; State Department, "Enabling Legislation for Program of Technical and Cultural Cooperation" (n.d.), NA, RG 353, Box 12, Folder Material in Support of Enabling Legislation; Olcott Denning to Rowena Rommel, April 29, 1947, NA, RG 353, Box 1, Folder History and Development.

74. "Orr Says That Only Food Can Bar WW III," *New York Times*, May 24, 1947; "Program to Meet the World Food Crisis," *New York Times*, November 9, 1947; "Population and Food," *New York Times*, September 12, 1948; "Osborne Stresses Farm Techniques," *New York Times*, May 4, 1949; "World Faces Bountiful Era, Scientists Forecast at U.N.," *New York Times*, August 27, 1949; "Savant Sees the Need to Restrain Births," *New York Times*, September 1, 1949; "Radioactivity Boosts Yields of Seeds," *New York Times*, January 25, 1948; "Chemicals May Aid Food Production," *New York Times*, April 30, 1948; "Noted Chemist Predicts Pottery Houses When Forests Are Gone," *New York Times*, May 10, 1949; "World Food Potential," *New York Times*, July 16, 1950; "Editorial: Gambling on Science," *New York Times*, September 11, 1949.

75. State Department, "Memorandum of Conversation," April 1, 1946; Messersmith to John Willard Carrigan, February 20, 1946, NA, RG 59, Decimal File 1945–1949, Box 5116; "Responsibilities of the Missions for the U.S. Government Program of Scientific and Technical Cooperation," August 9, 1946; Hewland W. Sargeant, "Policy regarding the Interdepartmental Committee on Scientific and Technical Cooperation," May 22, 1947, NA, RG 353, Box 1, Folder History and Development; Messersmith to Secretary of State, May 11, 1946.

76. Julius Matz to González Gallardo, October 19, 1946; E.W. Brandes to González Gallardo, September 13, 1949, AGN, Robles papers, Agriculture Section, Box 13, Exp. 110; Alonso Blackaller Valdés, "Breves nociones sobre leucinodes elegantalis gn." (n.d.), RFA, RG 1.2, Series 323, Box 20, Folder 125; Harrar to Mann, February 20, 1947, RFA, RG 1.1, Series 323, Box 2, Folder 14, RAC; Harrar, "Diary," October 23–24, 1947; J. Bernard Gibbs, "Citrus Fruit Report," November 7, 1945, NA, RG 166, Box 313; "Assignment of Mr. Herbert D. Smith by the USDA," April 5, 1948, and others, NA, RG 59, Decimal File 1945–1949, Box 5116; "Summary of Statement to Be Made by Colonel Arthur H. Harris for the Congressional Committee on Foreign Relations" (n.d.), NA, RG 59, Decimal File 1945–1949, Box 4357; Rafael de la Colina to Secretarío de Relaciones Exteriores, December 31, 1947, AHSRE, Exp. III-2397-9; Memorandum to Truman, "Actions to Implement the President's Proposal for Technical Assistance to Underdeveloped Areas," January 28, 1949; Jesse M. MacKnight to Rommel, August 17, 1947; "Report," September 1, 1949–August 31, 1950, p. 9; "Point IV," August 10, 1949; Interdepartmental Committee on Cultural and Scientific Cooperation, "1946 In-Service Training Program"; "Summary of Scientific and Cultural Activities, 1939–1949, Mexico," July 1, 1949, NA, RG 353, Boxes 1, 14, and 33, Folders History and Development, Circulars, and Summary of Scientific and Technical Activities by Country.

77. "Brannan Defends Point 4 Program," *New York Times*, September 29, 1949; John Boyd Orr, "The Food Problem," *Scientific American* 183, 8 (August 1950): 11–15; S.A. Du Pont, "Advertisement," *Tierra* 15 (December 1946): 860; Robert Brooks Taylor to Alemán, June 5, 1948, AGN, Alemán papers, Exp. 506.1/47; Jones-Hettelsater Construction Company to Carlos M. Cinta, November 17, 1947, AGN, Robles papers, Agriculture Section, Box 13, Exp. 110; Ch. Ramírez, "Conferencia"; Van Zwaluwenberg to Gonzalez Gallardo; "Emprestito a la Financiera," *El Universal*, February 27, 1947; "Respuestas al cuestionario"; Armour Research Foundation, "Proyecto"; "Memorandum para Alemán: La cera y otros subproductos del henequen," October 24, 1949, AGN, Robles papers, Agriculture Section, Box 2, Exp. 37; Advisory Committee . . . , "Annual Meeting," November 4–5, 1948, p. 7; Parker to Harrar, March 10 and

November 25, 1947, RFA, RG 6.13, Series 1.1, Box 49, Folder 555; González Gallardo to Harrar, October 30, 1948, RFA, RG 6.13, Series 1.1, Box 23, Folder 259; Weaver to Chester I. Barnard, Miller, LFK, and Harrar, May 6, 1949, RFA, RG 1.1, Series 323, Box 5, Folder 33; Stakman to Weaver, December 7, 1949; Harrar to Weaver, March 15, 1948; Weaver, "Diary on Trip to Colombia," September 17, 1949, RFA, RG 1.1, Series 323, Box 3, Folders 16 and 19; Miller to Harrar, August 3, 1949, RFA, RG 1.2, Series 323, Box 23, Folder 152; A. Arriola T. to Harrar, March 17, 1947, RFA, RG 1.1, Series 323, Box 2, Folder 14, RAC.

78. Raymond B. Fosdick, *The Rockefeller Foundation: A Review for 1945* (New York: The Rockefeller Foundation, 1945), p. 5; Fosdick, *The Rockefeller Foundation: A Review for 1946* (New York: The Rockefeller Foundation, 1946), pp. 9, 33; "Mexican Agriculture, Board Meeting," December 3, 1945; Emilio A. Yepes Y. to Miller, June 20, 1946; Bradfield to W.I. Myers, July 12, 1946; Weaver to Harrar, December 6, 1946; Harrar to Miller, April 3, 1947; Harrar to Miller, August 22, 1947; Harrar to Weaver, November 22, 1947, RFA, RG 1.1, Series 323, Box 2, Folders 10, 11, 12, 13, 14, and 15; "MAP," December 2–3, 1947, RFA, RG 1.2, Series 323, Box 17, Folder 103; "Annual Meeting of the Advisory Committee, MAP," October 30–31, 1947, RFA, RG 1.1, Series 323, Box 10, Folder 67, RAC; Stakman, "MAP," March 10–April 3, 1948.

79. Weaver, "Interview with C.A. Evans," September 9, 1948, RFA, RG. 1.1, Series 323, Box 3, Folder 16, RAC; Raymond B. Fosdick, *The Rockefeller Foundation: A Review for 1948* (New York: The Rockefeller Foundation, 1948), pp. 16–28; "Letter to the Editor by Guy Irving Burch," *New York Times*, July 21, 1951; "Scientists Promise More Food for All," *New York Times*, December 28, 1949; Perkins, "The Rockefeller Foundation," pp. 9–10.

80. "Population and Food," July 17, 1949, RFA, RG 6.13, Series 1.1, Box 42, Folder 474; W.I. Myers to Weaver, September 7, 1949, RFA, RG 1.1, Series 323, Box 3, Folder 19; "Annual Meeting of the Advisory Committee on Agriculture of the Rockefeller Foundation," October 13–14, 1949, p. 6, RFA, RG 1.1, Series 323, Box 10, Folder 67, RAC; Raymond B. Fosdick, *The Rockefeller Foundation: A Review for 1949* (New York: The Rockefeller Foundation, 1949), pp. 31–32.

81. Harrar, *Mexican*, pp. 35–36; "Blueprint for Hungry Nations," *New York Times*, January 1, 1950; Stakman, "Report of Trip to Mexico, Colombia, and Costa Rica for the Rockefeller Foundation," February 1950, RFA, RG 1.2, Series 323, Box 10, Folder 60, RAC.

82. Hewitt de Alcantara, *Modernizing*, p. 20; Bruce H. Jennings, *Foundations of International Agricultural Research, Science and Politics in Mexican Agriculture* (Boulder, Colo.: Westview Press, 1988), pp. 11, 39.

83. Flores, *Historias*, pp. 15–16; 353; Stakman, "MAP, Report," Septem-

ber 6–October 7, 1944, p. 7, RFA, RG 1.2, Series 323, Box 10, Folder 60, RAC.

84. "El agrónomo defensor de la tierra y apostol de la producción," *Tierra* 6, 5 (May 1951): 291–93.

85. "Resoluciones del V Consejo," p. 78.

86. Niblo, *War*, p. 267.

87. Peter Cleaves, *Professions and the State: The Mexican Case* (Tucson: University of Arizona Press, 1987), pp. 97–98; Hewitt de Alcantara, *Modernizing*, pp. 37–38; Harry M. Cleaver, "Origins of the Green Revolution" (unpublished Ph.D. dissertation, Stanford University, 1974), p. 315.

88. "Interview with Lorenzo Martínez Medina," in Jiménez Sánchez, *Las ciencias agrícolas*, p. 104; Nicolas Sánchez Durón, "La investigación agrícola y el desarrollo agropecuario de México," *Anales de la Sociedad Mexicana de Historia de la Ciencia y de la Tecnología* 2 (1970): 178; Mangelsdorf, "Interview," May 15, 1947, RFA, RG 1.1, Series 323, Box 2, Folder 15, RAC.

89. Peter H. Smith, *Talons of the Eagle: Dynamics of U.S.–Latin American Relations* (New York: Oxford University Press, 1996), pp. 147–49.

Chapter 6

The Mexican Revolution and the Green Revolution, 1950–1970

Although some scholars downplay the Green Revolution's impact on farm productivity, during the 1950s and 1960s Mexico was the first developing nation to experience it. By 1951 MAP varieties covered 70% of all land planted in wheat; in 1965, 80%; and in 1968, 90%. In 1952 Mexico imported significant quantities of wheat, corn, rice, and even garbanzos, but by 1956 wheat self-sufficiency was achieved. The MAP's corn did not attain the same acceptance. In 1968 improved varieties grew on only 20% of Mexico's *milpas* (maize land), but except for 1963, from the late 1950s through the 1960s Mexico was virtually self-sufficient in corn and wheat. From 1940 to 1955 insecticide imports rose from 432 to 30,526 metric tons, and chemical fertilizer use soared from 12,000 tons in 1950 to 379,000 tons in 1967. Whether or not the new technologies were the primary cause, during this period Mexico's agricultural output increased (4.6% per year) faster than the population (3.3% per year).[1]

The Green Revolution contributed to Mexico's "Economic Miracle." Farmers produced food for industrial workers and the un- or underemployed in growing cities, allowing others to grow export crops to pay for imported machinery. Presidents Adolfo Ruiz Cortines (1952–1958), Adolfo López Mateos (1958–1964), and Gustavo Díaz Ordaz (1964–1970) increased funding for irrigation projects, price supports, rural road construction, and research. The SAF, agricultural banks, and parastatals like *Guanos y Fertilizantes de Mexico* and PRONASE sold chemical fertilizers, pesticides, and hybrid seeds. Mexico's failed cultural revolution left a profoundly Catholic nation, and the Green Revolution averted food shortages, helping the State avoid the issue of birth control. Combined with land reform, which destroyed "feudal" obstacles to modernity, the new technologies modified the *agrónomos'* role in Mexico's revolutionary process and became their contribution to the Economic Miracle, the "specialized knowledge" that made them credible scientific professionals, and a success story that put a developing nation on the cutting edge of international science.[2]

Notwithstanding the legacy of revolution, Mexico joined the free world camp and contributed to the fight against Communism (and U.S. foreign policy) by acting as the base to extend the Green Revolution to other developing nations. During the 1950s and 1960s the Rockefeller Foundation began programs like the MAP in Latin America, Asia, and finally Africa, helping to contain Communism and advance U.S. economic interests by selling chemical fertilizers and pesticides. By the late 1960s the Green Revolution was a media event and an international triumph of U.S. science, but its shortcomings also became obvious when critics identified the economic and environmental problems that it caused. In 1970 Norman Borlaug won the Nobel Peace Prize for his hybrid wheat, but he received it in a tempest of controversy.

INTERNATIONALIZING AND PROFESSIONALIZING *AGRONOMÍA:* MEXICO'S GREEN REVOLUTION, THE ECONOMIC MIRACLE, AND THE COLD WAR

Some Mexican farmers fully or partly adopted the Green Revolution, and others rejected it, but it did attract much attention. Attendance at MAP and SAF demonstrations grew from tens to hundreds to 1,000 to 2,500 per meeting as the Green Revolution spread. Most were private farmers, but *ejidatarios* also listened to the *técnicos*. Commercial growers invited MAP scientists to their meetings, praised the government's agricultural policy, and called plant breeding the first example of its success. Alfonso Castillo, president of the *Confederación Nacional de Pequeños Propietarios*, often urged Ruiz Cortines to expand the reach of technical assistance, including to *ejidatarios*. In 1956 *rancheros* in Yerbanis, Durango, requested and received SAF help to try MAP maize.

Sonoran wheat growers supported Borlaug's research and offered funds to improve the El Yaqui station. Impressed by the yields of new dwarf wheats introduced in 1963, even skeptical private farmers planted SAF seeds. Organizations such as the *Asociación de Agricultures del Bolsón Mapimi*, the *Asociación de Agricultores del Rio Culiacán*, and the *Asociación Agrícola Local de Cafeticultores* of Bienvenido de Galeana, Puebla, praised the government's technical support. *Campesino* groups like the *Comité Regional Campesino* of Calkiai, Campeche, and the wheat, corn, and bean growers of Badiraguato, Sinaloa, asked the SAF to send *agrónomos* to teach them better farming practices. In 1962 farmers urged the agency to keep the Llera Station open.[3]

Like the private farmers who profited from the Green Revolution, some *ejidatarios* tried or adopted new seeds and agrochemicals. The Corn Commission occasionally produced hybrid seed in *ejidos*. In Tlaxcala farmers from the *Liga de Comunidades Agrarias* stopped growing barley for a local brewery because of low prices and grew "Supremo 211" and "Kentana" wheat successfully. In 1958 *ejidatarios* in Campeche doubled their yields by adopting one of the Corn Commission's varieties. In 1962 seven *ejidos* in Guerrero asked López Matoes to found a nursery to breed and distribute hybrid maize. Under *agrónomos'* supervision *campesinos* in Ameca, Jalisco and Tlalnepantla, Mexico State, grew hybrid corn and wheat successfully, and a Puebla farmer told López Mateos about his "magnificent harvest." In 1953 the *Asociación de Pequeños Propietarios Agrícolas* of Ramón Sánchez, Zacatecas, asked the SAF to send an *agrónomo* to supervise fertilizer applications. In 1954 the *ejidatarios* of Los Morales, Mexico State, reaped bountiful harvests of corn, beans, and wheat using *Guanomex* fertilizer provided by the BNCE. The *campesinos* of Hueyapan obtained higher yields from their fruit trees with regular applications and built a road to reach other markets. The independent *Campesinos Mexicanos Confederados* (Confederated Mexican Peasants) asked López Mateos to tell *Guanos y Fertilizantes* to hire more *agrónomos* to help peasants use chemical fertilizers. In 1960 *ejidatarios* in Chiapas and Veracruz and the farmers of Pabellón asked him to help them obtain agrochemicals, claiming that they were expensive and scarce. The *Asociación de Defensa Agrícola* of Papatlazolco, Puebla, wanted to try chemical fertilizers on their *milpas*, and some peasants in the State began to use them regularly. In 1963 *campesinos* and *pequeños agri-*

cultores (small agriculturalists) in Jalisco complained that *Guanos y Fertilizantes* did not satisfy their demand for ammonium sulfate. Peasants asked the SAF to teach them to use pesticides like BHC. *Campesinos* in Las Margaritas, Oaxaca, sprayed DDT on their *milpas* of *criollo* maize and began to use chemical fertilizers in the mid-1950s, when the BNCE forced loan recipients to buy them. Migrants to the Nuevo Cosoltepec colony in the Papaloapan basin prospered in the early 1960s, when they planted selected sesame and corn seed and used herbicides recommended by *agrónomos* at the SAF's Cotaxtla Station. Commercial and export crops like cotton and winter vegetables required heavy applications of fertilizers and pesticides, and bank financing or illegal renters of *ejidal parcelas* provided them. Supply shortages and the *campesinos'* poverty were the most significant factors limiting agrochemical use, not resistance to new technologies.[4]

It took time, water restrictions, and higher fixed prices for wheat to overcome farmers' reluctance to accept *agrónomos'* expertise, but the Green Revolution had the greatest impact in Sonora's irrigation districts. The new technologies made Hermosillo boom and fostered strong economic and cultural ties with the United States According to an *agrónomo*'s wife:

> If you fly from Tucson, Arizona to Hermosillo, what you are going to see is a type of agriculture that makes you rich, so rich. You will see homes like you've never seen in Mexico City, swimming pools and everything. They have such a lot of money. Why do you go to Europe, they ask? Tucson is much better. Tucson is new. Tucson is the capital of Hermosillo. The Tucson stores advertise in the local papers. They say, "this month you need give just 10 pesos for a dollar instead of 12." The ladies of these rich Mexican farmers like to save, so they get together and form a club and once a month they go to Tucson. Some saving![5]

The SAF and *agrónomos* working for agrochemical vendors gave organized private farmers effective technical assistance, and they used their purchasing power to buy insecticides, fertilizers, and other inputs at low cost, but *ejidatarios* often received obsolete varieties vulnerable to *chahuixtle*, impure seed, and no, expensive, or adulterated agrochemicals from the BNCE and PRONASE, as well as dictatorial treatment and little or bad advice from the *técnicos*.

During the 1960s Sonora's private farmers exported hybrid seed to India, Turkey, and Pakistan. The Green Revolution made *agrónomo* Luis Arturo Romo, a friend of Marte Gómez, rich when he bought irrigated land and started an International Harvester dealership. The *campesinos* of Quechuhueca, Sonora, and some *ejidos* in Sinaloa's Fuerte Valley used organizations created by *cardenistas* to get assistance and inputs to grow hybrid wheat, cotton, and sugar cane profitably, but most *ejidatarios* incurred mountains of debt and began to rent their *parcelas* illegally.[6] In the northwest the Green Revolution attained many of the MAP's architects' original objectives, but the modernization of Sonoran agriculture also demonstrates the abandonment of the wartime goal of using science to improve the lives of peasant farmers.

Many other farmers selectively adopted or rejected the Green Revolution. Some of San José de Gracia's *rancheros* used chemical fertilizers, and more bought modern farm implements, but the new seeds had little impact. Few *ejidos* got enough credit or technical support to use the technologies effectively. *Ejidatarios* in Temascalcingo rejected hybrid wheat and agrochemicals after they performed worse than traditional strains because it was not bred for the region. The *campesinos* of Las Margaritas sometimes mixed hybrid and *criollo* seed, but they preferred their own varieties because the ears had hairs that helped plants resist rot and feeding birds, as well as a larger volume that buyers found more attractive. They also favored organic fertilizers. By the mid-1960s only 7 of 100 peasant households in Ericuaro, Michoacán, supported themselves by farming; the rest lived off wages earned in a local packing plant or in the United States. In 1970 the *ejidatarios* and *pequeños propietarios* of Durango's Los Llanos region did not use improved seeds or fertilizers. Only a few *ejidatarios* in El Mangal, Veracruz, cooperated with the BNCE's campaign to introduce cotton, and those who did abandoned the crop after the small pumps that the bank supplied failed to raise water for the plants and insects ate the survivors. During the 1950s and 1960s *campesinos* who moved to agricultural colonies in humid tropical regions like the Papaloapan basin and the Candelaria River in Campeche struggled when Green Revolution technologies developed for temperate conditions failed in a very different environment. Many colonists could not afford agrochemicals, and pests consumed their crops. A legacy of the World War II *Hevea* program, diseases destroyed rubber plantations in Papaloapan, adding to tensions between the project directors and the *campesinos*.

Although he admitted that it took some time, Borlaug believed that the new seeds made farmers accept *agrónomos'* expertise. However, when Kenneth Wernimont worked at the U.S. Embassy during the 1950s, many farmers asked him to get seeds from the MAP because they did not trust the Corn Commission. *Ejidatarios* in Guanajuato felt that *agrónomos* preferred to work with private farmers. Many *campesinos* in Durango viewed an SAF campaign to promote sunflowers with skepticism. *Tierra* claimed that aside from a few progressive farmers, the *campesinos'* lack of confidence in new methods held back the modernization of agriculture, despite recent advances in research. Durán said that *ejidatarios* did not trust the BNCE. In an extreme example, *campesinos* shot *agrónomo* Guillermo Aguirre Morales when he tried to stop illegal logging near Guadalajara. In 1966, when *técnicos* began to relocate peasants in the Grijalva River Valley to build a large irrigation project, the *ejidatarios* of Santana attacked them. According to Armando Bartra, they had "no confidence in the actions of the government" and felt that "the Revolution always has defrauded us."[7] The Green Revolution gave some farmers' faith in *agrónomos* and the SAF, but many others still distrusted or disliked them.

The Green Revolution did not silence the *agrónomos'* enemies. The Right still called land reform an obstacle to economic development, and some leftists attacked the SAF and the MAP's approach to modernizing agriculture. In 1953 CNC leaders joined critics who claimed that *agrónomos'* research did not help poor farmers because they preferred working with the affluent and called them incompetent scientists whose jobs wasted government funds. In 1954 Enrique Navarro Palácios said that many of SAF branches "have no useful function" and provided jobs only for middle-class bureaucrats, and others called *agrónomos "doctores"* and *"licenciados"* whose knowledge reflected "the city applied to the land" and who could "learn from the experience, environment, and people who live there and suffer its problems."[8] Believing that *agrónomos* could not improve agriculture if they joined the bureaucracy right after graduation, Ignacio E. Zepeda and the *Asociación Nacional de Cosecheros* proposed a period of mandatory social service in rural areas. Critics urged the State to do more to help *campesinos* and *ejidatarios*, and a few *agrónomos* joined them.

The MAP's scientists knew that some *agrónomos* opposed their work. Many IIA researchers believed that Alemán favored the MAP and resented the lavish praise that it received, and others who conducted research before 1943 like Damién Correu felt demeaned when younger colleagues claimed that Mexican research began when the U.S. scientists arrived. The MAP, however, hired over 100 *técnicos* in 1951 and 1952, and its scientists produced new technologies faster than the IIA, breeding more hybrid corns and wheats and working on sorghum, soybeans, potatoes, soil science, pest and disease control, and other projects. Reflecting concern about *campesinos'* diet and poverty, in 1951 the MAP hired Ralph W. Richardson to run the *Huerto Familiar* (Home Vegetable Garden) program. To develop tropical agriculture, MAP plant breeders assisted two experimentation stations established by the Papaloapan River Basin Commission. By 1954 they created high-yield corn and rice varieties. To enhance *agrónomos'* credibility, the MAP gave them prominent roles in demonstrations of the new seeds. Analyzing the politics of Mexican agriculture, worried about the appearance of "favoritism" toward "the more intelligent large farmers" who "obviously . . . can help the experimental program more than the small growers," Stakman told the MAP's scientists to maintain "balance between their cooperative efforts with big agricultural operators and with the less favored groups . . . [and avoid] entangling alliances with certain industrial individuals and groups . . . especially those involved in politico-economic problems."[9] With jobs, new seeds, professional support, and wise politics, the MAP won the support of influential *agrónomos* such as Pascual Gutíerrez Roldán and Rodolfo Acosta of the Corn Commission, minister of agriculture Ricardo Flores Muñoz, and Ruiz Cortines.

Intransigent leftist Mesa A. was president of the Soviet–Mexican Cultural Relations Institute, visited China in 1950, and attended a symposium in Peking in 1955, but along with other forces disseminating U.S. culture in Mexico after World War II, the MAP gave *agronomía* a decidedly U.S. flavor. *Agrónomos* admitted that the United States supplanted Europe as their source of scientific inspiration. The ENA used English language texts and kept "Henry

Wallace" turkeys, named in honor of the American secretary of agriculture of the 1930s. Influential *agrónomos* like Gonzalo Robles, Fernández y Fernández, and Sánchez Colín, foundation fellows, and *Tierra*'s editors praised the MAP and its hybrids, attended U.S. universities, and worshiped U.S. science. For example, Emilio López Zamora and Alfonso Díaz del Pino told all farmers to grow hybrid corn, and José Luís de la Loma said that plant breeding was the first step in any project to increase farm productivity. Gilberto Mendoza Vargas praised the ENA's experimentation station, run by U.S. scientists, for teaching young *agrónomos* to conduct research. A few *agrónomos* tried to raise public awareness of the problems caused by soil erosion or advocated organic fertilizers, and the SAF began several biological pest control campaigns, some in collaboration with the USDA, but most supported the use of agrochemicals. Attending prestigious universities and meeting internationally famous scientists gave foundation fellows a positive impression of the United States.[10]

During the 1950s ENA students took pride in their professors' publications such as de la Loma's *Experimentación agrícola*. The press praised the MAP and *agrónomos'* contributions to the school's success. Guadalajara's daily *El Informador* claimed that the new seeds put Mexico on the front line in the fight against world hunger. *Agrónomos* proudly claimed that the acceptance of the Green Revolution proved that farmers "continue to acquire confidence in technical experts' advice."[11] In 1952 López Zamora said that Mexican plant breeders could develop better cotton varieties without foreign assistance, and he urged farmers to run their own experimentation stations to protect research from the ups and downs of SAF politics. By themselves or with the MAP's help, by the 1960s *agrónomos* developed new cacao, barley, sorghum, potato, sesame, and rice strains. Trained by the MAP's scientists and foundation fellowships, the *agrónomos* possessed a professional language that only specialists could comprehend. In 1954 Julián Rodríguez Adame told ENA students that "agriculture is a technical and complicated activity that requires the effective use of diverse knowledge, and that is why each day the farmer needs the advice of the agronomist and the specialist more"; the *Colegio de Ingenieros Agrónomos*, which promoted scientific professionalism, proclaimed that "*agrónomos* are complete agronomists . . . who work at a high scientific and technical level and represent

Mexico proudly among professionals from any part of the world," and de la Loma asserted that with many skilled plant breeders, "it would be an error to base projects to improve any aspect of our nation's agriculture on the results of foreign research."[12] Like Mexico's irrigation specialists, *agrónomos* were proud of their contributions to international science. Costa Rica's Inter-American Institute of Agricultural Sciences hired a Mexican Foundation fellow who studied corn breeding in the United States; the ENA and Monterrey's *Instituto Tecnológico* trained Central and South American students; *agrónomo* Efrain Hernández Xolocotzi praised the Mexican delegation's "important role" and "sound presentations" at a Latin American scientific congress, which enhanced "Mexico's social and cultural prestige in South America"; and Fernández y Fernández and Acosta said that science "had no frontiers," deeming scientific cooperation "not contrary to our nation's traditional zeal to protect its sovereignty."[13]

During the Green Revolution a career in agronomy attracted many young Mexicans. In 1959 the ENA founded the *Colegio de Postgraduados* to offer Masters degrees. By 1960 the MAP had employed 550 *agrónomos* and awarded 155 fellowships to Mexican students, and 52 earned Ph.D.s at U.S. universities. In 1954 about half the *agrónomos* who responded to a questionnaire came from large cities, one-quarter from "county seats" (*cabeceras municipales*), and one-quarter from "rural population centers" (*poblados rurales*). A few were children of *ejidatarios*, and even fewer were Native Americans, but *campesinos* who became *agrónomos* wanted to work in cities. Most came from middle-class families, and their fathers had jobs like customs inspectors and college professors. Many chose the career because local scholarships to agricultural schools offered their only chance to obtain professional training. During the 1950s the growth of the agrobureaucracy pleased the *agrónomos*, who rarely had to work outside the profession. Several thousand had government jobs, and they used their scientific successes to demand pay increases and more employment opportunities. New or larger agricultural parastatals created more jobs. In 1953 Ruiz Cortines chose de Dios Bojórquez to manage the flour and tortilla manufacturer MINSA (*Maiz Industrializada, S.A.*), resurrecting his career in government service; *Guanos y Fertilizantes* hired Gonzalo González H.; and PRONASE employed Daniel Rios Alcocer, who helped *zapatistas* in Morelos get *ejidos* during the

1920s. Some *agrónomos* moved to the private sector to take advantage of the Economic Miracle. Juan Ballesteros Acevedo bought a hotel and became president of the *Asociación Mexicana de Hoteles.* Orange promoter Genero Arzave founded *Química del Golfo,* and the spread of agrochemical farming made him a wealthy man. Marte Gómez was president of *Worthington de México,* a U.S. pump manufacturer, becoming a *prestanombre* (name-lender) to circumvent a law limiting foreign ownership of companies to 49%. The socialist López Zamora became a capitalist, selling motors and drilling machines.[14]

Reflecting the transition from *agraristas* to *técnicos,* some older *agrónomos* devoted their careers to land reform, and some worked in collective *ejidos,* like Francisco García Robledo, *Asesor Técnico* for the *Ingenio Los Mochis* in Sinaloa, but by the mid-1950s most, especially the young, avoided the CNA, which offered lower salaries than the SAF and jobs with "technical simplicity that does not attract professionals."[15] In the 1960s the agency could not find enough *agrónomos* to fill its posts. *Agrónomos* used *agrarista* rhetoric and discussed the *ejidos'* problems, but they refused to challenge politically powerful landowners who circumvented the Agrarian Codes. In 1953 Rodríguez Adame said that few farms were eligible for expropriation and advocated colonization projects to placate landless *campesinos.* In 1961 Edmundo Flores wanted to end land reform quickly. To preserve their alliance with a State that used *agrarismo* to claim legitimacy and validate their professional history, most tried to reconcile the Mexican and Green Revolutions. To find middle ground between rightists who wanted to eradicate the *ejido* and ardent *agraristas* who advocated expropriating all remaining large farms, including those owned by powerful Revolutionary Family members, *agrónomo* Manuel Marcue Pardiñas said that science was the only solution to the agrarian problem. José López Bermúdez proclaimed that "*agrónomos* know that if *ejidatarios' parcelas* cannot be enlarged, there is a new way to save them; to make production grow vertically . . . with higher yields, [then] . . . the *ejido* or *pequeña propiedad* will provide its own well-being and bread for all Mexicans."[16] Critics who said that *agrónomos* did not know how to make farmers more productive annoyed Luís L. León, who saw the MAP's seeds and training as the last phase of a rural transformation begun by the Mexican Revolution, in which his profession played, and should continue to play, a crucial role. He said that "the worst thing we could do now is agitate in rural areas and gravely wound production that is increasing and becoming better organized. . . . *agrónomos* should collaborate . . . , not with demagogic political or doctrinaire agitation, but rather with scientific and technical advice, by going to the fields to study and find concrete solutions to the problems of each *ejido* or region."[17]

In 1959 León and new minister of agriculture Rodríguez Adame urged ENA students to "make the goals consecrated in our agrarian reform a reality . . . [as] magnificent researchers, *técnicos* and farmers . . . [with] training to promote the scientific development of agriculture . . . [not as] politicians or agrarians."[18] *Agrónomos* often sounded like conservatives such as the National Confederation of Chambers of Commerce, who hired old Porfirian Julio Riquelme Inda and claimed that the only way to raise the *campesinos'* income was "to increase yields per hectare."[19]

In the 1950s *proyectismo* (development), interagency rivalry, job turnover, and changing bureaucratic goals continued to harm agricultural programs. Along with their collaboration with the MAP's scientists, *agrónomos'* class and professional prejudices influenced their projects. *Agronomos'* rigid control of colonists' farming practices in the Papaloapan basin led Thomas Poleman to compare the latter to plantation laborers. They asserted professional authority without studying local conditions and made many decisions that had ruinous consequences. De la Loma stressed the importance of field plot techniques in plant breeding but claimed that "capable workers under the constant supervision of the experiment's director" could use them.[20] Students did fieldwork in school, but many *agrónomos* saw working with their hands in fields alongside *campesinos* as demeaning. López Zamora said that BNCE *agrónomos* in La Laguna had "no interest in directing the *campesinos'* labors," and David Barkin found that few wanted to work in rural areas.[21] Many *agrónomos* criticized farmers who rejected the Green Revolution and often attributed this to character defects. Díaz del Pino lamented that some farmers had no faith in the MAP's hybrids. To Moises T. de la Peña, "ignorance and resistance" impeded the use of fungicides and insecticides on cacao plantations. López Zamora wanted *ejidatarios* to work collectively to introduce modern farm technologies, and he attributed all resistance to "cultural backwardness, especially among Native

Americans." Veteran *agrarista* Manuel Fitzmaurice saw an urgent need to "create a sense of responsibility in *ejidatarios*. . . . [because its deficiency] caused all the failures of . . . agricultural programs."[22] *Agrónomos* understood the logic of subsistence farming, but they still criticized corn-bean intercropping. *Tierra* lamented that "hundreds of thousands of *ejidatarios* and *pequēnos propietarios* almost completely lack the most basic knowledge of agronomy and only follow routine," and Fernández y Fernández said that "government paternalism" kept *ejidatarios* from developing a "mentality of impresarios with aspirations and initiative."[23]

The turmoil in the SAF subsided, but *agrónomos* still worried about their profession. The *Colegio de Ingenieros* fretted when critics said that they forgot the *campesinos*, admitted that some chose the career "more from opportunism than vocation and true interest in the field," and refuted those calling them "*bureaucratizada*" (committed to government employment) but confessed "without conceding the point" that the 407 who lived in Mexico City might be. One told an interviewer that ENA students often enrolled "just to seek government employment."[24] Despite their scientific accomplishments, many *agrónomos* fretted about their prestige. De la Loma urged students to use caution when telling farmers to change their practices, calling bad advice based on preliminary findings "dangerous . . . [because it] compromise[d] the credibility of agricultural science and scientists among the *campesinos*," and Blanco Macias urged *agrónomos* to wait for "solid results . . . because whatever advance that is followed by steps backward erodes farmers and *campesinos*' faith in scientific agriculture."[25] *Agrónomos* complained about low pay, lack of influence on agricultural policy, and the SAF's practice of hiring people without degrees. Marcué Pardiñas and Arturo García Campa felt that the SAM's political activities gave the public a negative impression, and Fábila lamented that some people saw *agrónomos* as a "plague upon agriculture."[26] Reflecting the legacy of *agrarismo*, *agrónomos* such as Mendoza Vargas believed that the SAF should solve economic and social problems and told their colleagues not to forget the *ejidos*. Fernández y Fernández worried about the population explosion and dared to urge his colleagues to promote contraception among the rural poor. Mesa A. lamented that although new technologies modernized farming in Sonora, *campesinos* lived like peons on nineteenth-century haciendas, and he chastised *agróno-*

mos who became cotton growers in the Rio Grande Valley for paying workers such low wages that soldiers had to watch them to prevent flight to the United States. In 1957 Felipe Salcedo Pérez urged the SAF to devote more effort to *de temporal* corn and bean farming, and *agrónomos* in the northwest asked Ruiz Cortines to let them work with the CNC to bring the Green Revolution to the *campesinos*. In 1960 José García y García of the CNA told López Mateos that the SAF's research failed to alter many farmers' "old, spent, useless practices," while agrochemicals exacerbated the *campesinos'* poverty. Durán praised the MAP's "important results" but called the work with corn a failure "despite scientific research."[27]

The SAF tried to mollify the critics. The MAP's scientists still bred hybrids, but in 1956 they agreed to devote some effort to open-pollinated strains that did not require the purchase of new seed every year to reach their potential. In 1957 the SAF admitted that its projects targeted irrigated lands and debuted a project to improve *de temporal* farming. In 1960 ENA scientists isolated a trait for drought resistance and crossed it with other corns to develop hybrids for rain-fed conditions. Responding to IIA director Taboada's complaints about funding, governors' demands for experimentation stations, and the MAP's privileged position and unwillingness to communicate, as well as his claim that peasant farmers would not use hybrids, Rodríguez Adame merged them, creating the *Instituto Nacional de Investigaciones Agrícolas* (National Agricultural Research Institute, hereafter INIA). Reflecting his colleagues' disdain for working with peasants in rural areas, he gave much of the responsibility for extension work to show the *campesinos'* Green Revolution technologies to the SEP, which organized teachers, *agrónomos*, and veterinarians in *Brigadas de Promoción Agropecuarias* (Agricultural Promotion Brigades). The CNI did extension work in the irrigation districts.[28]

These reforms did not bring peasant farmers a Green Revolution. Agricultural programs suffered from old problems like interagency rivalry. If farmers cooperated, bureaucrats' enrichment rather than science sometimes dictated what seeds they planted. When they worked with *campesinos*, *agrónomos* still tried to transform them or assert dictatorial control over their actions and culture. In 1961 Leobardo Jiménez S. did extension work in the *ejido* El Mangal, Veracruz. He began to admire the *ejidatarios'* farming practices, urged his colleagues to listen to *campesinos* to find the best solutions to their problems, and claimed that "the technology that is adapted to the conditions of the *ejidatario* [will be] most practical and efficient," but when he organized women to give them a voice in community development, he alienated many men, especially older males, and they refused to cooperate.[29] In 1963 *agrónomo* Atanasio Cuevas R. advised the *campesinos* of Salinas Victoria, Nuevo León, but the experience convinced him that bureaucrats and city people did not understand rural Mexico's problems. Denying that they preferred working with affluent and educated farmers, some *agrónomos* claimed that their ranks were too small to reach every *campesino* and urged the government to train and hire more *técnicos*. Díaz del Pino wanted to centralize all programs for *campesinos* in a single agency and, of course, to hire more of his colleagues to run them. The INIA bred open-pollinated corn, but Rodríguez Adame appreciated the high productivity of hybrids and supported their development. He chose ex-foundation fellow Rodolfo Peregrina of the MAP to run the INIA and two others as assistant directors and preserved the SAF's ties to U.S. plant breeders. In 1965 Díaz Ordaz announced a "fight to reduce corn growing areas to dedicate them to more remunerative crops," and the SAF planned to urge Mexicans to eat more wheat to consume the surplus that the new hybrids created.[30]

During the 1950s pundits fretted about rapid population growth and warned of impending famines. Along with distinguished scientists, Protestant and Catholic clergy, and powerful people like John D. Rockefeller III, the FAO labeled rural poverty and hunger as threats to world peace and pleaded for international scientific cooperation to eradicate them but worried that "ignorance, greed, strife, superstition, and blind adherence to tradition" would hinder the spread of new farm technologies in the Third World.[31] Proud of their contributions to the high productivity of U.S. farmers, scientists were ready for the task. Some believed that atomic energy could increase the world food supply dramatically by fostering rapid advances in genetics like giant "super corn" or plants that grew in the Arctic. In the short term, they wanted to introduce "modern agricultural techniques in the backward areas" to transform peasant farming into "a better system."[32] U.S. agribusinesses claimed that their technologies would feed the growing world population. Some scientists and activists raised the issue of birth control, such as physicist Sir Charles Darwin, grandson of the famous biologist, who said that "however much food is produced, there will always be too many mouths to feed," and Sir John Russell, who worried about Latin America's "Communist element" and advocated "a sound population policy . . . to ensure that the numbers do not outstrip the food resources."[33] Supported by advisers who saw food as a key weapon in the struggle between the free world and "International Communism" and bolstered by the Marshall Plan's success in rebuilding European agriculture and U.S. farmers' bumper crops, in June 1950 Truman began the Point IV Program, sending experts to the Third World and training its scientists as part of his containment policy. President Dwight D. Eisenhower recognized that increasing food production in the developing nations served U.S. security and economic interests and sidestepped the controversial issue of birth control. By 1961, 6,000 U.S. experts worked on Point IV projects in 58 countries.[34]

Abandoning the postwar goal of U.S.–Soviet cooperation, the Rockefeller Foundation wanted to use the MAP to confront the population explosion and the Communist menace. Favorable publicity in the Mexican, Latin American, and U.S. press pleased the foundation. The officers knew that many farmers supported the MAP, that new seeds sold U.S. farm machinery and agrochemicals, that training Latin American students made allies, and that scientific publications impressed the region's academics. To refute Malthusians and aid cold warriors, in 1950 Stakman told the Advisory Committee on Agriculture that "the western hemisphere may become the world's last refuge against Communism . . . [making it] of tremendous importance for the United States to demonstrate that the democratic system is capable of assisting backward people in raising their own standards of living." In 1951 the committee decided that "population pressures . . . [might cause] extensive famine . . . [creating] underfed . . . susceptible people . . . [and] world tensions" and presented the MAP as a way to avert Third World hunger, poverty, and "the reckless instability of people who have nothing to lose and perhaps something to gain by embracing political ideologies designed not to create individual freedom but to destroy it."[35]

Bolstered by the MAP's achievements and popularity, during the 1950s the foundation used Mexico as a base to extend the

Green Revolution in the Third World. The foundation's program in Colombia began in 1950 and trained scientists from Ecuador, Peru, Central America, and Chile. By 1952 Indian and Israeli farmers grew MAP wheat and corn. The directors of the Point IV Program and the State Department decided that the MAP was an excellent model for Third World agricultural assistance programs, and in 1956 the foundation began one in India. In 1957 the National Planning Association called the MAP an example of what private organizations could accomplish by providing technical assistance to Latin American agriculture. With Harrar's assessment that "rice fills the major food requirements for about half the world's population, yet there is less scientific knowledge about this cereal than any other major crop," in 1960 the foundation founded the International Rice Research Institute (IRRI) to bring the Green Revolution to East Asia.[36] He felt that U.S. scientists could handle the "immense challenge" of the population explosion but called it "meaningless" to believe that increasing Third World food production would end rural poverty and hunger without "intensive and extensive efforts in related disciplines" such as economics, education, and social and political reform, and Bradfield said that birth control was "wise" but also claimed that "we can delay the evil day when population will greatly exceed the food supply for some time by improving agriculture."[37]

Catholic, socialist, or pragmatist, *agrónomos* rejected birth control when Fernández y Fernández raised the issue in 1954. Mesa A. believed that more equitable wealth distribution would end the masses' poverty. Panfilo Rodríguez León said that contraception was contrary to Mexican culture and claimed that science and the government's economic program would bring prosperity to the poor. Blanco Macias, who studied plant breeding at Cornell during World War II and founded the soil conservation society *Amigos de la Tierra* (Friends of the Land) in 1951, was thoroughly satisfied with the MAP because it fed "the growing numbers added each year to the nation's dinner table." He praised its work on corn and family vegetable gardens, claimed that its 80 *agrónomos* worked "in complete harmony" with U.S. scientists, and thanked the SAF's directors for supporting it. In 1961 Flores used the Economic Miracle to refute critics of land reform, while Fernández y Fernández and Acosta claimed that agricultural research began to function effectively in 1943 and had already produced "truly astonishing" results.[38]

THE *AGRÓNOMOS* AND THE INTERNATIONAL GREEN REVOLUTION: SHINING AND BURNING IN THE SPOTLIGHT OF HIGH SCIENCE

In 1960 Eisenhower reacted to the popularity of Fidel Castro's revolution by expanding technical assistance and other programs in Latin America. To fight Communism in the Third World, in 1961 President John F. Kennedy founded the Agency for International Development (AID) and the Alliance for Progress to provide more aid. His "best and brightest," like Secretary of Agriculture Orville Freeman, proposed "an international contest, similar to the space race," and South Dakota Senator George McGovern understood the political benefits of exporting U.S. agricultural technology and recognized that it helped avoid the issue of birth control. To confront pessimists like FAO director Binay Rangan Sen, who lamented the prevalence of Third World hunger, and Raymond Ewell of the State University of New York at Buffalo, who forecast widespread famines, distinguished individuals like Sir Norman Wright, secretary of the British Association for the Advancement of Science, and scientific farmer Jonathan Garst, author of *No Need for Hunger* (1963), continued to make optimistic predictions about scientists' ability to feed the world's rapidly growing population. In 1965 President Lyndon B. Johnson's assistant secretary of agriculture Dorothy Jackson told the FAO that the solution to the world hunger problem lay "within our reach."[39]

In 1966 the threat of famine in India shattered the optimism. To avert disaster, the United States shipped surplus grain. Pointing to the urgent need for "a revolution in agriculture," the FAO warned of "disaster of an unprecedented magnitude" in 35 years unless leaders took "all possible measures" to "raise food production and halt the population explosion." William H. Draper, chairman of the Population Crisis Committee, proclaimed that "only birth control on a massive scale . . . [and] rapidly increased food production in the developing countries, can prevent the greatest worldwide catastrophe of modern times"; the *New York Times* agreed; and Ewell told Congress that "the food-population problem will in a few years dwarf and overshadow such political problems as Vietnam, Cuba, and the Congo."[40] In the Paddock brothers' 1967 book *Famine 1975*, William, an agricultural scientist, and Paul, a diplomat, said that the population explosion would cause catastrophic food shortages in many Third World nations. Influenced by biologist Garret Hardin's "Lifeboat Ethic," they urged the United States to aid nations that sold useful products, like Bolivia's tin, and let others that did not, like Paraguay, suffer famines because their politics and culture doomed agricultural assistance and birth control programs to failure. To avoid or postpone such politically explosive choices, Johnson and his advisers wanted to modify food aid policy under Public Law #480 to foster "greatly increased emphasis on international cooperation directed toward encouraging and assisting hungry nations to increase their own agricultural production and their ability to process and distribute food."[41] Echoing the FAO and prominent scientists who described "future Vietnams," in 1967 H.L. Bennett of the president's Panel on the Hunger Problem said that "to avoid a threat to the peace of the world as well as to our national security, we cannot afford to be too little and too late with our developmental assistance."[42]

Scientists again accepted the challenge to avert mass starvation. In 1966 Frederick Seitz, president of the National Academy of Sciences, discussed the MAP's achievements in Mexico at a symposium on world hunger and wished that "a similar pattern could be extended on a world wide scale with comparable success in the next quarter century."[43] To prevent famines that spark a World War III, Herbert J. Waters, head of AID's new War on Hunger Office, prominent scientists, and business leaders urged U.S. agribusiness to "encourage the expansion of private enterprise and private investment in the hungry nations."[44] With Johnson's support, U.S. capital formed the Agribusiness Council to solve Third World food problems. The Rockefeller Foundation's officers and scientists were pleased, claiming that the private sector could extend the Green Revolution more effectively than parastatals like PRONASE. William B. Murphy, president of Campbell's Soup, called the MAP "a classic example of what can be done by a well-directed research program."[45] Sterling Wortman, who bred plants for the MAP during the 1950s and in 1968 was the foundation's director of agricultural sciences, became vice president of the Agribusiness Council. To increase food production, Harrar told Third World leaders to have "a clear understanding . . . that agriculture is an industry and must be treated as such," and W. David Hopper advised them to "separate the goal of growth from the goals of social development and political participation . . . [because] their joint pursuit in unitary action programs is incompatible with the development of an effective strategy for abundance," thus admitting that the Green Revolution was not an anti-poverty program.[46]

By the mid-1960s the foundation thought that the MAP and the IRRI had made significant breakthroughs that would feed the hungry and advance U.S. foreign policy and economic interests in the Third World and presented their accomplishments to politicians, scientists, and the public. Borlaug acknowledged the eventual need for birth control, but in 1965 he told *Phytopathology* that "through a combination of improvements in conventional and non-conventional food production methods, man can feed the world's mushrooming human population for the next 100–200 years."[47] In 1966 the *President's Review* mentioned "agricultural revolutions," and the IRRI released its IR-8 hybrid, dubbed "miracle rice" by the press. Harrar told the *New York Times* that "it was not good enough to cure a man of yellow fever and then let him die of starvation" and told Congress that modernizing Third World agriculture with high-yield seeds and agrochemicals was a better foreign policy than food aid because it led to "social gains."[48] In 1967 Stakman, Mangelsdorf, and Bradfield published the official history of the MAP and the Green Revolution, *Campaigns against Hunger*. In a thinly veiled attack on *agrarismo*, which satisfied "the hunger of the landless for land, but was it satisfying their hunger for food?," they lauded Mexico's post-1940 agricultural "revolution" that gave "a better life" to "millions of human beings" and improved Mexicans' opinion of the United States.[49] Seeing no place for peasants in a modern society, its technologies "release many people no longer needed on farms for other needed services, as teachers, engineers, doctors, civil servants, and industrial workers."[50] Demonstrating their faith in social engineering, the authors boldly claimed that the MAP tried to "convert an unsophisticated society into a sophisticated one," by fostering "a long and happy step" "from back-breaking peasant farming to the intelligent business of farming."[51] Reflecting their view of the link between food, poverty and politics, of Colombia's "*La Violencia*" they wrote that "sunk in peasantry, they looked to political rather than scientific remedies to improve their lot."[52] Urging Asian nations to produce more rice, Borlaug pronounced that "what Mexico did, your country can also do, except that yours should do it in half the time."[53] Other foundation representatives made similar claims to Third World leaders, Congress, businessmen, and the U.S. foreign policy community. In 1968, to reach Africa, the foundation established the Institute for Tropical Agriculture in Nigeria.[54]

That year many First World leaders saw an agricultural revolution. Discussing IRRI rice, Freeman proclaimed that "the development of this seed is to agriculture what the steam engine was to industry."[55] Coining the name for the new technologies, AID director William Gaud told the Society for International Development that "record yields, harvests of incredible size and crops now in the ground demonstrate that throughout much of the developing world . . . we are on the verge of an agricultural revolution . . . I call it 'The Green Revolution.' "[56] The United States used "miracle rice" to fight the Vietnam War. Third World leaders praised the accomplishments of the foundation's scientists. At a celebration of the anniversary of the shah of Iran's inauguration, Ferdinand Marcos gave him a staff of "miracle rice." The *New York Times* called the Green Revolution a "Revolution that deserves the wholehearted support of even the most Conservative American."[57] The USDA's Lester Brown praised the "agricultural revolution" sweeping Asia but warned that it was "not the ultimate solution to the food-population problem" but bought only "some much needed additional time in which to mount effective family planning programs."[58] Similarly, some of the foundation's scientists felt that the media and politicians were too optimistic about the new seeds' potential. In 1968 the IRRI's Robert Chandler called IR-8 rice "a holding operation" in the struggle to contain the population explosion, and in 1969 he said that "there just isn't such a thing as a 'miracle rice.' No improved variety will solve the world food problem by itself."[59] When the foundation opened a research center in Colombia, David E. Bell told the press that he wanted to solve the *campesinos'* problems but refused to "promise any miracles."[60] Borlaug told a 1968 symposium, "I am hesitant to voice even cautious optimism, because complacency in any form is likely to prove disastrous. Unrestrained population increases are driving us into a corner from which it is difficult to emerge."[61] In 1969 Stakman proclaimed that science provides a way to produce more food and limit population growth "humanely and wisely."[62]

In 1969 President Richard Nixon's secretary of agriculture Clifford Hardin said that there was "no question about what the world's agricultural revolution is; it is a magnificent achievement . . . that offers new hope to the developing nations," and foundation researchers in Sonora claimed that a bumper harvest "should convince all skeptics that the Green Revolution has deep, strong roots."[63] Harrar urged Congress to use food aid to force developing nations to adopt the Green Revolution. He praised its socioeconomic impact: "The income of many populations is moving up substantially, enabling them to purchase a greater variety of foods and also to buy the products of urban based industry . . . [which] may also provide major and direct benefits to the developed nations, particularly in the expansion of cash markets for an ever broader range of industrial products."[64] The wave of optimism crested in 1970, when Borlaug won the Nobel Peace Prize. The Awards Committee said that he "made it possible for the developing countries to break away from hunger and poverty," and the FAO found "sufficient . . . evidence . . . to strongly suggest that a real technological breakthrough has been made."[65] The *New York Times* proclaimed that "from his work, many people in many countries will benefit in more abundant food for many years to come," and Boyd Orr told the press that "agriculture and engineering technology could eliminate hunger from the world within a decade."[66]

While the Green Revolution spread internationally, many *agrónomos* still criticized *campesinos* who rejected it. In 1964 PRONASE director Emilio Gutíerrez Roldán called "the cultural level of the people" the primary factor limiting the acceptance of the new seeds. Cuevas R. denounced the "impenetrable negligence, apathy, and inability to be sociable and cooperate" of Salinas Victoria's farmers, "making any plan, no matter how good, fail," traits that he attributed to their poor diet, lack of education, poverty, alcoholism, gambling, and "thievery."[67] To Díaz del Pino, misinformation made *campesinos* reluctant to try hybrid corn. In 1965 Díaz Ordaz praised *agrónomos* for attacking the "colonial practices" of most Mexican farmers. *Tierra* lamented that *campesinos* lacked education and funds to adopt new technologies and clung to their "antiquated methods" and claimed that their unwillingness to "follow *agrónomos'* instructions to increase their harvests" caused low yields in grain farming.[68]

Scientific achievements gave *agrónomos* ammunition to fight their enemies. During the 1960s they pointed with pride to Sonora's farmers, whose success with hybrid wheat reflected a "perfect union of technical experts and farmers," and *Agricultura Técnica en México* praised "our geneticists' creation of high yield maize hybrids and wheats resistant to *chahuixtle*," which "resolved . . . the urgent problem of feeding the people corn and wheat and our Government's need to import these grains each year."[69] *Tierra* lauded the new varieties' contributions to Mexico's Economic Miracle. Sugarcane experts helped the *ejidatarios* of Atencingo raise bumper crops, *agrónomos* at the Mexicali Valley experimentation station found ways to triple yields of cotton and wheat, and INIA researchers used their training and 1 million pesos to create new malt barleys that made Mexican breweries self-sufficient. López Zamora claimed that the prosperity achieved with better irrigation works, fertilizers, insecticides, and new crops in La Laguna restored the *campesinos'* faith in the government and its *técnicos*. In 1965 Minister of Agriculture Juan Gil Preciado proclaimed that the *agrónomos* had fulfilled their duty to the Revolution by using science and technology tobring prosperity to the *campesinos*. Other professionals praised the SAF's research. In 1968 foundation fellow *agrónomo* Oscar Brauer H. called his colleagues "researchers influenced by a developmentalist concept of agriculture and a humanistic philosophy who use science and technology to foster a more productive agriculture whose motivation is to improve the large *campesino* community."[70] *Agrónomos* basked in the light of international acclaim as Mexico exported disease-resistant potatoes to Costa Rica, Panama, Brazil, Colombia, Argentina, Haiti, and Thailand. Rodríguez Adame proclaimed, "We offer the people of Latin America, the new nations of Africa, and the world in general Mexico's agronomy, scientists, and seeds . . . to cooperate in the struggle against hunger and misery."[71]

The *agrónomos'* alliance with the State brought political rewards. Many *agrónomos* served as senators and federal deputies. In 1952 Sánchez Colín became governor of Mexico State. In 1965 Díaz Ordaz appointed Aaron Merino Fernández interim governor of

Puebla. In 1962 SAM's president, Senator José Rodríguez Elias, was the PRI gubernatorial candidate in Zacatecas. His successor, Alfonso Támez Mireles, acknowledged the *agrónomos'* debt to the State: "We are part of a generation benefitting from the social laws the Revolution produced."[72] For example, Jesús Garza Morales and his wife raised seven children in middle-class prosperity; two sons graduated from medical school, two were civil engineers, and one son and one daughter taught in normal schools; his other daughter studied languages and became a bilingual secretary. In 1964 López Mateos praised *agrónomos'* contributions to agricultural development. One *agrónomo* told an interviewer that he wanted to run for president. Wernimont claimed that the Green Revolution earned "entirely new recognition and respect" for *agrónomos,* but in 1968 Brauer H. lamented that "the opinions that other professionals and people in general have about [them] are quite low and poorly founded."[73] Nevertheless, by 1970 more than 7,000 had jobs related to agriculture in and outside the government, and they dominated the SAF and CNI. Economic reward and social status made agronomy an attractive career for young Mexicans.[74]

In 1964 French-trained *agrónomo* Jesús Uribe Ruiz denounced Mexico's *"intoxificación técnica"* (intoxication with technicality), the MAP, and its Green Revolution. Combining nationalism with professional pride, he compared Mexico to China during the Boxer Rebellion by claiming that the MAP "deformed and even worse de-Mexicanized agricultural research," ignoring Mexican farmers' needs to solve the problems of U.S. agriculture, like *chahuixtle.* The foundations's management of the ENA's experimentation station, "as if it was understood the school is incapable of doing it," insulted him. He told Mexican researchers not to follow "foreigners' " orders because it eroded their prestige. Calling the SAF "ineffective," he claimed that its scientists concentrated excessively on plant breeding and urged them to find other ways to raise productivity. Recognizing that peasants could not use hybrid corn, he claimed that the new seeds made rich farmers richer and poor ones poorer, joined critics who lamented the SAF's lack of interest in *de temporal* and tropical agriculture, and said that by promoting the Green Revolution, "each *campesino* becomes a living example of the mistakes of research and he denigrates it at every opportunity," eroding the profession's prestige. He lambasted critics who belittled *agrónomos'* expertise, and he called them "scientific dilettantes" who worried about "glory" rather than making "their knowledge serve the *campesinos.*" He said that jobs in "*la burocracía media mexicana*" (Mexican middle-class bureaucracy) co-opted and corrupted *agrónomos,* who were afraid to criticize the SAF. Before "appropriate technology" came into vogue, he claimed that his colleagues ignored the human aspect of rural development and wanted to create "simpler technologies" for peasant farmers. He thought that many peasant farming practices were primitive and admitted that *agrónomos* had problems working with *campesinos* because they gave orders like hacienda *mayordomos* supervising social inferiors, but he wanted to include them in the discourse of modernization: "One must hear, see, and understand the peasants to know their motivations, their opinions, their problems, and then translate this dialogue into technical language accessible to research workers, to find approaches that work in [their] language."[75] To train more *campesinos* as *agrónomos,* he urged the ENA to drop the *preparatorio* (preparatory) degree requirement because there were few of these schools in rural areas.

Like Uribe Ruiz, in 1964 *Tierra* praised *agrónomos* for increasing corn, bean, wheat, cotton, sugarcane, and coffee production, but noted that "these irrefutable triumphs were achieved in privileged regions with abundant water and credit and farmers with sufficient knowledge and initiative to use the latest advances in agricultural science. The rest of the country remains to be dealt with."[76] The editors even raised the issue of birth control. In 1967 Durán repeated this critique, admitting that the Green Revolution reached only more affluent farmers and bypassed the *campesinos.* He chastised the agrobureaucracy for ignoring the problems of peasant farmers, and he urged his colleagues to give the *campesinos* more control over agricultural development programs. He criticized the ENA for adopting a curriculum "almost exclusively oriented toward technical problems, detached from the social problems of organizing *campesinos.*"[77] That year, responding to an outburst of protests and land seizures in Puebla, the SAF launched an intensive program to bring the Green Revolution to the *campesinos* of the State (the *Plan Puebla*), but it failed. Nealticán's peasants preferred their traditional corn over the SAF's hybrid because the hybrid produced fewer kernels than their varieties, had a shorter stalk that provided less fodder for their livestock, and made tortillas with a taste and texture that they disliked. The hybrid was less resistant to corn worms, and they could not afford pesticides to control them. They did not trust the urban vendors of pesticides and hybrid seed. The hybrid's growing season required a uniform planting time, making it more vulnerable to late frosts than their traditional strains, a threat to their subsistence security. The project's directors could not find any improved variety that performed better than the local maize. The *agrónomos'* and the State's authoritarian treatment of the *campesinos*, such as forcing them to purchase crop insurance in groups, which they perceived as an insult to their autonomy, contributed to the program's failure. Agreeing with Adolfo Orive Alba of the CNI, *agrónomo* Carlos Manuel Castaños said that the INIA "did not give due attention to *de temporal* regions."[78]

By the late 1960s the Green Revolution was an international media event but also the target of vehement criticism. In 1968, as Díaz Ordaz prepared to show the world Mexico's progress by hosting the Olympic Games, the Economic Miracle unraveled. Some peasant producers held their own, like the *campesinos* of Las Margaritas, who attained some prosperity growing alfalfa, raising livestock, and selling cheese and milk, and the *ejidatarios* of Quechehueca and the Fuerte Valley, who adopted the Green Revolution successfully, but during the 1950s and 1960s unfavorable price trends, inflation, subsidies that favored urban consumers, and government neglect impoverished many *ejidos* and small farmers.[79] *Campesinos* who fled rural areas while the Green Revolution and the internationalization of Mexican agriculture proceeded did not get jobs as teachers, medical doctors, or even industrial workers as Stakman and his colleagues predicted; they struggled to find work and lived in growing shantytowns surrounding urban centers like Mexico City. The Revolutionary Family dominated politics, and the democracy that Francisco Madero sought in 1910 was as illusory as ever. When the army fired on student protesters in Tlatelolco, it was obvious that 20 years of reform followed by 25 of economic growth brought inequality and little social justice or democracy. The *agrónomos* and the SAF had to defend their role in the transformation of rural Mexico.

To many people, the Green Revolution was a triumph of international high science, but others pointed to problems that dimmed its luster. During the 1950s a few scientists worried that farmers adopted new pesticides so rapidly that they "outdistanced sound knowledge of the effect on human and animal health and on soil."[80] Rachel Carson's 1962 classic *Silent Spring* stimulated public interest in the environmental problems that pesticides caused. In 1966 Vogt told the National Audubon Society that the use of persistent pesticides in Latin America threatened migratory birds. By 1970 environmentalists blamed the Green Revolutionaries for encouraging Third World pesticide use. In 1963 scientists raised the issue of genetic erosion as farmers abandoned native strains to plant uniform scientifically created varieties. Experts discussed the problem at a 1967 conference in Rome and announced that the spread of Green Revolution strains threatened global food security and future advances in plant breeding. Other researchers found that many peasants preferred the taste of foods made from their traditional varieties.[81]

During the 1960s critics began to attack the economic effect of the Green Revolution. In 1964 the UN's World Health Organization warned that raising Third World farm productivity would drive peasants off the land. In 1967 agricultural economist W. Whitney Hicks said that the Green Revolution made some Mexican farmers prosper and gain political influence, especially in the northern irrigation districts, but it bypassed *campesinos* who grew *de temporal* corn. In 1969 Hicks' findings were confirmed by Donald K. Freebairn; Lena Dalrymple of AID, who called Mexico "hardly a complete model for other less developed nations"; and R. D'Shaw of the Overseas Development Council, who found that the Green Revolution did not create more rural employment.[82] Reporters and scholars who studied India's Green Revolution also found that it altered rural economies against the interests of peasant farmers, making them poorer rather than more prosperous. Dependency theorists put the Green Revolution in the context of prior introductions of modern technologies in the Third World, which exacerbated income distribution to the rural poor's detriment. To Samir Amin, it was "absurd . . . to suppose that the present-day Third World's agricultural revolution can yield the same results as those achieved in the west. The Green Revolution is no possible substitute for total, self-centered development, agricultural and industrial."[83] Anthropologist Gunnar Myrdal urged scientists to develop labor-intensive technologies "appropriate" for peasant farmers and to lobby for Third World land reform.

Indira Ghandi told the *New York Times* that "unless the Green Revolution is accompanied by a revolution based on social justice, the Green Revolution may not remain green."[84]

Eco-Marxists and other environmentalists saw the Green Revolution as another example of the domination of nature rooted in the capitalist market ethic. To protect the environment, they denounced the intensive use of agrochemicals and implored rural development planners to take ecological principles into account. Even a few U.S. agricultural scientists called the Green Revolution a fraud. William Paddock compared it to the Emerald City of Oz, which is visible only to people wearing the right glasses. "Take off the glasses," "and the Revolution proves to be an illusion—but devastating in the damage it can do to mankind's tardy efforts to limit the world's horrendous population growth."[85] Population biologist Paul Ehrlich joined Paddock, claiming that media optimism led to complacency on the issue of birth control and calling those who said that the Green Revolution would end world hunger "clowns" who must learn "some elementary biology, meteorology, agricultural economics, and anthropology."[86] Michael Allaby of *The Ecologist* lamented that "all our efforts to increase food availability for a higher population must be eventually self-defeating. It is worth remembering that after 30 years of aid, the developing countries are worse off than ever."[87] Nobel Prize winner George Wald warned that it was "utterly meaningless and bankrupt to believe that merely increasing food production is an answer to growing population."[88] Secretary of Defense Robert McNamara, the *New York Times*, and many scientists urged Third World nations to adopt birth control programs.[89]

The Mexican government denied that the Green Revolution had problems. In 1968 INIA director *agrónomo* Roberto Osoyo said that "most of goals that have been attained in self-sufficiency in food products have been reached chiefly by virtue of our systematic and coordinated research program," and in 1969 the SAF called the corn-breeding program a success, proclaiming that "now no one denies the advances our agricultural research experts achieved."[90] Díaz Ordaz planned to mint a 25-peso coin to commemorate the MAP's achievements, and he awarded Wellhausen the Aztec Eagle, the highest honor that non-citizens can receive. In 1970 Preciado called hybrid corn from Guanajuato's irrigated fields "a true triumph of research that serves the *campesinos*."[91] The

SAF proudly exported seeds to other countries and said that even the Indians of Sonora's Sierra Madre doubled their yields by using new varieties. Confronted with mounting criticism, some *agrónomos* denied that the Green Revolution had failed. Pointing to *ejidatarios* and *rancheros* from the outskirts of Mexico City, Díaz del Pino and other *agrónomos* claimed that many farmers used hybrid corn successfully and chastised those "who do not follow the government's advice."[92] *Agrónomos* proudly taught agricultural scientists from other Latin American nations to "feed their inhabitants better and fortify economic and social progress."[93] In 1969 Sánchez Durón told U.S. reporters that Mexico was "a nation of agricultural scientists" and boasted about the success of hybrid wheat in India and Pakistan, which "merited international acknowledgment for Mexican agricultural research."[94] In 1970 *Tierra* said that "Mexico proudly shares" Borlaug's prize because he was "surrounded by *agrónomos* and humble *campesinos*."[95]

Other *agrónomos* admitted that the Green Revolution had shortcomings. Using the criticism to address professional concerns, in 1968 *agrónomos* urged Díaz Ordaz to spend more money on research for peasant agriculture. López Zamora admitted that the Green Revolution did not bring prosperity and social justice to most farmers, but he argued that if more land reform was impossible and the population explosion inevitable, then organizing *ejidos* into single units of production to permit the use of its technologies was the only way to increase *campesinos'* income and feed the nation. Flores called hybrids "a major success" but extension programs a failure. Many *agrónomos* admitted that profitable modern farming existed alongside traditional production systems that perpetuated poverty. To make a more just Green Revolution, Orive Alba even proposed reducing the landholding limits for farms in irrigation districts and distributing the excess to the *campesinos*. *Tierra* praised the MAP but admitted that the SAF's research targeted the irrigation districts, that hybrid corn had limited acceptance, and that the Green Revolution did not bring prosperity to the *campesinos*, demanded a study to discover why, and urged the agency to study *de temporal* agriculture. The editors also said that farmers used too many chemical fertilizers and recommended more research on organic farming and pollution and soil erosion control. Recognizing the "pesticide cycle," Blanco Macias decried the harmful effects of DDT, calling pesticides "a terrible Frankenstein returning to destroy its creator."[96]

In spite of the criticism, the U.S. government lauded the Green Revolution as an example of what capitalist nations could do for the Third World. In 1969 Nixon praised the "miracle seeds" for averting "mass starvation" that "seemed possible" during the Indian food crisis. In 1970 AID's Delbert Myron admitted that in Mexico "a small number of the large private farmers have been the first ones to recognize the potential" but insisted that "modern agriculture" offered "an opportunity to break out of . . . traditional poverty."[97] Brown's *The Seeds of Change* lauded their "promise to improve the well-being of more people in a shorter time than any other single technological advance in human history. They are replacing disappointment and despair with hope; for literally hundreds of millions they can be the key to the door opening into the 20th Century."[98] Misinterpreting the MAP's objectives, Brown wrote, "Just as Henry Ford designed the Model T to meet the needs of middle class America, so the plant breeders designed the new wheat and rice varieties to meet the varied needs of peasant farmers in the Third World."[99]

The foundation's scientists challenged their critics. Joining many other agricultural scientists and representatives of agrochemical manufacturers, they denounced the attack on pesticides as a combination of bad science and naíveté. Reflecting a U.S. definition of a modern society, Robert Osler, deputy director of the *Centro Internacional del Mejoramiento de Maiz y Trigo* (International Center for Corn and Wheat Improvement, hereafter CIMMYT) in Sonora, said that the Green Revolution brought "economic improvement in the nation as a whole, not just improvement in yield," and his colleague Art Klatt claimed that its wheats "are grown over 95% of Mexico, so it must be nonsense to say they don't reach the small farmer."[100] Chandler admitted that the new technologies caused socioeconomic problems but asked, if there were so many poor and hungry peasants in nations like India, "how can it be wrong to increase the amount of food . . . [and] the incomes of farmers?"[101] In 1970 Harrar, now president of the foundation, claimed that the "Green Revolution cannot be made the scapegoat for the displacement of agricultural laborers and sharecroppers that has taken place for decades," denied that it led to complacency on birth control, and argued that it could bring beneficial social change to the rural Third World. He told the press the Green Revolution ensured that there was "no question that world food production could be enlarged to cope with the large population increases expected by the end of the century" but warned that "sharp increases in food production would be of little use in the long run unless human population can be brought under control."[102] Even Borlaug was less optimistic. Accepting his Nobel Prize, he said that "there can be no permanent progress in the battle against hunger until the agencies that fight for increased food production and those that fight for population control unite in a common effort," and he retreated from his 1965 prediction on science and the food supply, saying that the Green Revolution bought only 30 years before uncontrolled population growth would "destroy the species." He said that "millions and millions" of farmers grew his wheat, claimed that the Green Revolution could not "solve all the social problems which already existed," and chastised "biased" media reporting that made him "sick."[103]

Privately, Mangelsdorf and Wellhausen claimed that the primary reason that hybrid corn did not gain more acceptance in Mexico was government rather than private sector control over seed distribution, which fostered inefficiency and corruption, and they still believed that the *campesinos* would use it if "approached in the right way."[104] Mangelsdorf lamented that in 1962 "the lot of the small farmer, the *ejidatario,* had changed very little [since 1941]. In fact, relatively, he was worse off because his neighbors were better off, and he was standing still, economically speaking," a problem for which he had "no answer."[105] Fearing that *campesinos* would "form the slums of Mexico City and . . . Guadalajara and . . . Monterrey . . . , [become] people that cause riots in the streets . . . [and be] susceptible to agitation from outside," Wellhausen felt that something should be done about the problem and admitted that the MAP neglected "marginal" regions to concentrate "on the areas which are more productive from the standpoint of corn."[106] In 1967 Borlaug told an interviewer that *ejido* agriculture was not economically viable and gave farmers the chance to buy only "a bicycle . . . [or] transistor radio . . . [not] a house in which one would like to see citizens of all countries live . . . [or get] "a higher education" for their children, and opined that "maybe the time has come, but the politicians don't think so, to change this land for the masses and instead, get the people off the land."[107] The MAP's scientists were not oblivious to the profound social changes that their new seeds had wrought.

CONCLUSIONS

The Green Revolution was part of a profound change in rural Mexico's relationship to the Mexican Revolution. To increase farm productivity, the State followed the agricultural policies begun during the 1940s to their fruition. A few Mexican farmers, mostly in the private sector, reaped substantial gains from the Green Revolution, but many others, especially the *campesinos,* benefited little or not at all from the second revolution to sweep rural Mexico, despite making at least some effort to try the new technologies. Rather than a place to create social justice for the *campesinos* who struggled alongside Zapata, Villa, and other revolutionary leaders, with the Green Revolution rural Mexico served the needs of industrial and political elites and the urban middle class. If *cardenismo* was the pinnacle of *agrarismo,* by 1970 the agrarian movement was in major retreat. Both revolutions, agrarian and agricultural, did not bring prosperity to most *campesinos,* and a significant number fled the countryside for Mexico's cities or the United States. After creating a powerful tool for U.S. foreign policy and basking in the glow of a major triumph of international science, the Rockefeller Foundation and its scientists had to bear the burden of much criticism as the shortcomings of the Green Revolution became evident in Mexico and other nations. By hosting the MAP and becoming a base to extend the Green Revolution in the Third World, Mexico contributed mightily to the U.S. effort to use agricultural assistance to contain the spread of Communism. Mexico tried to construct an independent foreign policy, but the MAP's contribution to its industrialization project by reducing food imports and enabling the sale of subsidized foodstuffs to the urban working class overrode this concern. Further, the MAP helped Mexico become a regional leader in Latin America, an objective of the government's foreign policy for many years.

The Green Revolution helped move the *agrónomos,* an urban, middle-class group tied to the Revolutionary Family, away from *agrarismo,* but it also gave them the professional credibility that they had lacked but sought for many years. The spotlight of international high science shined on the *agrónomos*, an unusual role for scientists from the Third World. Most struggled to reconcile their contributions to the agrarian and Green revolutions, and some expressed profound denial as the latter developed. A few *agrónomos* criticized, rejected, or resisted the Green Revolution, especially the IIA's researchers, who became the MAP's professional rivals, and some called the MAP an example of Yankee imperialism. More *agrónomos* defected when the Green Revolution came under international attack, but even its critics appreciated the MAP's contribution to *agronomía* because it made the *agrónomos* scientific professionals, enhancing their position in the Revolutionary Family during a period when technical expertise was important for a successful bureaucratic career. A few *agrónomos* urged the State to implement land reform in the irrigation districts to make a more just Green Revolution, but they were not willing to challenge the Revolutionary Family by encouraging peasant land seizures or organizing agrarian petitioners as they did during the 1920s and 1930s. Jennings' assertion that the foundation's training program transformed the *agrónomos* into "a-political" scientists by defining standards of professional competence has some merit, but the State's agricultural and agrarian policy also dampened their enthusiasm for *agrarismo*.[108]

NOTES

1. Blanca Torres, *Hacia la utopía industrial* (México, D.F.: El Colegio de México, 1984), p. 82; Stephen R. Niblo, *War, Diplomacy, and Development: The United States and Mexico, 1938–1954* (Wilmington, Del.: Scholarly Resources, 1995), p. 186; Reed Hertford, *Sources of Change in Mexican Agricultural Production, 1940–1965* (Washington, D.C.: Economic Research Service, U.S. Department of Agriculture, 1971), pp. 2, 9, 20–23; Cynthia Hewitt de Alcantara, *Modernizing Mexican Agriculture: Socioeconomic Implications of Technological Change, 1940–1970* (Geneva: United Nations Research Institute for Social Development, 1976), p. 36; Folke Dovring, "Land Reform and Productivity: The Mexican Case, a Preliminary Analysis," unpublished paper, Department of Agricultural Economics, Agricultural Experimentation Station, University of Illinois, November 1966; Sergio Reyes Osorio et al., *Estructura agraria y desarrollo agrícola en México: Estudio sobre las relaciones entre la tenencia y el uso de la tierra y el desarrollo agrícola en México* (México, D.F.: Fondo de Cultura Económica, 1974), p. 121; David Barkin, "Food Production, Consumption, and Policy," in

Michael S. Werner, ed., *Encyclopedia of Mexico: History, Society and Culture* (Chicago: Fitzroy Dearborn, 1997), pp. 493–96.

2. Hewitt de Alcantara, *Modernizing*, p. 40; for *agrónomos* and the economic miracle see Edmundo Flores, *Tratado de economía agrícola* (México, D.F.: Fondo de Cultura Económica, 1961), pp. 38–39.

3. "Dos dias de demostración," *Agricultura Técnica en México* 4 (Summer 1957): 35; "Cotaxtla," *Agricultura Técnica en México* 11 (Winter–Summer 1961): 43–44; "Día del Agricultor en Sta. Elena, Méx.," *Tierra* 19, 11 (November 1964): 910; "V Consejo de la Asociación Nacional de Cosecheros," *Tierra* 7, 6 (June 1952): 379; "untitled document," October 4, 1954; Alfredo Brigada to Ruiz Cortines, November 11, 1956; Asociación de Agricultores del Rio Culiacán to Ruiz Cortines, July 12, 1956; Alfonso Castillo to Ruiz Cortines, February 16, 1953; Magdaleno Aguilar to Ruiz Cortines, June 21, 1954; Teofilo Álvarez Borboa to Ruiz Cortines, March 17, 1954, AGN, Ruiz Cortines papers, Exps. 506.1/38, 568.3/169, 506.1/162, 506.1/8, 151.3/773, and 501.1.122; Donato Miranda Fonseca to Francisco Cedillo, March 23, 1962; Elpidio Rincón López and others to López Mateos, July 14, 1960, AGN, Lopez Mateos papers, Exps. 506.1/41 and 506.1/20; Manuel Parra and R.D. Osler, "Primeros experimentos con maiz en Yerbanis, Durango," *Agricultura Técnica en México* 3 (Winter 1956–1957): 40; Elvin C. Stakman, "Report," April 16–May 14, 1954, p. 2, RFA, RG 1.2, Series 323, Box 10, Folder 60; Richard Bradfield, "Report on Mexico Trip," August 14–September 5, 1953; Alberto Salinas Ramos to MAP, April 15, 1952, RFA, RG 6.13, Series 1.1, Box 20, Folder 216, RAC; Hewitt de Alcantara, *Modernizing*, pp. 168–69.

4. Alfonso Díaz del Pino, "Diez toneladas de maiz por hectarea," *Tierra* 7, 1 (January 1952): 21–23, 53; "Editorial: Educación agrícola popular," *Tierra* 11, 7 (July 1956): 575; Reyes Osorio, *Estructura agraria*, pp. 733–35; Salvador Camelo Soler, "El maiz en la historia de México," February 18, 1958; Rubén C. Carrizosa to Adolfo Ruiz Cortines, February 28, 1953; Ignacio Gómez del Campo to DGA, January 11, 1954; Castillo to SAF, January 29, 1953, AGN, Ruiz Cortines papers, Exps. 521.8/124, 506.1/8, 506.1/83 and 503.11/116; Navarro Rogacino and others to López Mateos, March 8, 1962; Cruz Hernández Camarillo to López Mateos, April 14, 1961; Antonio Hidalgo to López Mateos, December 9, 1958, and others; Amado Robles Uzina to López Mateos, March 5, 1960; M. Hilario to López Mateos, November 26, 1962; Jesús Patiño Navarette to Ferrer Galvan Bourell, April 3, 1961; Raul Ortiz Mena to Galvan Bourell, May 17, 1963; Alberto Cruz Juárez to López Mateos, October 1, 1960; Luis Ramírez de Arellano to López Mateos, June 10, 1959; Dario L. Arrieta M. to SAF Agente General, June 18, 1962, AGN, López Mateos papers, Exps. 506.19/2, 506.1/32, 506.1/3, 506.1/16, 506.1/46, 506.1/33, 506.1/23, 506.1/25, 506.1/9, and 506.17/4; Christina Gladwin, "A View of the

Plan Puebla: An Application of Hierarchical Decision Models," *American Journal of Agricultural Economics* 58 (December 1976): 881; Barkin, "Agricultural Development in Mexico: A Case Study of Income Concentration," *Social Research* 37, 2 (Summer 1970): 314; Abraham Iszacvich, *Modernización campesina* (México, D.F.: Editorial Edicol, 1980), pp. 34, 42–43; Emilio López Zamora, *El agua, la tierra, los hombres de México* (México, D.F.: Fondo de Cultura Económica, 1977), p. 28; Hewitt de Alcantara, *Modernizing*, p. 81; Judith Friedlander, *Being Indian in Hueyapan: A Study of Forced Identity in Contemporary Mexico* (New York: St. Martin's Press, 1975), p. 64.

5. Stanley Johnson, *The Green Revolution* (London: Hamish Hamilton, 1972), p. 9; Hewitt de Alcantara, *Modernizing*, pp. 160–64.

6. Marte R. Gómez, *Biografías de agrónomos* (Chapingo, México: ENA, Colegio de Postgraduados, 1976), pp. 463–64; Hewitt de Alcantara, *Modernizing*, pp. 72, 173, 204–10; Manuel L. Carlos, *Politics and Development in Rural Mexico: A Study of Socioeconomic Modernization* (New York: Praeger, 1974), p. 15.

7. William C. Cobb, "Interview with Norman Borlaug," June 1967, p. 193; Cobb, "Interview with Kenneth Wernimont," March and May 1967, pp. 55, 66, RFA, RG 13, RAC; Luís González y González, *San José de Gracia: Mexican Village in Transition*, John Upton, trans. (Austin: University of Texas Press, 1974), p. 301; Billie R. DeWalt, "Inequalities of Wealth, Adoption of Technology and Production in a Mexican Ejido," *American Ethnologist* 2, 1 (February 1975): 154; Marilyn Gates, *In Default: Peasants, the Debt Crisis and the Agricultural Challenge in Mexico* (Boulder, Colo.: Westview Press, 1993), pp. 137–39; Leobardo Jiménez S., "El Mangal: Una población ejidal de la costa veracruzana," *Chapingo* 1, 1 etapa II (March 1961): 93–94; Román Badillo to López Mateos, April 23, 1960, AGN, López Mateos papers, Exp. 506.1/11; Carlos Manuel Castaños, *Testimonios de un agrónomo* (México, D.F.: Editorial Futura, S.A., 1981), pp. 48–49, 87; "Editorial: Los Clubes de Cinco Toneladas," *Tierra* 10, 3 (March 1955): 191; Tomás Aznar Álvarez to Ruiz Cortines, April 20, 1954, AGN, Ruiz Cortines papers, Exp. 541/331; Reyes Osorio, *Estructura agraria*, pp. 723–26; Marco Antonio Durán, *El agrarismo mexicano* (México, D.F.: Siglo Ventiuno, 1967), p. 152; Sergio Zendejas, "Appropriating Government Reforms: The Ejido as an Arena of Confrontation and Negotiation," in Zendejas and Pieter De Vries, eds., *Rural Transformations Seen from Below: Regional and Local Perspectives from Western Mexico* (La Jolla: Center for U.S.–Mexican Studies, University of California, San Diego, 1995), p. 35; Iszacvich, *Modernización*, pp. 32, 42–43; Armando Bartra, "Colectivización o proletarización: El caso del Plan Chontalpa," *Cuadernos Agrarios* 1, 4 (October–December 1976): 69.

8. Manuel Girault, "El ejido: Callejón sin salida," *Problemas agrícolas*

e industriales de México 5, 4 (1953): 1–25; Enrique Navarro Palácios, "El problema agrícola de México," April 30, 1954; Alberto M. Saavedra to Ruiz Cortines, June 10, 1954; Alfonso Santos Franco to Ruiz Cortines, February 11, 1953; Felipe Cortes G., "Plan para refaccionar al agricultor de México por un periodo de tres años a base de absoluta economía," September 25, 1954; Lorenzo Azua Torres, "Projecto de Pequeñas Granjas Tipo, que dentro de los más factible puedan superar el servicio social agrícola que nunca ha podido satisfacer los honorosos y bureaucráticos campos experimentales," September 30, 1953; Azua Torres, "Memorandum de la CNC al Presidente de la República sobre la creación de las Granjas Pilotos," September 30, 1953; Peregrino[?] to Enrique Rodríguez Cano, October 7, 1954, AGN, Ruiz Cortines papers, Exps. 506.1/38, 506.1/5, 506.1/15, and 506.1/381.

9. Ignacio E. Zepeda to Ruiz Cortines, May 21, 1954; "Generalidades sobre la situación agrícola" (n.d.); Constantino Guerrero to Ruiz Cortines, September 12, 1955; Enrique Cardeño Palácios to Ruiz Cortines, April 27, 1958, AGN, Ruiz Cortines papers, Exps. 506.1/38 and 545.3/409; Manuel Cavasos González, *La ciencia y la técnica aplicada a la agricultura y sus relaciones con el derecho agrario* (México, D.F.: Tipografía Ortega, 1953); Stakman, "Report," pp. 3–4, 12; Bradfield, "Report," p. 4; Bradfield, "Report on a Trip to Mexico," October 17–24, 1953, RFA, RG 1.2, Series 323, Box 10, Folder 61; Cobb, "Interview with Edwin J. Wellhausen," June 28, 1966, p. 179, RFA, RG 13; "Personal Técnico," 1951–1952, RFA, RG 6.13, Series 1.1, Box 42, Folder 535, RAC; David L. Clawson and Don R. Hoy, "Nealtican, Mexico: A Peasant Community That Rejected the 'Green Revolution,' " *American Journal of Economics and Sociology* 38, 4 (October 1979): 381; Thomas T. Poleman, *The Papaloapan Project: Agricultural Development in the Mexican Tropics* (Stanford, Calif.: Stanford University Press, 1964), p. 119; Rodolfo P. Peregrina, "La investigación agrícola en México," in Ateneo Nacional Agronómico, *Problemas agrícolas actuales* (México, D.F.: Ateneo Atenagro, 1955), pp. 49–50.

10. George A. Kneller, *The Education of the Mexican Nation* (New York: Columbia University Press, 1951), p. 197; Damién Correu Toledo, "Discusión del Primer Tema," in Ateneo . . . , *Problemas*, pp. 83–84; Ramón Fernández y Fernández, "La promoción agrícola," in Ateneo . . . , *Problemas*, p. 204; Karen and Olaf Wessberg to Ruiz Cortines, March 7, 1955, AGN, Ruiz Cortines papers, Exp. 506.1/3; Eleazar Jiménez J., "El empleo de enemigos naturales para el control de insectos que constituyen plagas agrícolas en la República Mexicana," *Chapingo* 12, 73 (January–February 1959): 191–208; López Zamora, *El agua*, pp. 157, 305; "Control biológico de las plagas en México," *Tierra* 19, 3 (March 1964): 157; Lane Simonian, *Defending the Land of the Jaguar: A History of Conservation in Mexico* (Austin: University of Texas Press, 1995), pp. 119–20; "Jojutla mejorado," *Agri-*

cultura Técnica en México 3 (Winter 1956–1957): 47; "Invitation to Meeting of Federación Mexicana de Organizaciones Agrícolas, Asociación Nacional de Cosecheros, y Unión de Productores de Maiz y Frijol," September 1959, AGN, López Mateos papers, Exp. 433/144; Fidel Márquez Sánchez, "Notas notararias," *Chapingo* 8, 54 (November–December 1955): 236–40; Comisión Nacional del Maiz, "La producción de semilla de registro para la formación de híbridos de maiz," *Chapingo* 9, 60 (November–December 1956): 232–38; "Interview with Efrain Hernández Xolocotzi," in Jiménez S., *Las ciencias agrícolas y sus protagonistas* (Chapingo, México: Colegio de Postgraduados, 1984), p. 211; Díaz del Pino, "Bate el record el Presan No. 1," *Tierra* 7, 3 (March 1952): 169; Díaz del Pino, "El maiz híbrido en México," *Chapingo* 8, 53 (September–October 1953): 200; Colegio de Ingenieros Agrónomos de México, *Directorio Agronómica Nacional* (México, D.F.: Talleres Gráficos de la Nación, 1954), pp. 428, 447; José Luís de la Loma, *Experimentación agrícola* (México, D.F.: Unión Tipográfica Editorial Hispano America, 1955), p. 12; Gonzalo Robles, "Agriculture and Livestock," March 23, 1954, AGN, Gonzalo Robles papers, Agriculture Section, Box 13, Exp. 115; "Nos habla el Dr. Wellhausen de los trabajos de la Oficina de Estudios Especiales," *Tierra* 7, 8 (August 1952): 534–39; Gilberto Mendoza Vargas, "La enseñanza y la extensión agrícola," February 1954, in Enseñanza Agrícola Superior, BCM Bound Volume #145–43, p. 21.

11. Gabriel Ítie Cantelue, "Breve comentario sobre un nuevo libro," *Chapingo* 9, 55 (January–February 1956): 15; "Editorial," *Chapingo* 9, 55 (January–February 1956): 3; "Editorial," *Chapingo* 12, 73 (January–February 1959): 189; "Editorial," *Agricultura Técnica en México* 3 (Winter 1956–1957): 1; "For the Welfare of Humanity," translation of article in *El Informador*, August 29, 1953, RFA, RG 1.2, Series 323, Box 23, Folder 155, RAC.

12. Moises T. de la Peña, *La producción y beneficio del cacao* (México, D.F.: 1953), p. 80; "Nuevas variedades de plantas cultivadas," *Tierra* 20, 12 (December 1965): 938; Baldomero Rodríguez M., "Cultivo del ajonjolí en el noroeste," *Tierra* 20, 1 (January 1965): 30, 83; "Selección del arroz," *Tierra* 19, 2 (February 1964): 98; "Mejoramiento de la papa," *Tierra* 19, 3 (March 1964): 237; "Mejoramiento del cacaotero en México," *Tierra* 19, 1 (January 1964): 48; Julián Rodríguez Adame, "Inauguración oficial de los cursos en la ENA," *Tierra* 9, 3 (March 1954): 268; López Zamora, *El agua*, pp. 122–23; Colegio . . . , *Directorio*, p. 479; de la Loma, *Experimentación*, p. 10–11.

13. Stakman, "Report," p. 5; Efrain Hernández Xolocotzi, "La participación de México de la Tercera Reunión Latino-Americano de Fitogenetistas, Fitopatólogos, Entomólogos, y Edafológicos," *Chapingo* 8, 53 (September–October 1955): 177; "The Reach of the Mexican Program,"

October 1950, p. 6, RFA, RG 1.2, Series 323, Box 23, Folder 155, RAC; Ramón Fernández y Fernández and Ricardo Acosta, *Política agrícola, ensayo sobre normas para México* (México, D.F.: Fondo de Cultura Económica, 1961), p. 98.

14. Mendoza Vargas, "La enseñanza," p. 5, in Enseñanza; Hertford, *Sources*, p. 9; Colegio . . . , *Directorio*, p. 464; Manuel Mesa A., "Segunda intervención," in Ateneo . . . , *Problemas*, p. 165; Hewitt de Alcantara, *Modernizing*, p. 231; Gobierno del Estado de México, *Campo Experimental Santa Elena, Informe Num. 1, 1952–1954* (Toluca: Dirección de Agricultura y Ganadería, 1954), p. 8; Peregrina, "La investigación," p. 79; Peter S. Cleaves, *Professions and the State: The Mexican Case* (Tucson: University of Arizona Press, 1987), p. 54; Oscar Brauer H., "La tecnología y la ciencia en el desarrollo de la agricultura," in Brauer H. et al., *Diagnóstico de las ciencias agrícolas en México* (México, D.F.: Ediciones Productividad, 1968), pp. 30–38; Wernimont, "Rockefeller Foundation Scholarships and the Mexican Revolution in Agricultural Science" (n.d.), pp. 8–9, RFA, RG 13, RAC; Martin H. Greenberg, *Bureaucracy and Development: A Mexican Case Study* (Lexington, Mass.: D.C. Heath, 1970); p. 101; David E. Lorey, *The University System and Economic Development in Mexico since 1929* (Stanford, Calif.: Stanford University Press, 1993), p. 31; Gómez, *Biografías*, pp. 35, 43, 126–27, 156, 247, 256, 343–45, 439.

15. Colegio . . . , *Directorio*, p. 461; Gómez, *Biografías*, 36–37, 109, 216.

16. Flores, "Agrarian Reform and Economic Development," *The Texas Quarterly* 2, 1 (Spring 1959): 109–12; Flores, *Tratado*, p. 397; Rodrigo García Trevino, "Agrarismo revolucionario y ejidalismo burocratico," *Problemas agrícolas e industriales de México* 5, 4 (1953): 29–32; Rodríguez Adame, "Antecendentes y realizaciones de la Reforma Agraria," *Problemas agrícolas e industriales de México* 5, 4 (1953): 81–82; Manuel Marcue Pardiñas, "Intervención," in Ateneo . . . , *Problemas*, pp. 251–57; López Zamora, *El agua*, pp. 210–12; Fernández y Fernández and Acosta, *Política agrícola*, p. 35; José López Bermúdez to Fernández y Fernández, June 6, 1956, ARFF, Box 180.

17. Luis L. León, "La actuación del gremio agronómico en la reforma agraria," *Boletín de la Sociedad Mexicana de Geografía y Estadística* 78 (1954): 66–68; Gómez, *Biografías*, p. 45.

18. Ramón Fernández y Fernández, *Chapingo hace 50 años* (Chapingo, México: ENA, Colegio de Postgraduados, 1976), p. 61; Juan de Dios Bojórquez, *Los agrónomos de México* (México, D.F.: Ediciones del Boi, 1959), p. 7; Armando Narvaez López, "Inauguración de cursos," *Chapingo* 12, 73 (January–February 1959): 209–10; "Inaguaración oficial de los cursos en la ENA, Chapingo, Méx.," *Tierra* 14, 3 (March 1959): 266.

19. "Debe dar más rendimiento nuestro campo" (n.d.), AGN, Ruiz Cortines papers, Exp. 545.3/41; Gómez, *Biografías*, pp. 440–41.

20. Wernimont, "Rockefeller Foundation," pp. 15, 19; Gómez, *Biografías*, p. 385; Poleman, *The Papaloapan Project*, pp. 9, 119, 139–41; de la Loma, *Experimentación*, p. 14.

21. "Nos habla Leonel Robles, del Departamento de Agronomía del Tecnológico de Monterrey, N.L.," *Tierra* 7, 5 (May 1952): 314–18; López Zamora, *El agua*, p. 140; Barkin, "Agricultural Development," p. 316.

22. Díaz del Pino, "¿Es buen negocio el cultivo del maiz?" *Tierra* 6, 4 (April 1951): 222; de la Peña, *La producción*, p. 53; Manuel Fitzmaurice, "Intervención," in Ateneo . . . , *Problemas*, p. 285; López Zamora, *El agua*, p. 135.

23. Emilio Alanis Patiño, López Bermúdez, and Mesa A., "Problemas de tenencia y aprovechamiento de la tierra en México," *Problemas agrícolas e industriales de México* 5, 4 (1953): 149–50; "Editorial: Educación," p. 575; Peregrina, "La investigación," p. 44; "Plan Agrícola," pp. 941–44; Fernández y Fernández, "El caso de 'El Mangal,' " *Chapingo* (epocha II) 1, 1 (March 1961): 105–10.

24. Colegio . . . , *Directorio*, pp. 436, 443, 452, 470–78; Wernimont, "Rockefeller Foundation."

25. De la Loma, *Experimentación*, pp. 4–5; Blanco Macias, "El Programa Agrícola," p. 219.

26. "Otra imposición más, ahora, en la SAM," *Excelsior*, May 3, 1956; Blanco Macias, "La agricultura nacional y los agrónomos," *Chapingo* 7, 50 (March–April 1955): 60–62; Rogelio González to Ruiz Cortines, January 30, 1955; Mariano Parra Hernández and others to Ruiz Cortines, December 3, 1952, AGN, Ruiz Cortines papers, Exps. 702.2/33 and 702.2/103, p. 14; Gilberto Fábila, "Primera intervención," in Ateneo . . . , *Problemas*, p. 161.

27. Mendoza Vargas, "La enseñanza y extensión agrícola," in Ateneo . . . , *Problemas*, p. 127; Fernández y Fernández, "La promoción agrícola," in Ateneo . . . , *Problemas*, p. 200; Mesa A., "La cuestión agraria actual," in Ateneo . . . , *Problemas*, p. 248; Felipe Salcedo Pérez, "Agricultura de temporal: Fuente de alimentos populares," *El Universal*, November 19, 1957; López Bermúdez to Ruiz Cortines, March 22, 1957, AGN, Ruiz Cortines papers, Exp. 433/75; Narvaez López, "Breve resumen de conferencias sustentadas en la ENA," *Chapingo* 12, 75 (May–June 1959): 308–13; José García y García to López Mateos, September 11, 1960, AGN, López Mateos papers. Exp. 506.1/24; Durán, "La reforma agraria en México," *Chapingo* 13, 81 (May–June 1960): 113–18.

28. "Annual Report of the Corn and Sorghum Project," July 1955–June 1956, RFA, RG 6.13, Series 1.1, Box 12, Folder 142, RAC; "Editorial," *Agricultura Técnica en México* 5 (Winter 1957–1958): 1; "Campo 'El Horno,' Chapingo, México," *Agricultura Técnica en México* 10 (Summer 1960): 612;

"Interview with Rodríguez Adame," in Jiménez S., *Las ciencias agrícolas,* pp. 13–15; "Editorial: La educación agrícola cambia de rumbo," *Tierra* 14, 5 (May 1959): 405; Reyes Osorio, *Estructura agraria,* pp. 661–65.

29. Greenberg, *Bureaucracy and Development,* p. 93; Jiménez S., "El Mangal," pp. 94–101.

30. Atanasio Cuevas R., "Un año de trabajo de extensión en Salinas Victoria, N.L," *Tierra* 19, 2 (February 1964): 122; Díaz del Pino, "El cultivo del maiz," *Tierra* 19, 9 (September 1964): 769; "Editorial: Educación," p. 525; "Editorial: Técnica y técnicos," *Tierra* 20, 9 (September 1965): 699; "Interview with Julián Rodríguez A.," pp. 15–16; Jesús Uribe Ruiz, *Problemas y soluciones en el desarrollo agrícola de México* (México, D.F.: Academía Mexicana de Ciencias Agrícolas, 1964), pp. 72–73; "Informe Presidencial," *Tierra* 20, 9 (September 1965): 703; "Plan Agrícola Nacional, primera etapa 1965–1966," *Tierra* 20, 12 (December 1965): 942–43; Raymond Wilkie, *San Miguel: A Mexican Collective Ejido* (Stanford, Calif.: Stanford University Press, 1971), p. 163.

31. "Science Notes World's Future Food Supply," *New York Times,* December 30, 1956; "Attack on Famine Urged by U.N. Unit," *New York Times,* January 20, 1952; "Big Gains Reported by U.N. Food Group," *New York Times,* May 9, 1952; "U.N. Sees Population of 4 Billion by '84 Causing World Food Crisis," *New York Times,* May 3, 1954; "Dramatic Strides in Food Forecast," *New York Times,* February 21, 1953; "Farm Revolution Shows No Signs of Slowing," *New York Times,* January 11, 1960; "Food Output Tops Population Rise," *New York Times,* September 6, 1960.

32. "Food Output Rise Seen in Science," *New York Times,* February 16, 1952; "Atoms May Unlock Arctic for Farms," *New York Times,* August 16, 1955; "Scientists Warned on Point Four Plan," *New York Times,* December 27, 1950; Sir John Russell, *World Population and World Food Supplies* (London: George Allen & Unwin, 1954), p. 470.

33. "Race with Famine Seen," *New York Times,* September 10, 1957; "World Food Gains Held Insufficient," *New York Times,* November 8, 1958; "Parenthood Unit Scores U.S. Stand," *New York Times,* December 4, 1959; "Birth Control Held World's Best Hope," *New York Times,* December 3, 1959; "Birth Rate Break Called Necessity," *New York Times,* November 6, 1958; Russell, *World Population,* pp. 7, 467.

34. "Brannan Says Food Is Part of Defense," *New York Times,* November 21, 1951; John Kerry King, "Rice Politics," *Foreign Affairs* 31, 3 (April 1953): 453–60; "No Food Shortage Found in Europe," *New York Times,* July 30, 1951; "World Corn Crop Put Near Record," *New York Times,* October 27, 1952; "Point Four to Urge Capitalist Farms," *New York Times,* January 8, 1951; "Eisenhower Asks World Wide War against Hunger," *New York Times,* December 12, 1949; "President Bans U.S. Help for Birth

Control Abroad," *New York Times*, December 3, 1959; Patricia L. Kutzner, *World Hunger: A Reference Handbook* (Santa Barbara, Calif.: ABC-CLIO, 1994, p. 67.

35. "The Reach," pp. 1–4; Gobierno del Estado . . . , *Campo Experimental*, p. 8; Norman Borlaug, "The Impact of Agricultural Research on Mexican Wheat Production," *Transactions of the New York Academy of Science* 20, 3 (January 1958): 288; "Minutes of the Annual Meeting of the Advisory Committee for Agricultural Activities," October 26, 1950, RFA, RG 6.13, Series 1.1, Box 32, Folder 361; "Agriculture and the Rockefeller Foundation" (n.d.), p. 2–3. RFA, RG 6.13, Series 1.1, Box 25, Folder 280; "The World Food Problem, Agriculture, and the Rockefeller Foundation," June 14, 1951, pp. 1–2, RFA, RG 6.13, Series 1.1, Box 32, Folder 361, RAC; John H. Perkins, "The Rockefeller Foundation and the Green Revolution, 1941–1956," *Agriculture and Human Values* 26 (Summer–Fall 1990): 11.

36. "Editorial: Mexico's Private Point Four," *New York Times*, October 11, 1952; "A U.S. Foundation Aids Latin Farms," *New York Times*, December 9, 1956; Arthur T. Mosher, *Technical Cooperation in Latin American Agriculture* (Chicago: University of Chicago Press, 1957), pp. 107–10; Chester I. Barnard, *The Rockefeller Foundation: A Review for 1950 and 1951* (New York: The Rockefeller Foundation, 1952), pp. 10, 42; "Rockefeller Unit Widens Mexico Aid," *New York Times*, October 10, 1952; Elvin C. Stakman, Richard Bradfield, and Paul C. Mangelsdorf, *Campaigns against Hunger* (Cambridge, Mass.: Harvard University Press, 1967), pp. 219–30; "Rice Yield Stirs Farm Scientists," *New York Times*, March 9, 1958; Kutzner, *World Hunger*, p. 70; Perkins, "The Rockefeller Foundation," pp. 10–13.

37. "Big Need for Food Seen in Decades," *New York Times*, August 21, 1958; J. George Harrar, "Technologic Revolution in Agriculture," text of talk delivered December 8, 1960, at the Food Protection Symposium, Washington, D.C., in Harrar, *Strategy toward the Conquest of Hunger*, selected papers of J. George Harrar (New York: The Rockefeller Foundation, 1967), pp. 175–76; "Soil Experts Get Production Plea," *New York Times*, August 16, 1960.

38. Mesa A., "Intervención," in Ateneo . . . , *Problemas*, p. 223; Panfilo Rodríguez León, "Intervención," in Ateneo . . . , *Problemas*, p. 225; Flores, *Tratado*, pp. 110, 377; Blanco Macias, "El Programa Agrícola," pp. 216–19; Fernández y Fernández and Acosta, *Política agrícola*, p. 61; Simonian, *Defending the Land*, p. 140. For Mexican reactions to birth control before 1940 see Anna Macías, *Against All Odds: The Feminist Movement in Mexico to 1940* (Westport, Conn.: Greenwood Press, 1982), pp. 92–94.

39. Stephen G. Rabe, *Eisenhower and Latin America: The Foreign Policy of Anticommunism* (Chapel Hill: University of North Carolina Press, 1988),

pp. 134–52; "Stabilizing of World's Population Is Advocated by Rockefeller III," *New York Times*, November 7, 1961; "Famine Envisioned in Population Rise," *New York Times*, May 11, 1963; "Editorial: Malthus Specter," *New York Times*, June 30, 1963; "Food Output Gain Cited in Report," *New York Times*, December 3, 1959; "Soil Experts Get Production Plea," *New York Times*, August 16, 1960; "Aid Needy Lands, Churches Urged," *New York Times*, April 29, "Editorial: The World's Food," *New York Times*, September 11, 1960; "Fight on Hunger Urged," *New York Times*, October 26, 1961; "Science Debated at Food Congress," *New York Times*, June 6, 1963; "Food for Peace Goals," *New York Times*, February 26, 1961; "U.N. Aide Says Famine Is Likely before 1980," *New York Times*, June 4, 1963; "Scientist Foresees Widespread Famine," *New York Times*, September 2, 1964; "Briton Sees Man Able to Solve Food Problems," *New York Times*, November 15, 1964; "Review of *No Need for Hunger* by Jonathan Garst," *New York Times*, January 26, 1964; Howard W. Mattson, "Food for the World," *International Science and Technology* 48 (December 1965): 28–39; "U.S. Calls on FAO to Conquer Hunger," *New York Times*, November 21, 1965.

40. "U.N. Told World Faces Food Crisis," *New York Times*, March 23, 1965; "Agriculture Revolution Urged," *New York Times*, July 6, 1966; "Widespread Starvation in a Decade Is Seen," *New York Times*, December 3, 1965; "Editorial: The Malthusian Specter," *New York Times*, October 14, 1966; *World War on Hunger*, Hearings before the Committee on Agriculture, House of Representatives, 89th Congress, 2nd Session (Washington, D.C.: U.S. Government Printing Office, 1966), p. 139.

41. William Paddock and Paul Paddock, *Famine, 1975! America's Decision: Who Shall Survive*? (New York: Little, Brown & Company, 1967), pp. 9–11, 65–78, 201; "Johnson May Revise Food Aid Policy," *New York Times*, November 8, 1965; "Peace Panel Asks Foreign Aid Rise," *New York Times*, November 30, 1965; Orville Freeman, *World without Hunger* (New York: Praeger, 1968), p. 167.

42. "FAO Urges Radical Change to Close Food-Population Gap," *New York Times*, October 12, 1967; "World Aid Urged to Avert Starvation," *New York Times*, June 18, 1967; "Food Experts See Future Vietnams," *New York Times*, November 15, 1967; "U.S. Hybrid Corn Called of Little Value Abroad," *New York Times*, August 10, 1967; "2 New U.S. Agencies Proposed to Aid Poor Lands Food Output," *New York Times*, December 29, 1967.

43. "Letter to the Editor by Richard K. Manoff," *New York Times*, December 9, 1966; Frederick Seitz, "Foreword," in *Prospects of the World Food Supply*, Proceedings of a Symposium during the 103rd Annual Meeting of the National Academy of Sciences (Washington, D.C.: National Academy of Sciences, 1966), n.p.

44. "AID Official Sees War Threatening World Hunger," *New York Times*, December 7, 1966; "Libertyville, Illinois: Growth Science Center Sees Economic Aid," *New York Times*, October 16, 1966; "U.S. Industry Set to Feed Poor," *New York Times*, August 13, 1967; "Experts Meet Here on World Hunger," *New York Times*, September 12, 1967; "Private Food Aid Gets New Stress," *New York Times*, September 14, 1967; Lester R. Brown, "The Stork Outruns the Plow," *Columbia Journal of World Business* 2, 1 (January–February 1967): 16–18.

45. William B. Murphy, "Probe into the World of Ideas," *Columbia Journal of World Business* 2, 1 (January–February 1967): 94.

46. Sterling Wortman, "Toward the Conquest of Hunger," in *Symposium on the Strategy for the Conquest of Hunger*, Proceedings of a Symposium Convened by the Rockefeller Foundation, April 1 and 2, 1968, at the Rockefeller University (New York: The Rockefeller Foundation, 1968), p. 2; W. David Hopper, "Investments in Agriculture: The Essentials for Payoff," in *Symposium*, p. 105; "Self Help Urged for Needy Lands," *New York Times*, October 10, 1967; "Big Industry to Aid the Farmer," *New York Times*, November 20, 1967; Stakman et al., *Campaigns against Hunger*, p. 60; Rockefeller Foundation, *President's Review, 1966* (New York: The Rockefeller Foundation, 1966), p. 3; Cobb, "Interview with Mangelsdorf, p. 109, RFA, RG 13, RAC.

47. Borlaug, "Wheat, Rust, and People," *Phytopathology* 55, 10 (October 1965): 1097.

48. Rockefeller Foundation, *President's Review, 1966*, pp. 17–31; "A Malay Farmer Looks to New Rice," *New York Times*, December 10, 1967; "Ford, Rockefeller Foundations Adjusting to a Changing World," *New York Times*, January 30, 1967; *World War on Hunger*, pp. 82–84.

49. Stakman et al., *Campaigns against Hunger*, pp. 1, 9, 93, 129.

50. Stakman et al., *Campaigns against Hunger*, p. 321.

51. Stakman et al., *Campaigns against Hunger*, pp. x, 24.

52. Stakman et al., *Campaigns against Hunger*, p. 218.

53. Stakman et al., *Campaigns against Hunger*, p. 283.

54. "A Research Farm Opens in Nigeria," *New York Times*, April 26, 1970; Kutzner, *World Hunger*, p. 77.

55. "Farming Revolution in Poorer Lands Is Held Near," *New York Times*, May 14, 1968; "Freeman Foresees Saigon Export Rice Again," *New York Times*, July 7, 1968.

56. William Paddock and Paul Paddock, *Time of Famines* (New York: Little, Brown & Company, 1976), p. xvii.

57. "South Vietnam Expects Bumper Rice Crop," *New York Times*, November 9, 1969; Carlos P. Romulo, "Strategy for the Conquest of Hunger," in *Symposium on the Strategy*, 1968, pp. 116–22; "Editorial: The Green Revolution," *New York Times*, May 21, 1968.

58. "Changes in Asia: Growth of Cities and Food Output," *New York Times*, July 31, 1968.

59. "Short Plant Rice May Double Crop," *New York Times*, May 27, 1968; Hubertus Zu Lowenstein, "The Story of a Sophisticated Breed," *Ceres* 2, 1 (January–February 1969): 22.

60. "Tropical Crops Studied," *New York Times*, November 18, 1968.

61. Borlaug, "National Production Campaigns," in *Symposium on the Strategy*, p. 101.

62. C.M. Christensen, *E.C. Stakman, Statesman of Science* (St. Paul, Minn.: The American Phytopathological Society, 1984), p. 125.

63. *Symposium on Science and Foreign Policy—The Green Revolution*, Proceedings before the Subcommittee on National Security Policy and Scientific Developments, House of Representatives, 91st Congress, December 5, 1969 (Washington, D.C.: U.S. Government Printing Office, 1970), n.p.; "Developers of High-Yield Grains Report a Victory over Drought and Disease," *New York Times*, September 16, 1969.

64. *Symposium on Science*, p. 20.

65. "U.S. Agronomist Gets Nobel Peace Prize," *New York Times*, October 22, 1970; FAO, "Provisional Indicative World Plan for Agricultural Development, Summary and Conclusions," 1970, in Antonio Barreto, *A Study of the Social and Economic Implications of the Large Scale Introduction of High-Yielding Varieties of Food Grains: A Selection of Readings* (Geneva: United Nations Research Institute for Social Development, Report No. 71–6, 1971), pp. 91–92.

66. "Scientist of the Fields Norman Ernest Borlaug," *New York Times*, October 22, 1970; Boyd Orr, "Food Enough for Everyone," *New York Times*, December 17, 1970.

67. "Interview with Emilio Gutíerrez Roldán," *Tierra* 19, 11 (November 1964): 884; Cuevas R., "Un año," pp. 122–23.

68. Diaz del Pino, "El cultivo del maiz: El maiz hibrido," *Tierra* 19, 1 (January 1964): 26; "Editorial: Técnica," p. 699; "Editorial: Progreso en el campo," *Tierra* 19, 2 (February 1964): 107; "Informe Presidencial," *Tierra* 20, 9 (September 1965): 702; "Origen e importancia del cultivo del maiz en México," *Tierra* 24, 6 (June 1969): 418.

69. "Editorial: Progreso," *Tierra* 19, 5 (May 1964): 365; Brauer H., "La tecnología," pp. 34, 56; "Editorial: Los resultados concretos que abren el camino a la agricultura tecnificada," *Agricultura Técnica en México* 12 (Winter 1961–1962): 1; Julián Rodríguez Adame, *La Secretaría de Agricultura y Ganadería y su función en el desarrollo agropecuario nacional* (México, D.F.: Guanos y Fertilizantes de México, 1964), pp. 13–14.

70. "Editorial: Semillas," *Tierra* 19, 4 (April 1964): 279; David Ronfeldt, *Atencingo: The Politics of Agrarian Struggle in a Mexican Ejido* (Stanford, Calif.: Stanford University Press, 1973), p. 180; López Zamora, *El*

agua, pp. 35, 184–87, 211; Reyes Osorio, *Estructura agraria*, p. 908; "Se inauguron en la ENA los cursos del año de 1965," *Tierra* 20, 2 (February 1965): 101–3; Brauer H., "La tecnología," p. 28; Vicente Cedillo Vázquez, *Principales factores del desarrollo agricola en México 1925–1960* (México, D.F.: UNAM, 1964).

71. Orive Alba, *La irrigación*, p. 11; Rodríguez Adame, *La Secretaría*, p. 23; "Nuevas variedades de plantas cultivadas," *Tierra* 20, 12 (December 1965): 938.

72. Sebastián Hernando Castilla, "Campo Experimental de Santa Elena, Méx.," *Tierra* 14, 9 (September 1959): 791–93; Donato Miranda Fonseca to Gustavo Díaz Ordaz, 21 June 1962; Alfonso Támez Mireles, "Discurso," April 23, 1962, AGN, López Mateos papers, Exps. 437.3/85 and 437.3/144; Ronfeldt, *Atencingo*, p. 197; Wernimont, "Rockefeller Foundation Scholarships," p. 17.

73. "Editorial: Un sexenio ventuoso," *Tierra* 19, 9 (September 1964): 705; Brauer H., "La tecnología," p. 51; Wernimont, "Rockefeller Foundation Scholarships," p. 6; Gómez, *Biografías*, p. 229–30.

74. "Urgencia de la reforma agraria y sus perspectivas en el decenio 1970–1980," *Tierra* 25, 10 (October 1970): 780; Roderic A. Camp, "A Middle Level Technocrat in Mexico," *Journal of Developing Areas* 6, 4 (July 1972): 576; Greenberg, *Bureaucracy and Development*, p. 99.

75. Uribe Ruiz, *Problemas y soluciones*, pp. 68–70, 71–87, 115, 160–62, 167, 205, 222–27, 236–38, 264, 411, 477, 481–85.

76. Uribe Ruiz, *Problemas y soluciones*, pp. 58–60, 94–96, 115–18, 165, 182–87, 222–28; "Editorial: Mirando hacia el porvenir," *Tierra* 19, 11 (November 1964): 881.

77. Durán, *El agrarismo*, pp. 18–20, 77, 108–17.

78. Clawson and Hoy, "Nealtican," pp. 379–81; Guillermo González López, "Lecciones del Plan Puebla para el cambio tecnológico en el campo Mexicano," *Comercio Exterior* 40, 10 (October 1990): 962–67; CIMMYT, "The Puebla Project, 1967–1969 (Progress Report of a Program to Rapidly Increase Corn Yields on Small Holdings)," Mexico, 1969, in Barretto, ed., *A Study*, pp. 110–11; Andrew Pearse, *Seeds of Plenty, Seeds of Want* (New York: Oxford University Press, 1980), pp. 186–90; Adolfo Orive Alba, *La irrigación en México* (México, D.F.: Editorial Grijalva, 1970), p. 223; Castaños, *Testimonios*, p. 60; Gladwin, "A View of the Plan Puebla," p. 881.

79. Iszacvich, *Modernización campesina*, pp. 154–60; Hewitt de Alcantara, *Modernizing*, pp. 214–34; Wilkie, *San Miguel*, pp. 140–41; Frans J. Schryer, *The Rancheros of Pisaflores: The History of a Peasant Bourgeoisie in Twentieth-Century Mexico* (Toronto: University of Toronto Press, 1980), p. 23.

80. "Food Output Rise Seen in Science," *New York Times*, February 16, 1952.

81. Rachel Carson, *Silent Spring* (Cambridge, Mass.: Riverside Press, 1962); "Bird Peril Seen in Latin Nations," *New York Times*, November 13, 1966; "India Has Problem in a Bumper Crop," *New York Times*, June 26, 1968; "Roundtable Discussion with Sir Otto Frankel, W.K. Agble, J.B. Harlan and Ezra Bennet on the Genetic Dangers of the Green Revolution," *Ceres* 2, 5 (September–October 1969): 35–38; Judith Miller, "Genetic Erosion: Crop Plants Threatened by Government Neglect," *Science* 182, 4118 (December 21, 1973): 1231–33; Chadd Graham, "Hidden Peril of the Green Revolution," *New Scientist*, October 22, 1970, pp. 171–73; "Scientists Hoard Old Grain Seeds," *New York Times*, October 4, 1970.

82. "U.N. Group Warns on Push to Cities," *New York Times*, July 21, 1964; W. Whitney Hicks, "Agricultural Development in Northern Mexico, 1940–1960," *Land Economics* 43, 4 (November 1967): 393–402; Donald K. Freebairn, "The Dichotomy of Prosperity and Poverty in Mexican Agriculture," *Land Economics* 45, 1 (February 1969): 31–42; Lena G. Dalrymple, "New Cereal Varieties: Wheat and Corn in Mexico," *AID Spring Review*, May 1969, in Barreto, ed., *A Study*, p. 94; R. D'Shaw, "The Employment Implications of the Green Revolution" (Washington, D.C.: Overseas Development Council [mimeograph], June 1970).

83. "Green Revolution Transforming Indian Farming, but It Has a Long Way to Go," *New York Times*, May 28, 1969; Wolf Ladejinsky, "The Ironies of India's Green Revolution," *Foreign Affairs* 48, 4 (July 1970): 758–68; "Editorial: A Green Revolution Turns Red," *New York Times*, January 8, 1970; Harald Monthe-Kaas, "Green and Red Revolutions," *Far Eastern Economic Review* 67, 12 (March 19, 1970): 21–24; Solon Barraclough and Jacobo Schatan, "Technological Policy and Agricultural Development," May 1970, in Barreto, ed., *A Study*, pp. 29–30; Roger Loya, "The Green Revolution: Some Comments on the Roles of Rural Institutions," November 1970, in Bareto, ed., *A Study*, pp. 85–86; Samir Amin, "The Limits of the Green Revolution," *Ceres* 3, 4 (July–August 1970): 15.

84. "World Food Parley Warned of Short Food Supplies," *New York Times*, August 11, 1970; "The Green Revolution Has Sharply Increased Grain Yields but May Cause Problems," *New York Times*, October 22, 1970.

85. Murray Bookchin, "Toward an Ecological Solution," *Ramparts* 8, 11 (May 1970): 7–15; Robert Allen, "Review of *The Subversive Science: Essays toward an Ecology of Man*," *The Ecologist* 1, 3 (September 1970): 43–44; William Paddock, "How Green Is the Revolution?" *Bioscience* 20, 16 (August 1970): 897.

86. Paul Erlich and John P. Holden, "Population and Panaceas: A

Technological Perspective," *Bioscience* 19, 12 (December 1969): 1065; "Canadian Urges Nature Revival," *New York Times*, March 16, 1969.

87. Michael Allaby, "One Jump Ahead of Malthus," *The Ecologist* 1, 1 (July 1970): 28.

88. "The End of Civilization Feared by Biochemist," *New York Times*, November 19, 1970.

89. "Foe of Pollution Sees Lack of Time," *New York Times*, August 10, 1969; "McNamara Appeals for Population Curb in Notre Dame Talk," *New York Times*, May 2, 1969; "Editorial: Malthus Revisted," *New York Times*, May 9, 1969; "World Botanical Congress to Meet in Soviet in 1975," *New York Times*, September 7, 1969; W.L. Klarman, "Letter to the 'Symposium on Science and Foreign Policy—The Green Revolution," December 5, 1969, in Barretto, ed., *A Study*, p. 26.

90. Roberto Osoyo, "Mexico: From Food Deficits to Self-Sufficiency," in *Symposium on the Strategy*, p. 10; DGA, "Avances logrados por nuestros técnicos de agricultura," *Tierra* 24, 6 (January 1969): 462.

91. "Coin Column," *New York Times*, October 19, 1969; "Juan Gil Preciado contesta que las condiciones generales son más favorables," *Tierra* 25, 8 (August 1970): 581; "Interview with Edwin J. Wellhausen," in Jiménez Sánchez, *Las ciencias agrícolas*, p. 290.

92. "Algunos resultados de la técnica agrícola en México," *Tierra* 24, 7 (July 1969): 504; SAF, "La enorme importancia de la investigación agrícola," *Tierra* 25, 6 (June 1970): 431; "Origen y importancia," pp. 417–18; "Siembra del maiz," *Tierra* 24, 3 (March 1969): 184.

93. DGA, "Avances," p. 462.

94. "Inauguración del Congreso de la Asociación de Editores de Revistas Agropecuarias de Estados Unidos de Norteamérica," *Tierra* 24, 8 (August 1969): 579.

95. "Editorial: Norman Ernest Borlaugh [*sic*]," *Tierra* 25, 11 (November 1970): 815.

96. "Pensemos en un México mejor," *Tierra* 24, 2 (February 1969): 106; López Zamora, *El agua*, pp. 233–37; Edmundo Flores, *Vieja revolución, nuevas problemas* (México, D.F.: Cuaderno de Joaquín Mortiz, 1970), pp. 79–83; Hewitt de Alcantara, *Modernizing*, p. 45; Orive Alba, *La irrigación*, p. 229; "Editorial: C.I.M.M.Y.T.," *Tierra* 25, 8 (August 1970): 575; "El maiz híbrido," *Tierra* 24, 6 (June 1969): 432, 477–79; "Editorial: Agricultura paradogica," *Tierra* 24, 11 (November 1969): 815; "Editorial: Hacia una nueva política agrícola," *Tierra* 25, 7 (July 1970): 495; "Editorial: Agricultural de temporal," *Tierra* 25, 4 (April 1970): 255; "Editorial: Materia orgánica," *Tierra* 25, 6 (June 1970): 415; "Editorial: El espejismo de la Revolución Verde," *Tierra* 25, 2 (February 1970): 95; Blanco Macias, "DDT, el homicido silencioso," *Tierra* 25, 1 (January 1970): 44.

97. "Text of Nixon Message to Congress on Foreign Aid," *New York*

Times, May 29, 1969; Delbert T. Myron, "Integrating the Rural Market into the National Economy of Mexico," *Development Digest (USAID)* 8, 4 (October 1970).

98. Lester R. Brown, *The Seeds of Change: The Green Revolution and Development in the 1970s* (New York: Praeger, 1970), p. 196.

99. Brown, *The Seeds*, p. 16.

100. "Report Hits Book by Rachel Carson," *New York Times*, April 20, 1965; Johnson, *The Green Revolution*, pp. 4, 14–15.

101. "Rice Boom in Asia Raises Doubts," *New York Times*, April 6, 1970.

102. Rockefeller Foundation, *The President's Review and Annual Report, 1970* (New York: The Rockefeller Foundation, 1970), pp. xxi–xxiii; "Scientist Sees Man's Activities Ruling Climate by 2000," *New York Times*, April 30, 1970.

103. "Population Control Urged," *New York Times*, December 11, 1970; Johnson, *The Green Revolution*, pp. 6–7; "Prize to Aid Research," *New York Times*, October 22, 1970.

104. "Interview with Mangelsdorf," pp. 109–11; "Interview with Wellhausen," pp. 102, 109–10.

105. "Interview with Mangelsdorf," p. 110.

106. "Interview with Wellhausen," pp. 56, 171.

107. Cobb, "Interview with Borlaug," p. 72.

108. Camp, "A Middle Level Technocrat," pp. 577–79; Bruce H. Jennings, *Foundations of International Agricultural Research: Science and Politics in Mexican Agriculture* (Boulder, Colo.: Westview Press, 1988), pp. 63, 102; Cleaves, *Professions*, p. 68; Friedrich E. Schuler, *Mexico between Hitler and Roosevelt: Mexican Foreign Relations in the Age of Lázaro Cárdenas* (Albuquerque: University of New Mexico Press, 1998), pp. 34–35.

Chapter 7

The Aftermath of the First Green Revolution and the Birth of a Second? 1970–2002

The shortcomings of the Green Revolution and the energy crisis of the 1970s stimulated interest in new approaches to agricultural research and development. The Green Revolution was only a short-term solution to Mexico's food supply problems. During the 1960s Mexico began to import basic foodstuffs, and in the 1970s, 15 to 20% of the cereal grains that the people consumed came from abroad. Even Borlaug's wheat provided only a temporary respite as deficits began in 1974, but the chemical fertilizer and pesticide industries that grew with the Green Revolution continued to expand as commercial farmers diversified into winter vegetables and sorghum. The Green Revolution internationalized *agronomía,* and new ideas from abroad had some impact in Mexico, but the legacies of the Rockefeller Foundation's training program and the *agrónomos'* professionalization combined with the realities of Mexican politics to preserve the dominance of a top-down structured and plant-breeding and agrochemical-based approach to rural modernization.

Faced with the political debacle of Tlatelolco and criticism of the Green Revolution, politicians developed projects to assist peasant farmers and eliminate the "dualistic" structure of Mexican agriculture.[1] President Luis Echeverría (1970–1976) evoked *cardenismo* by reviving *agrarismo,* even expropriating prime farmland in the Yaqui Valley, establishing new collective *ejidos,* and beginning the *Proyecto Internacional de Desarrollo Rural* (International Rural

Development Project, hereafter PIDER) to modernize *de temporal* and tropical agriculture. Inspired by the anti-gringo tradition of Mexican nationalism and the optimism generated by the short-lived oil boom, in 1980 President José López Portillo (1976–1982) promoted food self-sufficiency to avoid trading non-renewable fossil fuels for U.S. cereal grains. He spent more money to modernize peasant farming by founding the *Sistema Alimentario Mexicano*, which subsidized Green Revolution inputs, raised fixed prices for basic foodstuffs, and created more comprehensive crop insurance for *de temporal* producers. He also brought an agricultural mission from Hungary to challenge the dominance of U.S. science.

In 1982 this project went bust when oil prices collapsed, inflation skyrocketed, and a serious recession began. The 1980s, called the "lost decade," put more stress on the *campesinos*. With the rising cost of agrochemicals and seeds, usurious interest rates, bribes demanded by corrupt officials in the agricultural banks, and low prices for crops grown for the internal market, especially corn and beans, small-scale food production fetched little to no income. More peasants fled to the United States or overpopulated and polluted urban areas like Mexico City.[2] President Roberto de la Madrid (1982–1988) experimented with neoliberal economic policies, reducing subsidies and joining the General Agreement on Tariffs and Trade in 1986 to establish a free marketplace and open the door for more foreign investment. In 1988, after being declared winner of a hotly contested election, PRI presidential candidate Carlos Salinas de Gortari embarked on an extensive neoliberal project, privatizing many parastatals. Rejecting the State's agrarian policy since the Mexican Revolution, Salinas terminated the land reform and began to grant *ejidatarios* titles to their *parcelas*. The North American Free Trade Agreement (NAFTA) put some previously protected rural producers in direct competition with U.S. farmers but also opened the market to more Mexican farm exports. By "inserting the rural village into the global economy," Salinas hoped to use neoliberalism to begin a second Green Revolution by inviting foreign agribusinesses to transform the *campesinos* into prosperous farmers.

THE FADING LUSTER OF THE GREEN REVOLUTION AND THE EMERGENCE OF NEW APPROACHES TO RURAL DEVELOPMENT

In the early 1970s criticism of the Green Revolution continued to mount. The UN worried about Third World famines and land concentration, biologist J.N. Black attacked the use of non-renewable fossil fuels to grow food, and anti-hunger activist Francis Moore Lappe's best-seller *Diet for a Small Planet* linked Third World famines to the developed nations' consumption of resource-intensive foods like meat. Organic farmers denounced agrochemicals, and environmentalists chastised the industry's public relations campaign against the movement. Anthropologist Marvin Harris asked:

> Can we continue to believe that the creation of high yielding rice was inspired by a vision of miracles for the peasants of Asia? I think not. For the peasants, these seeds . . . are anti-miracles, veritable seeds of their own destruction. The precise objective of the managers of the Green Revolution is to wipe out the class of small farmers and to replace them with efficient agribusiness men who will be heavily dependent upon industrial products and world markets.[3]

In his 1972 book *The Green Revolution*, Stanley Johnson described a heroic struggle of scientists working to avert catastrophic famines, but he warned that without birth control, the "population crunch" would be even worse with a larger world population.[4]

The 1973 energy crisis drove the price of agrochemicals higher, and pundits worried about fertilizer shortages causing Third World famines. David Pimentel and other distinguished scientists warned that "a careful assessment should be made of the benefits, costs, and risks of high energy demand Green Revolution agriculture . . . to be certain that this program will not aggravate the already serious world food situation," providing mainstream scientific support to organic farming and appropriate technology advocates.[5] Dismayed by Asian and African rice shortages, the *New York Times* said that "charts plotting production since World War II offer little evidence that a 'Green Revolution' exists" and compared experts who used expressions like "miracle rice" to those who saw "light at the end of the tunnel" in Vietnam.[6] In 1974 the threat of famine in India made the world food situation look even worse. Believing that world events since 1967 validated their predictions, the Paddock brothers confidently reissued *Famine 1975*, changing the title to *Time of Famines*. They chastised the Rockefeller Foundation and other agencies for spreading biased information about the Green Revolution's potential to solve world hunger and claimed that it fostered complacency in developing birth control programs. In 1976 Michigan State University President Clifton R. Wharton Jr. told the World Food Conference that millions of people in the Third World faced starvation within a decade. Evidence that the Green Revolution made rich farmers richer and poor ones poorer mounted, and some critics like French agronomist Rene Dumont and Lappe took a radical stance by proposing revolutionary change or the reduction of First World standards of living as the only solution to the food/population problem. In 1976 anthropologist Hewitt de Alcantara said that the MAP's scientists ignored the social, economic, and political forces that affected peasant farmers and failed to consider whether or not they could use their new technologies. Other critics claimed that the Green Revolution helped Third World leaders dodge politically charged issues like land reform, the redistribution of wealth, and birth control.[7]

During the 1970s many scientists still rejected organic farming. For example, Daniel L. Arrom told the American Association for the Advancement of Science that "reading the literature of the organic food movement, one is astonished by their ignorance of chemistry and their wishful thinking."[8] A few, however, such as Hardy Vogtmann of the University of Alberta and Mark Cowan of Great Britain's Dye Research Center joined alternative agriculture enthusiasts like University of Kentucky literature professor Wendell Berry and environmentalists like John J. DuPont to advocate organic farming practices and chastise colleagues who ignored them. To reduce pesticide use, scientists at 18 universities used computers to build models of crop ecosystems. In 1978 S.H. Wittner, director of the Michigan State Agricultural Experimentation Station, praised "non-polluting . . . sparing of capital, management, and non-renewable resources" and "scale neutral" technologies, and said that the United States "could make its greatest contribution" to the Third World by spending more money to develop them. In 1979 the USDA began to study organic farming, claiming that "the challenge to agriculture in this decade [the 1980s] will be to develop farming methods that produce the necessary quantity and quality of food . . . without adversely affecting our soil resources and environment. It is likely that organic technology . . . will play an important role in meeting that challenge."[9] In 1980 Robert Rodale, denied an audience by Alemán's SAF, said that organic farming was the only way to prevent soil depletion, the scarcity of petrochemicals, and population growth from causing a catastrophic famine by 2000.[10]

The issues raised by the environmental movement, the organic farmers, Marxists, and the Green Revolution's critics led some scientists to seek radical solutions. Building on British economist E.F. Shumacher's "Small is Beautiful" concept and Wendell Berry's ecosystems approach to farming, scientists at the New Alchemy Institute at MIT's Woods Hole Oceanographic Institute renounced the ethic of continually increasing production by establishing more control over nature and lamented that large corporations and the military determined the agenda of scientific research. To retain both human and biological diversity, they proposed a "whole systems" approach to develop "appropriate technologies" for peasant farmers to keep them on the land without damaging the environment or fostering greater dependence on nonrenewable resources, and they urged Third World nations to adopt land reform and other policies to redistribute wealth to the poor. They studied wind generators, fish ponds, household sewage purification, and intensive vegetable cultivation to create integrated agricultural systems. With low-cost decentralized research, scientists would seek "ways of living which will help alleviate oppression at all levels, against the earth as well as against people," and peasants would contribute to their modernization.[11] Worried about the energy crisis and rural unemployment, anthropologist Billie R. DeWalt urged scientists to develop "appropriate technologies." During the late 1970s and 1980s some foreign academics and scientists advocated interdisciplinary research, including social and agricultural experts "rooted in an understanding of the rationale underlying traditional production systems . . . [to create] a more labor-intensive land use system that can achieve higher productivity and absorb more labor."[12] Development agencies promised to include peasant farmers in planning and implementing projects, but in practice their input was often superficial or ignored.[13]

Many scientists still thought that new technologies like farming the ocean or irrigating deserts with desalinated seawater would dramatically increase world food production, continued to defend the Green Revolution, and supported the agrochemical industry's campaign against the environmentalists. Joined by Third World leaders like Peruvian Pedro G. Beltrán, they implored the United States to bring the Green Revolution to the rest of the Third World, claiming that this was the only way to stop famine and Communist revolutions. In 1975 World Bank director Warren G. Baum told the press that "Green Revolution must be repeated" on small farms, and *Farm Journal* columnist Roe Black thought that "peasant farmers will adopt the new seeds and farming methods as rapidly as American farmers, when given a chance."[14] University of Chicago economist Theodore Schultz said that environmentalism was not based on "firmly established knowledge." He chastised critics who thought that "agricultural research must be designed to benefit small farmers, as if it were possible for . . . scientists to invent a high-yielding variety of wheat or rice that would not be productive on large farms." Agricultural economist Keith O. Campbell labeled organic and appropriate technology advocates "idealistic and impractical" for urging farmers "to return to their grandfather's methods for the sake of the environment and . . . curtailing fossil fuel use."[15]

Before the energy crisis the Rockefeller Foundation was quite optimistic about the Green Revolution. For example, in 1971 Borlaug said that "millions of farmers" used the new technologies and would eventually prosper, but the true responsibility for the latter fell to "economists, sociologists, and politicians," not agricultural scientists, and claimed that he "was convinced" that peasants would use birth control "once the technique is available . . . as they adopted the change in agricultural methods that gave rise to the so-called Green Revolution."[16] An ardent critic of the environmental movement, Borlaug called *Silent Spring* "a diabolical, vitriolic, bitter, one-sided attack" and warned that "unless this vicious, shortsighted propaganda campaign is stopped, it will bring disaster to the world."[17] Similarly, Lewis Roberts told the press that "the spread of the Green Revolution could double cereal production in the next 25 years," CIMMYT and IRRI scientists Randolph Barker and Don Winkelmen said that "social conditions" would improve as more farmers used high-yield strains, and D.S. Athwal, IRRI associate director, warned that "either you apply the new technology or you starve."[18] After 1973 some Green Revolutionaries were still optimistic about its potential. In 1975 foundation vice president Sterling Wortman continued to speak of "miracle" rice and wheat. In 1975 Borlaug said that politicians caused the Green Revolution's negative economic consequences, not its technologies, and he urged Third World leaders to bring it to peasant farmers. In 1976 foundation scientist Peter R. Jennings admitted that the Green Revolution did "not solve the problem of the world's food supply, but rather . . . demonstrated an approach to a solution" and maintained that "recent developments in agricultural research give reason for optimism," and Chandler told reporters that the spread of Green Revolution varieties would avert world famine, at least until the end of the century.[19]

The energy crisis and criticism of the Green Revolution did lead to changes in the Rockefeller Foundation's approach to agricultural development. Recognizing that traditional farming systems include non-resource-intensive methods of controlling pests and maintaining soil fertility, foundation scientists began to study them. D.J. Greenland, director of the Tropical Agricultural Center in Nigeria, advocated intercropping tree and food crops and labor-intensive rather than mechanized cultivation, and the IRRI tested rice intercropping. Reflecting the "era of limits" philosophy, in 1975 President John H. Knowles wrote, "Worshiping mindlessly at the altar of an ideology based on endlessly growing production and consumption will become increasingly expensive to us and our fellow riders on this planet of ours."[20] To address concerns about genetic erosion, CIMMYT began to mix seeds of *chahuixtle*-resistant wheats. The Social Sciences Division studied Mexico's Green Revolution and confirmed that its benefits were distributed unevenly by region and social class. In 1976 Wellhausen told *Scientific American* that the Green Revolution reached "larger, more commercial farmers, who were in a better position to afford fertilizer and other inputs," and said that Mexico, "having accomplished one agricultural revolution . . . finds itself sorely in need of another."[21] To devise better programs for peasant farmers, he urged scientists to study the social sciences to understand traditional agriculture, but he still believed that *campesinos* could use Green Revolution technologies profitably with proper technical support. Wortman admitted that "the introduction into developing countries of Western-style, large scale, mechanized farming is not a solution," and he urged Third World nations to adopt birth control programs.[22] By 1979 Borlaug was making dire predictions about agricultural research and the population explosion. He implored Third World leaders to enact rural poverty programs and convince religious leaders to support birth control programs immediately. In 1978 Green Revolution booster-turned-Worldwatch Institute environmentalist Lester Brown advocated an ecological approach to rural development, denounced agrochemical farming, and urged Third World nations to "reform elitist social and political structures." In 1980, advocating lifestyle changes in the First World, he told *U.S. News and World Report* that "we are at the outer limits of what the Green Revolution can deliver."[23]

By 1980 the threat of world famine and the publicity surrounding the Green Revolution diminished, but some experts still worried about the loss of genetic stock due to the domination of a few multinational corporations in the seed market, and others urged scientists to create new technologies so farm productivity would keep pace with population growth. During the 1980s many Third World nations made impressive gains in food production, discrediting the alarmists. India, China, and even Bangladesh attained food self-sufficiency, and the energy crisis became the oil glut. In 1989 the International Food Policy Institute claimed that the increase in Third World food production proved that Green Revolution technologies were viable on small farms. New varieties contributed to the gains, which diminished the market for U.S. grain exports. To protect U.S. farmers, in 1986 Bob Meyer, ex-president Jimmy Carter's assistant secretary of agriculture, wanted to stop "giving away" farm technology, and even proposed reducing funding for research to prevent more technological breakthroughs that might put them out of business. Ironically, the Green Revolution also provided the setting for James C. Scott's classic 1985 monograph on peasant resistance, *Weapons of the Weak*.[24]

During the 1980s some experts claimed that a second Green Revolution based on genetic engineering would let food production keep pace with population growth and make peasants prosper, but others worried that the results of the first one would be repeated: large farmers and agribusiness would get richer while small and medium-scale producers could not afford the new technologies. Like the first Green Revolution, the social, political, and economic context of genetic engineering research will put a strong imprint on the new technologies. Scientists may end malnutrition by creating more nutritious cereal grains like high-protein rice or make profitable agriculture possible in arid regions by developing drought-resistant varieties. Some new technologies like nitrogen-fixing corn or wheat that make their own insecticides might reduce the use of agrochemicals, but techniques like cloning pose a new threat to the genetic diversity of crop plants. Suggesting the probable outcome of a biotechnology-based second Green Revolution, which demands costly and complicated research with substantial funding and assistance from the governments and universities, transnational agribusiness firms based in the First World used patent laws and control over research to develop technologies that sustain demand for agrochemicals and dependence on other industrial inputs. Reflecting continued disregard for *campesino* agriculture, at the behest of agribusiness the Third World's germ plasm became the "common heritage of mankind," but firms retained proprietary control of new strains developed by genetic engineering.[25]

During the 1980s and 1990s scientists supported by the Rockefeller Foundation and other government and non-governmental organizations continued to develop new crop varieties and study the lessons of the Green Revolution. For example, after 15 years of crossbreeding, the International Potato Center in Lima, Peru, perfected the "hairy potato," which has hairs on its stalks that contain a substance that constipates the Colorado beetle, damaging its reproductive organs, and traps smaller insects like aphids, reducing the need for pesticides. In 1982 Wellhausen admitted that Green Revolution technologies did not meet the needs of peasant farmers. He supported researchers' efforts to modify them and urged Mexico to undertake an all-out campaign to make *campesinos* more productive. Reflecting the recognition that critics of the Green Revolution had merit, during the 1980s some agencies adopted "farming systems research" to foster political empowerment, open and democratic communication between experts and peasants, and mutual respect for modern and traditional knowledge.

Bolstered by the 1987 report of the UN World Commission and Environment and Development, during the 1990s First World advocates of sustainable development, such as the Oklahoma-based World Neighbors Aid, worked with peasant farmers in Third World nations like Honduras and Kenya to find environmentally sound methods of increasing their income, such as raising rainforest butterflies. In 1988 the World Bank, which managed the 13 research institutes that the MAP spawned, admitted that in some situations, like regions of Africa with abundant and inexpensive land, traditional, extensive methods are more cost-effective than labor- and input-intensive Green Revolution farming. In 1991 they promised to "take into account farmers' needs and knowledge" and avert environmental problems by studying soil management and biological pest control.[26] In Mexico CIMMYT began to work on sustainable wheat and corn strains for poor farmers, using genetic engineering to add built-in tolerances to pests, diseases, and environmental stresses, and also tried to attain one of the *cardenistas'* objectives: creating a high-quality protein-content maize. In 1994 Worldwatch sounded another warning about the earth's capacity to produce more food: "The world's farmers can no longer be counted on to feed the projected additions to our numbers. . . . As the nineties unfold, the world is facing a day of reckoning."[27] Serious famines occurred in several Third World nations, where many rural people remain impoverished and malnourished, but by 2002 the worldwide catastrophe forecast by the Paddocks and others thankfully had not occurred.

THE AFTERMATH OF THE GREEN REVOLUTION AND THE *AGRÓNOMOS*

The international debate surrounding the Green Revolution influenced the *agrónomos,* as did political changes in the Revolutionary Family after Tlateloloco. In 1970 Mexico's quest for scientific self-sufficiency and autonomy led to the founding of the *Consejo Nacional de Ciencia y Technología* (National Council of Science and Technology, CONACYT), which supports research and education. Echeverría increased undergraduate agronomy programs from 4 to more than 30 and doubled the SAF's research budget. In 1976, 10 schools offered Masters degrees. In 1971 the SAM praised him for creating new jobs in the agrobureaucracy. *Agrónomos* campaigned for candidates and carefully watched the ebb and flow of PRI politics to learn "where they will be when jobs are distributed" during the next administration. When ministers of agriculture took office, *agrónomos* knew that a new *patrón* brought many *clientes* seeking prime jobs with him. *Agrónomos* continued to prosper in politics. Echeverría appointed Brauer H. minister of agriculture, and in 1975 rumors about his presidential aspirations circulated. In the 1970s and early 1980s the demand for *agrónomos* was strong. The profession attracted more young Mexicans, including some *campesino* youths like Pedro Ariziaga, who saw professional training as his only chance for upward socioeconomic mobility. By 1980, 40,000 Mexicans studied agronomy, and most who finished their degrees found jobs as the State continued to increase expenditures on research and extension. For example, de la Madrid decentralized agricultural programs to facilitate adaptation to local conditions. In 1984 the SAF employed 1,806 *investigadores* (researchers) and some 18,000 extension agents, over 3,000 more than the USDA, but the economic crisis of the 1980s forced the State to retrench. By 1988 new *agrónomos* faced an uncertain future.[28]

Taking their cue from Echeverría's leftist rhetoric and his appeal to nationalism by conducting an independent foreign policy, *agrónomos* discussed land reform, denounced politically powerful farmers who circumvented the Agrarian Codes, and criticized the Green Revolution's economic shortcomings. Scholars acknowledged rural Mexico's dualistic structure and lamented the effects of technological colonialism. The Rockefeller Foundation and the Green Revolution became rather "politically incorrect" in Mexico. In 1976, when Fernández y Fernández wrote a history of the ENA, he felt compelled to say that "the Cárdenas administration's agricultural research program "continued in its infancy. In 1940 the Rockefeller Foundation had to come to put it on sound foundations. But, with the ideas in vogue today, acknowledging this is seen negatively as is even more the least expression of gratitude."[29] Similarly, in 1971 López Zamora claimed that *ejidatarios* did not benefit from Calles' irrigation projects because foreign experts worked in CNI. Sociologist Jerjes Aguirre Avellaneda lamented that research was "directed at irrigated export crops, with a deep foreign [mainly North American] influence that fostered a divorce between technological and social issues and projects that almost completely ignore the subsistence *de temporal* farming that feeds approximately 10 million Mexicans."[30] These attitudes persisted under López Portillo. For example, in 1977 *agrónomo* Rafael Ortega Paczka said that foreign-trained professors had too much influence at the ENA, claimed that his overspecialized colleagues ignored the big picture of rural development, and lamented that they failed to recognize the problems that Green Revolution agriculture caused. Similarly, in 1981 Castaños admitted that "the problems of Mexican agriculture are more than technical problems, they are social problems, and they should not be analyzed only in the cold context of science," and he lamented that research institutes like the Construction Industry Association of New Orleans (CIANO) solved the problems of other nations' farmers.[31]

Criticism of the Green Revolution had some impact on Mexican agronomy. Influenced by Marxist critiques of technological development, the example of Mao Tse-tung's China, environmentalists, appropriate technology advocates, and the ideas of Argentine scientist Oscar Varsavsky, a few intellectuals even questioned whether Mexico should replicate the industry and technology-based economic model of the developed nations. Anthropologist Judith Friedlander found one young ENA student, an *agrónomo*'s son, who joined the *indigenista* group Confederated Movement for the Restoration of Anauak. Along with other students of rural Mexico who acknowledged the failure of prior attempts to introduce Green Revolution agriculture, some *agrónomos* found wisdom in the *campesinos'* farming practices. For example, INIA *agrónomo* Rogelio López Ildefonso claimed that intercropped corn and beans generated more income than separate cultivation and advocated research to make the system more productive. When the government began an agricultural program in Campeche, *agrónomos* promised to learn from the local Maya. Dr. Juan M. Mattieu warned that "in the future the technological development of agriculture should balance the agroecosystem without completely transforming the environment."[32] Calling the MAP's training program for *agrónomos* a legacy of "intellectual colonialism," ENA professor and former foundation fellow Hernández Xolocotzi wanted to "Mexicanize" research. He advocated a broad socioeconomic approach to rural development, began to study traditional farming systems like corn-bean intercropping, and trained students to extend his work. Other *agrónomos* told their colleagues to study peasant farming systems before trying to modify them. Castaños urged directors of agricultural programs to include rural power brokers like *comerciantes libres* (businessmen who buy from peasant farmers), *caciques* (village leaders), and even priests to facilitate *campesino* cooperation by working with instead of against the culture in which they lived and farmed.[33]

Reacting to the criticism of the Green Revolution, during the 1980s *agrónomos* like Rodríguez Adame and Taboada said that farmers and *campesinos* should provide input to the planning stage of agricultural programs. The ENA's *Colegio de Postgraduados* ran appropriate technology projects in *ejidos* in Puebla and Veracruz. The National Institute for Biotic Research in Jalapa and the Quintana Roo Research Institute helped *ejidatarios* learn sustainable forestry, recycling of biomass, fish farming, and small-scale livestock production. The *Secretaría de Pesca* (Ministry of Fisheries) promoted aquaculture in *ejidos* to provide "fish for the *ejidatario*'s table."[34] In El Guamuchil, Jalisco, young irrigation engineers used the *ejidatarios'* methods of removing heavy stones and worked alongside them after they realized that science offered no economically feasible way to solve their problems. Gledhill found that the *agrónomos* who worked for PRONASE in Briseñas, Michoacán, "talked to rather than at" the *campesinos* and spent some time with them in the fields.[35] In 1987 Rodríguez Adame joined conservationist Enrique Beltrán on the board of the *Instituto Mexicano de Recursos Naturales Renovables* (Mexican Institute of Renewable Natural Resources). Alanis Patiño proposed a national campaign to stop soil erosion, family planning, and an intensive research and extension program to make *campesinos* more productive farmers. In 1988 the SAF's Carlos Vidali urged the government to support genetic engineering research, but he also admitted that "biotechnology is not a panacea that lets us abandon the utilization and improvement of traditional production technologies."[36] However, the appropriate technology advocates represented a minority of the *agrónomos*.

The Green Revolution and the MAP's training put a profound stamp on Mexican agronomy. Many *agrónomos* of the revolutionary era like Luis León, who in 1971 said that the Green Revolution was the final phase of the agrarian reform, thought that it was a total success. In 1984 de la Loma called criticism of the Green

Revolution "absurd." Jesús Alarcón Moreno, who attended the ENA in the 1920s, thought that the Rockefeller Foundation's work brought great benefits to Mexico. *Tierra* praised MAP scientists like Wellhausen and thanked them for increasing corn production. Even critics of the government's agricultural policy like economist Sergio Reyes Osorio lauded the achievements of Mexico's plant breeders and soil scientists. *Agrónomos* and parastatals maintained close ties with U.S. researchers and institutes and used them as sources of knowledge for Mexico's programs. Agricultural journals urged the State to hire more *agrónomos* to "liberate" the *campesinos* with Green Revolution technologies. Most *agrónomos* believed in scientific agriculture and denigrated peasant farming systems. The *agrónomos* who designed Echeverría's program to modernize *campesinos'* "primitive technologies" concluded that they did not adopt the Green Revolution because they could not afford it. To solve the problem, the *Compania Nacional de Subsistencias Populares* (National Company of Popular Subsistences, hereafter CONASUPO) adopted various measures to increase their purchasing power to help them buy fertilizers, pesticides, and improved seeds, leading to a dramatic increase in the agency's budget. The SEP's 260 secondary agricultural schools (*Escuelas Técnicas Agrícolas*) taught agrochemical farming, the use of hybrid seed, and other principles of the Green Revolution to "change the ancestral customs of *campesino* farming."[37]

During the 1970s and 1980s the SAF's research institutions continued to develop new hybrids, including for tropical and *de temporal* agriculture. In Campeche SAF and agricultural bank *agrónomos* forced peasant borrowers to use hybrid seed. Recognizing that hybrids were less drought-resistant than local strains, the *campesinos* tried to breed varieties adapted to their conditions, but some lost the seed stock that they had used for many years. The government and *agrónomos* like Norberto Aguirre appreciated the international prestige earned by creating and exporting improved plant varieties to countries like India, Iraq, Iran, Pakistan, and Syria. Many *agrónomos* believed in agrochemicals, and some had a callous attitude toward farmworkers whose health was harmed by them. During the 1980s an SAF colonization project in Yohaltún, Campeche, caused an environmental disaster after several years of rice monoculture made the soil incapable of absorbing water, all but destroying its ability to sustain agriculture of any type. With their career tied to the support of politically influential strawberry farmers, CNI employees in Zamora, Michoacán, helped them violate irrigation laws by looking the other way when they exceeded water allotments. To build political support for the *Sistema Alimentario Mexicano*, *agrónomos* again claimed that hybrid seeds, agrochemicals, and farm machinery offered a technological solution to the *campesinos'* poverty and inefficient farming systems. Edmundo Flores, who participated in the *agrónomos'* transformation from *agraristas* to *técnicos*, claimed that the Rockefeller Foundation's assistance led to "the genetic improvement of corn and wheat that made the Green Revolution possible . . . [and denounced] 'small is beautiful,' and appropriate technologies . . . [as] either nonsense, if taken seriously, or political diversions to attain goals of a different nature."[38]

Agrónomos still saw *campesinos* and their culture as impediments to the success of their projects. For example, Castaños said that their "fatalism" and obedience to priests, who told peasants that sunflowers came from "atheistic . . . Communist" Russia, impeded a 1971 campaign to establish the crop in Durango. The directors of the collective *ejidos* in the *Plan Chontalpa* irrigation project and the *agrónomo* who ran the *Plan Maiz* in Temascalcingo used coercion to force *campesinos* to cooperate. Rejecting peasants' knowledge of the region's climate and geography as inferior to their "scientific" training, *agrónomos* who ran the *Plan Chontalpa* forced them to plant corn in a flood-prone area, claiming that the drains they built would solve the problem, but heavy rains drowned the entire crop. Similarly, after ignoring *campesinos'* insistence that they planted too late in the growing season, weeds, drought, and finally floods destroyed the improved rice that *agrónomos* introduced. Under Echeverría, when CONASUPO's Field Coordination Program gave *campesinos* a voice in planning community development projects and responded to their requests by acting as a broker with other agencies, state and local politicians and *caciques* resisted projects that undermined their wealth or power, and the project ended when he left office. In Jalisco, the media lauded the triumph of Hungarian technology when the *ejido* Lomas de Tejada produced a bumper corn crop for the *Sistema Alimentario Mexicano*, forgetting that the *campesinos'* decision to ignore the foreign *técnicos* and plant in March to benefit from an early rainy season was the main reason that they succeeded. Many of the bureaucrats whom Marilyn Gates and Alberto Arce interviewed during their research on agricultural programs in Campeche and Jalisco claimed that *campesinos'* apathy, laziness, irresponsibility, unwillingness to work together, and resistance made them fail. The bureaucrats who ran the *Sistema Alimentario Mexicano* did not consult peasants when they planned and implemented the program. Contrary to Arce, many *agrónomos* did not "abandon . . . traditional bureaucratic stereotypes about rainfed peasants resisting change because of ignorance or traditionalism."[39]

During the 1980s, with the encouragement of senior *agrónomos* like Rodríguez Adame, some Mexican scientists entered the field of biotechnology. In 1982 the *Instituto de Mejoramiento de la Producción de Azúcar,* operated by the parastatal sugar monopoly *Azúcar, S.A.,* began an ambitious project by using genetic engineering to develop new cane varieties but decided to abandon it in 1984. By 1987 some 100 Mexican scientists conducted basic biotechnology research related to crop plants, including two Ph.D.s and six technicians from the Biotechnology Laboratory of the Genetics Center at the ENA's *Colegio de Postgraduados,* who worked on export crops like strawberries and pineapples, as well as other tropical fruits like papayas, agaves, nopal cactus, and forest trees, but only researchers at the *Centro de Investigación y de Estudios Avanzados* (Center of Investigation and Advanced Studies, CINVESTAV) in Irapuato tried to develop new basic food crop varieties. Antonio Turrent Fernández complained that "the curricula [at most agricultural schools] are oriented toward modern, capital intensive agriculture or livestock operations."[40]

Some scholars and MAP scientists claim that the Rockefeller Foundation's training program established the *agrónomos'* credibility as scientific professionals and ended their aversion to "getting their hands dirty" conducting experiments in the fields, but other academics and observers paint a different picture. Stakman thought that the MAP eradicated *proyectismo,* but Gates says that it still affects Mexico's agricultural programs, as does "the dominant technocratic mentality of many agricultural development agencies" that treat *campesinos* like "a resource to be managed."[41] The Green Revolution did not establish the *agrónomos'* professional credibility with *campesinos* who did not adopt it or even with some farmers who did. Some of Sonora's wheat growers suspected that the seed that PRONASE sold caused a 1977 outbreak of *chahuixtle,* one that *agrónomo* Carlos Castaños called "a parasite infesting public agencies . . . [with an] atrophied brain," and the local press called the SAF incompetent for failing to solve the problem.[42] Echeverría's revival of *agrarismo* made *pequeños propietarios* distrust and even detest the agrobureaucracy and the *agrónomos* who worked in it. De la Loma thought that many Mexicans believed that the ENA trained Communists. The legacy of *agrarismo* continued to diminish the *agrónomos'* credibility as scientific professionals.

When he began his career in 1971, University of Guadalajara agronomy graduate "Sr. López" easily found work as an SAF extension agent because *agrónomos* from older and more prestigious schools like the ENA competed for administrative jobs in Mexico City. Confirming the findings of students of Mexican higher education, Castaños and de la Loma claimed that the proliferation of agricultural schools diminished the quality of scientific training because few good teachers were available to staff them. They also said that their younger colleagues preferred urban, middle-class comforts over social service to the *campesinos*, like the attitude of fictional *agrónomo* Cruz López in playwright Felipe Santander's prizewinning 1978 work *El extensionista* (The Extension Agent). *Campesino* youths who became *agrónomos* tended to stay in the cities where they went to school, and sometimes their parents left the countryside with them. In 1984 Álvaro Echeverría Zuno lamented that too many SAF employees, especially the most experienced, hold urban administrative positions, rather than work with *campesinos* in rural areas. That year journalist Alan Riding observed that:

> government agricultural research and extension agents contribute little to the ejidos. Thousands of agronomists and extension agents graduate every year, but more prefer to work in offices rather than in the field. Such research stations as do exist frequently ignore the local needs of farmers; the result is minimal communication between the researcher, the extension agent, and the peasant.[43]

In the 1970s DeWalt found that most *ejidatarios* in Temascalcingo rarely talked to government experts, and those whom Arce interviewed in the 1980s in Nextipac, Jalisco, said that *técnicos* never saw their crops. Veteran *agrarista* Nícolas Domínguez chastised SAF extension agents for their lax sense of duty and unwillingness to go to the fields (they preferred giving instructions at town hall meetings), and Gledhill said that Guaracha's *campesinos* had poor relations with SAF *técnicos* who "seem loath to visit the *parcelas*."[44] The majority of SAF field-workers in Zapopán lived in Guadalajara and used a vehicle purchased with an agency loan to drive 30 to 45 minutes to work. In 1988 the SAF admitted that too many agronomy students came from cities and that because the institutions that offer degrees are located there, those from the country prefer working as urban bureaucrats after they graduate and "have no interest in establishing roots and living together with the people of rural areas."[45]

Other typical problems of the past such as job instability, changes in objectives with each new presidential administration, insufficient resources to conduct effective extension programs, and the bureaucratic imperative of numerous reports continued to limit the effectiveness of SAF programs. Some researchers had no interest in solving the problems of the farmers in their districts, and other agrobureaucrats used their jobs for personal enrichment, such as the SAF officials in Jalisco who profited by developing the export melon industry. Internal power struggles and turf wars with other branches of the agrobureaucracy made apparently viable projects fail. For example, the SAF's 1971 sunflower campaign in Durango earned more money for peasants than corn or beans, but competition from the BNCE, the political downfall of Minister of Agriculture Manuel Bernardo Aguirre, and the inability to comprehend *campesino* culture contributed to the program's demise. Unlike Greenberg's assessment of the CNI, the Green Revolution failed to internalize an "ideology of efficiency" in the SAF. In the early 1970s the agency had a bad reputation that led to funding cutbacks. In 1976 López Portillo tried to streamline the agrobureaucracy by merging the SAF with the CNI, but *agrónomos* from the latter attained dominance in the new agency, and irrigated agriculture continued to receive more and better support. Ex-SAF extension agents took control after several years of conflict with former CNI employees, beginning a brief period when *de temporal* farming received some attention under the *Sistema Alimentario Mexicano*, but the victors' main objective was to obtain more political influence. Reflecting the short institutional memory inherent in the *sexenio* (six-year) system, in 1979 the SAF promoted sunflowers in Hidalgo, but conditions that also prevailed in Durango produced the same negative results. Similarly, during the 1980s the agency began new rice projects in Campeche, forgetting disasters like Bonafil and Yohaltún. The neglect of timely supervision of irrigated cornfields in El Guamuchil allowed insects to consume the crop. In 1981 Mesa A. denounced rampant "corruption and treason" in the CNA. Reflecting the opinion of many other *agrónomos* and the *campesinos* in Santander's fictional "Tenochtlén," Castaños lamented that "the people see us as part of the government and the *campesinos* do not distinguish between the PRI, the Agrarian Reform Ministry, the Bank, or CONASUPO; all are the government and in one way or another branches of this government have let them down on many occasions."[46] In 1990, to explain the persistence of rural poverty, Guillermo Knockenhauer criticized bureaucrats and denounced "the paternalism, bureaucratic centralization, and corruption in public agricultural institutions."[47]

THE AFTERMATH OF THE GREEN REVOLUTION AND MEXICAN FARMERS

In an ironic twist of fate, the ostentatious lifestyle of Sonora's wheat growers drove many into bankruptcy in the early 1970s. Some *campesinos* eventually used Green Revolution technologies but surrendered control over farming decisions to the agricultural banks or other agencies that supplied seeds and agrochemicals. In the 1970s most *ejidatarios* in Temascalcingo cooperated with an SAF campaign to introduce chemical fertilizers, despite their poor relations with the *agrónomo* who ran the program. Although some eventually withdrew, the *ejido*'s poorest *campesinos* continued to use them to increase their subsistence food supply. In Oaxaca, Mixtec farmer Alejandro Santiago Salazar reaped bumper harvests of corn when he began to use chemical fertilizers, pesticides, and improved varieties. The INIA's demonstrations of new technologies often attracted thousands of farmers. Studies sponsored by the ENA, parastatals, and the SAF during the 1980s found that many peasant farmers used some Green Revolution technologies such as chemical fertilizers without adopting the whole package like hybrid corn and pesticides. For example, wealthier *campesinos* in western Chiapas with PRI connections regularly bought fertilizers and pesticides. The *Sistema Alimentario Mexicano* gave generous subsidies to get farmers to use Green Revolution technologies to produce basic food crops. From 1979 to 1981

PRONASE doubled its distribution of improved seeds, and *campesinos* in many regions tried them. For example, in 1981 Mesa A. found peasants in northern Guerrero planting improved maize varieties, and Arce discovered that after 1980 most cultivators in Zapopán grew high-yield corn and used chemical fertilizers, but better-off farmers were much more likely to plant more corn, wheat, and beans to take advantage of higher subsidized prices. In 1988 the SAF said that improved varieties produced 29% of Mexico's corn crop. A 1989 study of maize growers with top-quality lands found that 35% used high-yield strains. Thanks to the State's post-1970 programs, many *campesinos* were familiar with hybrids and agrochemicals and had experience interacting with *agrónomos*, but Green Revolution technologies did not bring prosperity to most peasant food producers, despite many attempts to diffuse them more widely.[48]

Similar stories of *campesino* rejection of the Green Revolution transcended the cultural differences of Mexico's diverse rural regions. Continuing a trend that emerged at the height of the 1930s reforms, Nugent found that in Namiquipa mestizo peasants' access to technical assistance and new technologies was often determined by their willingness to cooperate, that is, to submit to agrobureaucrats' authority. To retain autonomy and control over the farming process, these *campesinos* "renounce technologically innovative or capital-intensive methods of working the land to free themselves from the bank and break the cycle of perpetual indebtedness."[49] Sanderson discovered that the Nahua peasants of Amatlán also avoided dependence on external providers of technology. Without access or cash to purchase agrochemicals, peasants in remote villages found that traditional varieties produced more corn than hybrids. The *campesinos* of Nealticán decided their organic fertilizers worked better than the agrochemicals that the government's experts promoted, and they avoided dependence on the often unscrupulous vendors of these products. *Ejidal* politics also influenced technology use because *ejido* presidents and powerful families sometimes provided access to water, seed, or fertilizers only to their allies. The failure of programs like the *Huerto Familiar*, which forced peasants to buy vegetable seeds that did not grow as the experts had promised because of climatic variations, made *campesinos* think that *agrónomos* were incompetent. When soil erosion, broken machinery, drought and torrential rains, a Johnson grass plague introduced in contaminated rice seed, crop failures, and *parcelas* smaller than promised turned the initially successful Bonfil colonization project into a seething cauldron of dissent, the *ejidatarios* ran the SAF's *agrónomos* out of town. In Jalisco, *campesinos* threatened the lives of *Sistema Alimentario Mexicano* fieldworkers who invoked an "idle land law" to force them to grow more corn. Some peasants innovated without *agrónomos'* help, such as Arcadio Amézquita of Lagos de Moreno, Jalisco, who grew lentils successfully. *Campesinos* resented experts who said that they refused to cooperate and *agrónomos* who denigrated their agricultural knowledge, and many viewed with disdain the State, "which never keep[s] . . . promises" and runs programs "based on lies and deceptions."[50]

Some entrepreneurial *ejidatarios* benefited from the State's massive expenditures on agricultural programs from 1970 to 1982, in some cases by reinvesting their short-term earnings in non-farm ventures like trucks or small shops, but the 1980s debt crisis left most peasant farmers in dire straits. *Campesino* food producers bypassed by the Green Revolution remained very poor, and they still ate corn and beans with every meal, adding chiles, tomatoes, onions, eggs, fresh or canned milk, and bread to their diets when they had disposable cash to buy them, but many also spent their money on less healthful processed foods or snacks like *Sabritas* chips (Mexico's Frito-Lay's) or Coca-Cola. Abandoned by the State when the *Sistema Alimentario Mexicano* ended, their prospects for the future seemed bleak as they could not farm profitably or be absorbed by the stagnant economy of the overpopulated cities. For example, in 1986 Benito Juárez Rosas sold his 22-acre *de temporal* sesame farm in Michoacán, moved to the sprawling shantytown of Nezahualcoyotl, and bought four sewing machines for a home tailor shop, but competition from other migrants limited his earnings to around four dollars a day. *Campesinos* who moved to humid tropical regions like Campeche to take advantage of colonization projects struggled to survive while farming systems developed for temperate-zone agriculture failed and degraded the environment. As the millennium ended, the UN claimed that desertification drove many *campesinos* off the land each year. In some regions, such as Santa Ana del Valle, Oaxaca, many peasants abandoned traditional farming to make textiles for tourists, work in overcrowded cities like the State capital or Mexico City, or find jobs in southern California restaurants. Similarly, modernization bypassed Hidalgo's Sierra Alta, causing "an absolute economic decline."[51]

The population of Mexico City and the suburbs and shantytowns of Mexico state passed 20 million during the 1990s. The multitude of street vendors, children performing circus acts on crowded boulevards, and beggars made it obvious that un- and underemployment would be serious problems for the future. Noxious air pollution and toxic substances in the water and air provide further evidence that life for the urban poor will be difficult. Discussing a plague of rats in the city in 1985, an official with the State's environmental protection agency called Mexico City "a massive accumulation of errors in the history of humanity."[52] The flourishing *maquiladora* factories of border cities like Tijuana and Ciudad Juárez also attracted many migrants, but they earn low wages, live in substandard housing, and suffer the consequences of lax enforcement of environmental regulations.[53]

During the 1980s U.S. agribusinesses like Castle & Cook and General Foods took advantage of lower production costs by investing substantial funds to develop export fruit and vegetable production in Mexico. The companies helped finance the opening of new agricultural regions such as the southern desert of Baja California, where tomatoes are grown for the U.S. market at less than half the cost of raising them in California. In 1985 Mexico sent nearly $1.5 billion worth of food to the United States. Staying in rural areas to work for foreign agribusiness also offered a bleak future for the *campesinos*. Reporters, doctors, and social and environmental scientists have documented the deleterious health effects of constant exposure to pesticides, not only to farmworkers but also to their young children. Fertilizer and pesticide runoff mix with urban and industrial waste, including heavy metals like mercury and lead, to pollute fresh waters like the Rio Lerma and Lake Chapala and brackish ones like the coastal lagoons of Sinaloa. The environment of tropical regions like Campeche suffered from degradation after forests were cleared for agricultural projects and then turned over to pasture when these failed. Further, rural wages remained low to keep exports such as vegetables and fruits competitive in international markets. In 1991, according to the *Christian Science Monitor*, 70% of the 17 million Mexicans in "extreme poverty" lived in rural areas.[54]

By 1990 many Mexicans agreed that the Green Revolution made life worse for peasant farmers. Acknowledging the *campesinos'* poverty and other problems, some experts defended them. They told *agrónomos* not to assume that their knowledge was superior to the *campesinos'* traditions and to take their opinions into account when they design and direct agricultural programs. Others urged *campesinos* to form independent organizations to break the power of "elites and political bosses willing to bow to established economic and social power,"[55] but some peasants had already begun to renegotiate the discourse of rural development. For example, by the late 1970s Bonfil's *ejidatarios* hired lawyers to demand attention from the agrobureaucracy, and they had some success. During the 1980s some *campesinos* such as the 1,000 coffee farmers from 17 Tzeltal and Tzotzil Mayan communities in Los Altos, Chiapas, who created the *Unión de Ejidos y Comunidades de Cafeticultores del Beneficio Majomut* formed independent organizations to promote "a system of cultivation inherited from our Mayan ancestors," including intercropping with fruit trees, food crops, and "ritual plants," and marketed their own produce, asserting their right to choose how they farmed.[56] In 1983 Protestant missionaries in the Isthmus of Tehuantepec helped *campesinos* found the *Unión de Comunidades Indigenas de la Región del Istmo* to grow coffee organically and sell it in Germany and Holland. In 1987 organic produce growers organized the *Asociación Mexicana de Agricultores Ecológicos* to establish quality and production standards. An unintended consequence of many ineffective programs, the "bureaucratization of agriculture" taught many *campesinos* how to negotiate successfully with external actors like bureaucrats and agribusiness representatives, including *agrónomos*.[57]

AGRONOMÍA NEOLIBERAL

A symptom of Mexico's malaise during the debt crisis, rural development programs after the Green Revolution contributed to "capital loss, stifled social initiative, and environmental devastation."[58] At the suggestion of the World Bank, which did not introduce the notion to Mexican agricultural policy, Salinas wanted farmers to cultivate high-value crops to earn export revenue, some of which could be used to import food. The bank also urged him to eliminate agrochemical and hybrid seed subsidies because they promoted inefficiency and the cultivation of unprofitable food crops. Sounding like the MAP's scientists in the 1960s, in 1989 Salinas accepted the bank's recommendation to "separate policies of increasing productivity from those of alleviating poverty" and concentrate technical support in regions with the potential for high productivity, abandoning *de temporal* maize farming.[59] As part of his privatization of parastatal firms, the government sold its interest in *Guanos y Fertilizantes de México*. To capitalize rural Mexico by giving *ejidatarios* title to their land so they could secure private loans and by ending land tenure insecurity that hindered investment and innovation, Salinas also terminated the agrarian reform and began to privatize the *ejido*, deconstructing an important pillar of the PRI's revolutionary legitimacy, which many of his critics and old *agraristas* recognized. Seeking his place in history as "the President who reinvented the revolution 'without myth,' " Salinas said that this was the only way to alleviate "the poverty and misery of millions of *campesinos*," and the SAF promised to close "the widening enormous gap between the traditional and modern sectors in the countryside."[60] Co-opted for many years, CNC leaders supported Salinas' policies.

Salinas took the *campesinistas'* ideas into account and promised to modernize the *ejido* "without altering the *campesino* way of life, or alienating *campesinos* from their land and their communities."[61] He knew that research and higher education were crucial for the progress of agriculture and proposed reforms to improve them. Reflecting the international acceptance of these ideas, the SAF admitted that its research benefited only commercial farmers and promised to study *campesinos'* traditional knowledge and devote more effort to *de temporal* farming. Reacting to the failure of many top-down structured programs, but also partly responding to the scarcity of resources, the *salinistas* wanted to help *campesinos* find their own solutions to their problems. Jaime de la Mora Gómez, director general of the bank *Banrural*, promised to "eliminate all forms of paternalism," and other officials claimed that they would include *ejidatarios* in the "discourse of modernization" by promoting bi-directional communication with *agrónomos* who ran programs.[62] Some scientists now devote their careers to solving peasant farmers' problems. For example, biologist Victor M. Trujillo works on technologies appropriate for the farming conditions of *campesinos* in Chiapas. Researchers at several institutions are studying traditional practices, such as Professor Alejandro Angulo of the University of Colima, who acknowledged that "indigenous technologies contain much knowledge of appropriate management of natural resources," and the ENA is training specialists in "agroecology" who try to improve them with appropriate technologies. Scientists at UNAM and the *Instituto Politécnico Nacional* work on biotechnologies, such as using hormones to increase the root growth of corn, to make *campesinos* more productive farmers.[63]

Some *campesinos* took advantage of opportunities offered by alternative approaches to agricultural development. Inspired by local teachers and advised by *agrónomos* from the *Centro de Investigaciones Ecológicos del Sureste* in Tapachula, in 1989 over 1,000 Mayan *ejidatarios* in Chiapas' Sierra Madre de Motozintla formed the *Union de Ejidos "Professor Otilio Montaño"* "to find an alternative path of totally autonomous and independent development to guarantee long term well-being and progress . . . [that is] not destructive and agrees with our culture, natural environment, and socioeconomic reality." The organization began a reforestation project and grew coffee, corn, wheat, vegetables, beans, and squash without agrochemicals. To end dependence on coffee monoculture, they raised edible mushrooms in the processing wastes and planted apple trees. In 1988 some 50,000 coffee growers from 70 independent peasant leagues in eight states united to form the *Coordinadora Nacional de Organizaciones Cafetaleras* and used the organization to market their crop in the United States and Europe by establishing their own firm, *Promotora Comercial de Cafes Suaves Mexicanos, S.A.* With traditional farming systems, the help of *agrónomos* trained in agroecology, and contacts with the International Federation of Organic Agriculture Movements and environmentalists in Germany and the United States, these *campesinos* entered the organic coffee market. Their organization is committed to "democratic practices, autonomy, and political pluralism."[64]

Neoliberalism offered another way to modernize peasant agriculture: globalization. To "insert the rural community into global society," Salinas facilitated contracts between *ejidos* and foreign and national agribusiness firms to bring state-of-the-art technologies to Mexico. For example, *campesinos* who belonged to the Popular Farmers and Workers' General Union led by *agrarista*

Margarito Montes signed a contract with Canada-based Ault Foods, which invested $100 million in a dairy farm project. Some experts worry that the continued globalization of Mexican agriculture will lead to further environmental degradation, but occasionally it had more promising outcomes. In Chiapas, the Elan Coffee Company of San Diego worked with *ejidatario* coffee growers, teaching them to use organic methods to obtain high prices from buyers who satisfy the U.S. demand for "natural foods." Mexican organic coffee growers also organized the firm Aztec Harvests to establish marketing contacts with an Emeryville, California, buyer. According to minister of agriculture Carlos "Hank" González, a month after the privatization of *ejidos* began in March 1992, *ejidatarios* had made over 100 deals with agribusinesses. Foreign companies like those that developed export fruits and vegetables in El Llano Grande, Jalisco, usually hire or consult U.S., Japanese, or Israeli experts, not *agrónomos*.[65]

In 1988, reflecting the rise of new ideas about the State's role in the economy, *agrónomo* and Ph.D. Rolando de Lasse, president of the *Colegio de Ingenieros Agrónomos*, urged agronomy schools to admit fewer students and adopt curriculum reforms to train professionals for private sector jobs or a practice in addition to service in the bureaucracy, but neoliberalism could not erase the history of Mexican agronomy. To solve the problems of corn growers in the humid tropics, the SAF continued to develop new drought-resistant maize hybrids, rejecting traditional crop strains and methods of obtaining seed. Founded in 1993, the *Procampo* program offered discounted agrochemicals to peasants who voted correctly, ignoring the severe soil degradation that fertilizers had wrought. SAF spokesman Pascacio Taboada said that "the government has reduced pesticide use and promoted organic farming wherever possible . . . [but] the bottom line is production."[66] In 1990 an *agrónomo* working in Campeche confessed that "today's bureaucracy lacks humanity and understanding of *campesino* culture," and Fernando E. Vega and Javier Trujillo Arriaga said that the study of traditional farming systems "has not received meaningful support." In 1992 Angulo chastised the SAF's "ignorance of the objectives of organic agriculture."[67] Under neoliberalism Mexico's rural environment continued to decay.

The end of price supports brought economic ruin to *ejidatarios* like Faustino Salinas of Tlaxcala, who during the 1980s switched from barley to wheat in response to high, subsidized prices, then fell into debt when Salinas removed them and the downward adjustment to his insurance failed to cover his loans when frost destroyed his crop. Government studies still called *de temporal* cultivation of corn and beans unprofitable, and many Mexican producers could not compete in an open market. Gates found that many *agrónomos* favor the cultivation of export crops and believe that food self-sufficiency is misguided. Some still want Mexicans to eat less corn and beans. In 1992 Undersecretary of Agricultural Planning Luis Tellez Kuenzer lamented that gains in cereal grain farming that made Mexico self-sufficient in corn and beans that year came "at a very high cost," and announced that the government now intended "to change production to more value-added crops, such as fruits, vegetables, and cattle," but like many *ejidatarios*, Faustino Salinas' *de temporal* land cannot grow crops that foreign or domestic agribusinesses want to buy.[68]

Procampo also promoted improved corn varieties, and some better-off farmers took advantage of continued price subsidies by planting them in the fall-winter season, but overall from 1990 to 1994 the government reduced its technical assistance for *ejidatarios*, and most used less fertilizer and hybrid seed after the State privatized these industries. In 1993 ENA rector Carlos Orozco Alam announced that science and technology could not transform the *campesinos* who farmed some 8 million hectares of *de temporal* lands into competitive producers and urged them to find jobs in the industrial sector. In 1994 Adrián Lozano Baenz, president of Mexico State's National Chamber of the Transformation Industries, called local "corn monoculture . . . unsustainable." Corruption, conflicts, and the sheer vastness of the task slowed the confirmation of land titles to a snail's pace through the end of the Ernesto Zedillo administration (1994–2000). Reflecting the transition to neoliberalism and recognition that many *campesinos* would lose their subsistence autonomy, the firm *Grupo Maseca* began to develop nutritionally enhanced corn flour.[69] Another cycle of rhetoric, promises, ill-conceived projects, and retreat had seemed to run its course.

CONCLUSIONS

During the 1970s critics of the Green Revolution raised difficult political issues like the redistribution of wealth, land reform, and a fundamental transformation of the world economy. In the 1980s and especially in the 1990s, however, politicians rejected these strategies when widespread famines failed to occur, Communism collapsed, and governments contracted welfare states. Today experts want to use the global market to alleviate poverty, rather than distributing existing wealth. These trends were reflected in Mexico, which saw a shift toward *cardenismo* under Echeverría; prosperity, an activist State, and retreat with the oil boom and bust; neoliberalism in the face of economic crisis under de la Madrid; and the deconstruction of the Revolution by Salinas. As members of the Revolutionary Family, *agrónomos* were tied too closely to the State to mobilize peasants to demand reform. Indeed, reflecting their professional identity as *técnicos* committed to scientific and technological solutions to rural problems, Arce found that a middle-level SAF bureaucrat punished a young *agrónomo* who helped *ejidatarios* who agreed to spend less money on alcohol build better homes, telling him to confine his work to providing advice about farming. Contrary to Arce, the transition from PIDER to the *Sistema Alimentario Mexicano* to neoliberalism was not caused by changes in the agrobureaucracy, nor was it the first time that "agronomists became the dominant actors operating within a framework of a notion of 'efficiency' "; rather, like the eras of revolutionary reform and the Green Revolution, *agrónomos'* efforts to ally with the State and their identity as middle-class urban professionals shaped their approach to rural development and their relations with *campesinos* throughout this period.[70]

In 1994 the wave of optimism generated by neoliberal-induced economic growth and Salinas' limited democratic reforms crashed with the EZLN Rebellion, the assassination of PRI presidential candidate Luis Donaldo Colosio, renewed inflation, and the scandal that engulfed the president and his brother, driving the former into exile and the latter into prison. Many Mexicans called Salinas the "*chupacabra*" (literally, goat sucker, a vampire-like creature). As the new millennium approached and the world's population surpassed 6 billion, rural Mexico's problems were still unresolved. Helped by corrupt politicians and police, in some regions businessmen destroy the environment and use violence and legal trickery to crush *campesinos* who challenge their control over resource use. Some peasants continue to flee to the overcrowded cities or cross the border illegally to find a better life in "*el norte*," while others languish in poverty in the rural areas."[71] The preservation of the *milpa* and the peasant community requires democratic reform and new approaches to agricultural development. The picture appears bleak, but recent events in Mexico offer some reason for optimism.

NOTES

1. Edwin J. Wellhausen, "The Agriculture of Mexico," *Scientific American* 235, 3 (September 1976): 129; Roe C. Black, "The Real Answer to Starvation," *Farm Journal* 99, 2 (February 1075): 33–35, 45; Manuel Moreno Sánchez, *México, 1968–1972: Crisis y perspectiva* (Austin: University of Texas Press, 1973), pp. 28–29; Sergio Reyes Osorio and Salomon Eckstein, "El desarrollo polarizado de la agricultura mexicana," in Miguel Wionczek et al., *¿Crecimiento o desarrollo económico?* (México, D.F.: SepSetentas, 1971), pp. 28–35; Sergio Reyes Osorio et al., *Estructura agraria y desarrollo agrícola en México: Estudio sobre las relaciones entre la tenencia y el uso de la tierra y el desarrollo agrícola en México* (México, D.F.: Fondo de Cultura Económica, 1974), p. 149.

2. Alberto Arce, *Negotiating Agricultural Development: Entanglements of Bureaucrats and Rural Producers in Western Mexico* (Wageningen, the Netherlands: Agricultural University Wageningen, 1993), p. 148; Daniel Nugent, *Spent Cartridges of Revolution: An Anthropological History of Namiquipa, Chihuahua* (Chicago: University of Chicago Press, 1993), p. 26.

3. "Farm Output up in North America," *New York Times*, September 18, 1971; "Food Production Gains in Far East," *New York Times*, August 23, 1971; "U.N. Group Calls for Farm Reform," *New York Times*, November 21, 1971; "Poorer Countries Lag Further in Food Production," *New York Times*, October 10, 1972; "On the Imminence of Disaster," *New York Times*, March 14, 1972; J.N. Black, "Energy Relations in Crop Production—A Preliminary Survey," *Annals of Applied Biology* 67, 2 (March 1971): 272–77; Patricia L. Kutzner, *World Hunger: A Reference Handbook* (Santa Barbara, Calif.: ABC-CLIO, 1994), p. 82; Robert Rodale, "The Responsibility of Being Organic," *Organic Gardening and Farming* 18, 3 (March 1971): 34–38; Frank Graham, "DDT Is Too Much with Us," *New York Times*, February 20, 1971; "Environmental Movement Scores Some Important Gains in Three Years," *New York Times*, April 9, 1973; Marvin Harris, "How Green the Revolution?" *Natural History* 81, 4 (June–July 1972): 30.

4. Stanley Johnson, *The Green Revolution* (London: Hamish Hamilton, 1972), p. 3.

5. David Pimentel, L.E. Hurd, A.C. Bellotti, M.J. Forster, I.N. Oka, D.D. Sholes, and R.R. Whitman, "Food Production and the Energy Cri-

sis," *Science* 182, 4111 (November 2, 1973): 443–49; Robert Rodale, "1974—The Year Organic Methods Intruded in National and International Affairs," *Organic Gardening and Farming* 21, 12 (December 1974): 4; "Letter to the Editor by Elizabeth Rutledge," *New York Times*, November 17, 1974.

6. "FAO Sees Food Threat in Fertilizer Shortage," *New York Times*, December 13, 1973; "FAO Sees Third World Menaced by Fertilizer Cut," *New York Times*, March 10, 1974; "Editorial: Vanishing Glut," *New York Times*, January 2, 1973; "The Green Revolution Hasn't Ended Hunger," *New York Times*, April 13, 1973; "Persistant Rice Plants Sought for Asia," *New York Times*, October 11, 1973; "Acute Food Shortage Found around Globe," *New York Times*, April 13, 1973; A.G. Leeks, "Rice in Crisis—Again," *Ceres, FAO Review on Development* 6, 5 (September–October 1973): 17.

7. "Editorial by James P. Brown," *New York Times*, May 30, 1973; Nicolas Wade, "Green Revolution (II): Problems of Adapting a Western Technology," *Science* 186, 4170 (December 27, 1974): 1191; "New Thought for Food Is Urged," *New York Times*, December 8, 1973; "Food: A Crisis for All," *New York Times*, September 19, 1974; "Millions in Food Short Nations Held in Danger," *New York Times*, June 29, 1976; William Paddock and Paul Paddock, *Time of Famines: America and the World Food Crisis* (Boston: Little, Brown & Company, 1967), pp. xii–xiv, xxiii, xxx, 252; Keith Griffin, *The Political Economy of Agrarian Change: An Essay on the Green Revolution* (Cambridge, Mass.: Harvard University Press, 1974); "U.S. Needy Found Poorer, Hungrier than 4 Years Ago," *New York Times*, June 20, 1974; Harry M. Cleaver, "The Origins of the Green Revolution" (unpublished Ph.D. dissertation, Stanford University, 1974), pp. 18–19, 224; "Letter to the Editor by Assistant Professor Mark D. Dornstretch," *New York Times*, December 9, 1974; Rene Dumont, "World Hunger: Crisis or Scandal," *UNESCO Courier* (May 1975): 16–23; Cynthia Hewitt de Alcantara, *Modernizing Mexican Agriculture: Socioeconomic Implications of Technological Change, 1940–1970* (Geneva: United Nations Research Institute for Social Development, 1976), p. 23; "Combating the World Food Crisis," *Intellect* 104, 2371 (January 1976): 278; "A Sociologist Doubts Rich Will Do Enough to Abolish Hunger," *New York Times*, June 30, 1976; Frances Moore Lappe and Joseph Collins, *Food First: Beyond the Myth of Scarcity* (New York: Ballantine Books, 1977), pp. 128, 153–54; "A Growing Clamor for a Well Fed World," *New York Times*, October 19, 1977; "U.N. Food Conference Opens with a Gloomy Report," *New York Times*, September 5, 1979; Andrew Pearse, *Seeds of Plenty, Seeds of Want* (New York: Oxford University Press, 1980), pp. 16, 172–77.

8. "Organic Food Myth Dispelled," *New York Times*, March 10, 1974.

9. Susan Carhart, "Return to Organic Farming," *Chemistry* 46, 9 (Oc-

tober 1973): 18–21; "Green Revolution's Salad Days Are Over," *Chemistry* 49, 4 (May 1976): 22–23; Jerome Goldstein, "Adversary Scientists and Agricultural Renegades," *Organic Gardening and Farming* 18, 4 (April 1971): 77–80; Goldstein, "Researchers Take a New Look at Organic Farming," *Organic Gardening and Farming* 24, 6 (June 1977): 68–71; John J. DuPont, "Editorial: Hunger in the Midst of Plenty," *The Conservationist* 32, 2 (September–October 1977): 1; S.H. Wittner, "The Next Generation of Agricultural Research," *Science* 199, 4327 (January 27, 1978): 375; Rodale, "Stocking Up on Future Food," *Organic Gardening and Farming* 27, 8 (August 1980): 22–28.

10. Robert Rodale, "Plan Your Food Future Now," *Organic Gardening and Farming* 27, 11 (November 1980): 26–30.

11. E.F. Schumaker, *Small Is Beautiful: A Study of Economics as If People Mattered* (London: Blond & Briggs, 1973), pp. 142–48; Wendell Berry, "Where Cities and Farms Come Together," in Richard Merrill, ed., *Radical Agriculture* (New York: New York University Press, 1976), pp. 14–24; Murray Bookchin, "Radical Agriculture," in Merrill, ed., *Radical Agriculture,* pp. 3–13; Merrill, "Preface," in Merrill, ed., *Radical Agriculture,* p. xviii; John Todd, "A Modest Proposal: Science for the People," in Merrill, ed., *Radical Agriculture,* pp. 261, 268; Michael Perelman, "The Green Revolution: American Agriculture in the Third World," in Merrill, ed., *Radical Agriculture,* pp. 114–20; Wade Green, "The New Alchemists," *New York Times,* August 8, 1976.

12. David L. Clawson and Don R. Hoy, "Nealtican, Mexico: A Peasant Community That Rejected the 'Green Revolution,'" *American Journal of Economics and Sociology* 38, 4 (October 1979): 384–85; Pearce, *Seeds of Plenty,* pp. 230–47; Billie R. DeWalt, *Modernization in a Mexican Ejido: A Study in Economic Adaptation* (Cambridge, Mass.: Cambridge University Press, 1979), pp. 263–64; David Winder and Deborah Eade, "Agricultural Issues in the United States and Mexico: Views from a Third Country," in Bruce F. Johnston, Cassio Luiselli, Celso Cartas Contreras, and Roger D. Norton, eds., *U.S.–Mexico Relations: Agriculture and Rural Development* (Stanford, Calif.: Stanford University Press, 1987), p. 376.

13. Anthony Hall, "Community Participation and Development Policy: A Sociological Perspective," in Hall and James Midgley, eds., *Development Policies: Sociological Perspectives* (Manchester: Manchester University Press, 1988), pp. 92–95.

14. John H. Franklin, "In 100 Years There'll Be 30 Billion of Us," *New York Times,* August 22, 1971; "Homo Sapiens: The Manna Maker," *New York Times,* August 9, 1975; Eugene Rabinowitz, "Man on Earth," *New York Times,* April 15, 1972; "Editorial: Food for the Americas," *New York Times,* October 14, 1972; Nicolas Wade, "Agriculture: Rise to Prominence at Home and Abroad," *Science* 182, 4119 (December 28, 1973): 1322;

Wade, "Green Revolution (I): A Just Technology, Often Unjust in Use," *Science* 186, 4169 (December 20, 1974): 1093–96; Wade, "Green Revolution (II)," p. 1191; Lester R. Brown, *By Bread Alone* (New York: Praeger Publishers for the Overseas Development Council, 1974); "Nations Demand Agricultural Aid," *New York Times*, August 3, 1975; Black, "The Real Answer," p. 33.

15. Theodore W. Schultz, "The Hungry, Crowded, Competitive World," *Bulletin of the Atomic Scientists* 33, 8 (October 1977): 30–31; Keith O. Campbell, *Food for the Future: How Agriculture Can Meet the Challenge* (Lincoln: University of Nebraska Press, 1979), pp. ix, 24, 81–83, 146.

16. "Expert Predicts Improved Grains," *New York Times*, January 12, 1971; Norman Borlaug, "The Green Revolution: For Peace and Bread," *Bulletin of the Atomic Scientists* 27, 6 (June 1971): 6–7, 42–45; Borlaug, "In Defense of DDT and Other Pesticides," *UNESCO Courier* (February 1972): 6; Paddock and Paddock, *Time of Famines*, p. xxiii.

17. "Borlaug Condemns Ban on Pesticides," *New York Times*, November 3, 1971; "D.D.T.'s Use Backed by Nobel Prize Winner," *New York Times*, November 9, 1971; Borlaug, "D.D.T., the First Domino," *New York Times*, November 21, 1971; "Agricultural Science and the Public," *Hortscience* 8, 5 (October 1973): 352.

18. "Poor Lands Do Little on Protein Deficiency," *New York Times*, April 19, 1973; "The Green Revolution Hasn't Ended Hunger," *New York Times*, April 13, 1973; "Research Institute Seeks Green Revolution in Nigeria," *New York Times*, September 13, 1972; Borlaug, "Agricultural Science and the Public," p. 352.

19. "Scientist Fears World Food Shortage," *New York Times*, February 28, 1974; "Food: A Crisis for All," *New York Times*, September 19, 1974; Wade, "Green Revolution: Creators Still Quite Hopeful on World Food," *Science* 185, 4154 (September 6, 1974): 844–45; Black, "The Real Answer," p. 33; Alan Anderson, "The Green Revolution Lives," *New York Times Magazine*, April 27, 1975, p. 80; Peter R. Jennings, "The Amplification of Agricultural Production," *Scientific American* 235, 3 (September 1976): 190; "Sociologist Doubts Rich Will Do Enough to Abolish Hunger," *New York Times*, June 30, 1976.

20. "The Traditional Farmer Is Africa's Hope," *New York Times*, December 16, 1974; D.J. Greenland, "Bringing the Green Revolution to the Shifting Cultivator," *Science* 190, 4217 (November 28, 1975): 841–44; Robert Allen, "New Strategy for the Green Revolution," *New Scientist* 63, 909 (August 8, 1974): 320–21; Rockefeller Foundation, *The President's Review and Annual Report* (New York: Rockefeller Foundation, 1975), p. 2.

21. "The Green Revolution Lives," *New York Times*, April 27, 1975; Barbara H. Tuckman, "The Green Revolution and the Distribution of

Agricultural Income in Mexico," *World Development* 4 (1976): 17–24; Wellhausen, "The Agriculture of Mexico," p. 134.

22. Wortman, "Food and Agriculture," *Scientific American* 235, 3 (September 1976): 31–40; Wellhausen, "The Agriculture of Mexico," p. 147; "Interview with Edwin J. Wellhausen," in Leobardo Jiménez Sánchez, *Las ciencias agrícolas y sus protagonistas* (Chapingo, México: Colegio de Postgraduados, 1984), p. 314.

23. "Mondale Hears Warning by Food Expert in Mexico," *New York Times*, January 9, 1978; Borlaug, "Civilization Will Depend More upon Flourishing Crops than on Flowery Rhetoric," Alfred M. Landon Lecture Series on Public Issues, March 20, 1979, Kansas State University, Manhattan, Kansas, pp. 1, 10–12; "Food and Population, Wary Optimism," *New York Times*, February 5, 1978; "Crisis in Food," *New York Times*, June 16, 1979; Lester R. Brown, *The Twenty Ninth Day* (New York: W.W. Norton & Company, 1978), pp. 1–6, 152; James N. Wallace, "Green Revolution Hits Double Trouble," *U.S. News and World Report* 89, 4 (July 28, 1980): 37-38. "Renewable Resources on Decline," *New York Times*, February 5, 1980.

24. "Botanists Warn of Plant Misuse," *New York Times*, June 17, 1980; Clifton R. Wharton Jr., "Food, the Hidden Crisis," *Science* 208, 4451 (June 27, 1980): 1415; "The World Doesn't Need Our Famers," *Washington Post*, December 29, 1985; John W. Mellor and Frank Z. Riely, "Expanding the Green Revolution," *Issues in Science and Technology* (Fall 1989): 67-68; "America the Bountiful Has Been Farmers' Undoing," *Los Angeles Times*, May 13, 1986; James C. Scott, *Weapons of the Weak: Everyday Forms of Peasant Resistance* (New Haven, Conn.: Yale University Press, 1985).

25. "Agricultural High-Tech," *New York Times*, October 25, 1981; "Vast Agricultural Gains Seen in Plant Genetics," *New York Times*, December 19, 1983; Susantha Goonatilake, *Aborted Discovery: Science and Creativity in the Third World* (London: Zed Books, 1984), p. 134; Ahmed Iftikahr, "Advanced Agricultural Biotechnology, Some Empirical Findings on Their Social Impact," *International Labor Review* 128, 5 (1989): 553–70; Gerd Junne and Annemieke Roobeek, "Social and Economic Impact of Biotechnology," *Biofutur* (November 1989): 48–50; Fawler C. Lachkovics et al., "Bio-farms: The End of the End. Third World Farmers and the New Plant Genetics," *Development Dialogue* 1–2 (1988): 61; "New Technology Called Harmful to Small Farms," *San Jose Mercury News*, March 18, 1986; "New 'Green Revolution?' " *Christian Science Monitor*, September 26, 1986; N. Patrick Peritore, "Biotechnology: Political Economy and Environmental Impacts," in Peritore and Ana Karina Galve-Peritore, eds., *Biotechnology in Latin America: Politics, Impacts, and Risks* (Wilmington, Del.: Scholarly Resources, 1995), p. 17.

26. "Beware the Green Revolution," *The Economist*, November 5, 1988, p. 75; "Spunky Spud a Star in 'Green Revolution,' " *Santa Barbara News Press*, October 27, 1992; John Walsh, "The Greening of the Green Revolution," *Science* 252 (November 5, 1991): 26; "Interview with Edwin J. Wellhausen," in Jiménez Sánchez, *Las ciencias agrícolas*, p. 324; Hall, "Community Participation," p. 101; Arce, *Negotiating*, pp. 3–4.

27. Rosa Elvia Barajas, "Biotecnología y Revolución Verde, especificidades y divergencias," *Sociología* 6, 10 (May–August 1991): 73; "Learning to Give as Much as We Take from Earth," *Los Angeles Times*, January 18, 1994.

28. Flores, "Science and Technology," p. 1398; John Gledhill, *Casi Nada: A Study of Agrarian Reform in the Homeland of Cardenismo* (Albany: State University of New York, Institute for Mesoamerican Studies, 1991), p. 220; David E. Lorey, *The University System and Economic Development in Mexico since 1929* (Stanford, Calif.: Stanford University Press, 1993), p. 34; Reyes Osorio, *Estructura agraria*, pp. 909, 912; Mario García Sordo, "Desempleados o subempleados, más de 90 mil agrónomos," *El Financiero*, October 5, 1988, p. 39; Arce, "The Social Construction of Agrarian Development: A Case Study of Producer–Bureaucrat Relations in an Irrigation Unit in Western Mexico," in Norman Long, ed., *Encounters at the Interface: A Perspective on Social Discontinuities in Rural Development* (Wageningen, the Netherlands: Agricultural University Wageningen, 1989), p. 25; Jiménez Sánchez, "Prologo," in Jiménez Sánchez, *Las ciencias agrícolas*, p. xxiv; Peter S. Cleaves, *Professions and the State: The Mexican Case* (Tucson: University of Arizona Press, 1987), pp. 50–56, 68; Arroyo and Waissbluth, "Desarrollo biotecnológico," p. 62; Castaños, *Testimonios*, pp. 28–29, 171; Merilee Grindle, "Power, Expertise, and the Tecnico: Suggestions from a Mexican Case Study," *Journal of Politics* 39, 2 (May 1977): 407; Grindle, *Bureaucrats, Politicians, and Peasants in Mexico: A Case Study in Public Policy* (Berkeley: University of California Press, 1977), p. 101.

29. Reyes Osorio, *Estructura agraria*, p. 910; Ramón Fernández y Fernández, *Chapingo hace 50 años* (Chapingo, México: ENA, Colegio de Postgraduados, 1976), p. 136; Emilio López Zamora, *El agua, la tierra, los hombres de México* (México, D.F.: Fondo de Cultura Económica, 1977), p. 243–45.

30. "Urgencia de la reforma agraria y sus perspectivas en el decenio 1970–1980," *Tierra* 25, 10 (October 1970): 780; Dirección General de Divulgación, "Telegramas de felicitación," *Tierra* 26, 9 (September 1971): 646; Jerjes Aguirre Avellaneda, *La política ejidal en México* (México, D.F.: Instituto Mexicano de Sociología, 1976), pp. 108–9.

31. Rafael Ortega Paczka, "Reorganización del mejoramiento genetico del maiz en el INIA," in Efrain Hernández Xolocotzi, ed., *Agroecosistemas*

de México: Contribuciones a la enseñanza, investigación, y divulgación agrícola (Chapingo, México: ENA, Colegio de Postgraduados, 1977), p. 370; Castaños, *Testimonios*, pp. 75, 220.

32. Moreno Sánchez, *México, 1968–1972*, pp. 42–46; Reyes Osorio et al., *Estructura agraria*, pp. 396–397; Julio Boltvinik, "Estrategia de desarrollo rural, economía campesina e innovación tecnológica en México," *Comercio Exterior* 26, 7 (July 1976): 813–26; Judith Friedlander, *Being Indian in Hueyapan: A Study of Forced Identity in Contemporary Mexico* (New York: St. Martin's Press, 1975), pp. 170–73; Rogelio López Ildefonso, "Asociación de cultivos maiz-frijol," *Agricultura Técnica en México* 3, 3 (July 1971): 98–101; López Ildefonso, "Asociación maiz-frijol como frijol-ganancias," *Tierra* 28, 9 (September 1973): 704; Marilyn Gates, *In Default: Peasants, the Debt Crisis and the Agricultural Challenge in Mexico* (Boulder, Colo.: Westview Press, 1993), p. 118; Juan M. Mattieu, "Contaminantes agroquímicos y su impacto en el brota," *Agronomía* 151–52 (September–October and November–December 1973): 31.

33. Efrain Hernández Xolocotzi, "El agroecosistema, concepto central en el análisis de la enseñanza agrícola en México," in Hernández Xolocotzi, ed., *Agroecosistemas*, pp. xv–xvi; Alberto Ramos Rodríguez and Efrain Hernández Xolocotzi, "Reflexiones sobre el concepto de agroecosistema," in Hernández Xolocotzi, ed., *Agroecosistemas*, pp. 531–38; Castaños, *Testimonios*, p. 51; "Interview with Efrain Hernández Xolocotzi," in Jiménez Sánchez, *Las ciencias agrícolas*, pp. 221–25.

34. "Interview with Julián Rodríguez Adame," in Jiménez Sánchez, *Las ciencias agrícolas*, p. 25; "Interview with Edmundo Taboada Ramírez," in Jiménez Sánchez, *Las ciencias agrícolas*, p. 140; Antonio Turrent Fernandez, "Research and Technology for Mexico's Small Farmers," in Johnston et al., *U.S.–Mexico Relations*, p. 318; Winder and Eade, "Agricultural Issues," p. 368; Pedro Ojeda Paullada, *Política Pesquera 1983–1985* (México, D.F.: Secretaría de Pesca, 1987), p. 32.

35. Arce, "The Social Construction," pp. 41–42; Gledhill, *Casi Nada*, p. 318.

36. Instituto Mexicano de Recursos Naturales Renovables, *Trigesimo quinto y trigesimo sexto informe anual* (México, D.F.: 1988); Carlos Vidali, "Apuntes sobre la modernización del sector agropecuario" (part II), *El Día*, August 23, 1988; "Interview with Emilio Alanis Patiño," in Jiménez Sánchez, *Las ciencias agrícolas*, pp. 362–63.

37. Secretaría de Agricultura y Ganadería, "Voto formal de los agrónomos de México," *Tierra* 26, 4 (April 1971): 261; "Agradecidos por sus 28 años de actividad Doctor Edwin J. Wellhausen," *Tierra* 27, 1 (January 1972): 152–53; "Productora Nacional de Semillas," *Tierra* 27, 9 (September 1972): 674; "Se presentara un informe completo de importante actividad agrícola," *Tierra* 25, 9 (September 1970): 707; Reyes Osorio, *Estructura*

agraria, pp. 906–7; Castaños, *Testimonios*, pp. 98–99; "Editorial," *Agricultura Técnica en México*, 3, 5 (July 1972): 161–62; SEP, Dirección General de Educación Tecnológica Agropecuaria, "La enseñanza de la agricultura en las Escuelas Técnicas Agropecuarias," *Tierra* 28, 2 (February 1973): 42–43, 157; "Interview with José Luís de la Loma y de Oteyza," in Jiménez Sánchez, *Las ciencias agrícolas*, p. 178; "Interview with Jesús Alarcón Moreno," in Jiménez Sánchez, *Las ciencias agrícolas*, p. 197; Grindle, *Bureaucrats, Politicians, and Peasants*, pp. 80–90; Gates, *In Default*, pp. 108–9.

38. Pedro Reyes Castañeda, "Mejoramiento de los híbridos de maiz (Zea mays L.): NLH 1, NLH 2, y NLH 3 para clima caliente seco," *Agronomía* 155–56 (May and July 1974): 82–86; Reyes Castañeda, "Mejores híbridos de maiz por el cruzamiento de germoplasmo seleccionado en clima caliente húmedo y caliente seco," *Agronomía* 169–70 (September–October and November–December 1976): 34–46; "Brillantes resultados de la formación de variedades de hibridos," *Tierra* 25, 11 (November 1970): 860; "Semillas para siembras Mexicanas," *Tierra* 26, 8 (August 1971): 570; Centro de Investigaciones Agrícolas del Golfo Central, INIA, SARH, *Manual de producción de arroz de temporal en el Estado de Veracruz* (Veracruz: Campo Agrícola Experimental Cotaxtla, 1982); Federico López de Alba, "Food Production, Environmental Protection, and Health Effects in Mexican Agriculture," *American Journal of Industrial Medicine* 18 (1990): 495–98; Ernest Feder, *Strawberry Imperialism: An Inquiry into the Mechanisms of Dependency in Mexican Agriculture* (The Hague, the Netherlands: Institute of Social Studies, 1977), pp. 123–24; Edmundo Flores et al., *La ciencia y la tecnología en México* (México, D.F.: CONACYT, 1982), p. 53; Flores et al., "Science and Technology in Mexico: Toward Self-Determination," *Science* 219 (1983): 1398; "Interview with Norberto Aguirre Palancares," in Jiménez Sánchez, *Las ciencias agrícolas*, p. 55; Gates, *In Default*, pp. 117–18, 166–68; Arce, *Negotiating*, pp. 24, 76.

39. Castaños, *Testimonios*, pp. 9, 87, 102; Armando Bartra, "Colectivización o proletarización: El caso del Plan Chontalpa," *Cuadernos Agrarios* 1, 4 (October–December 1976): 57, 87; Grindle, *Bureaucrats, Politicians, and Peasants*, pp. 142–63; Gates, *In Default*, pp. 146, 224–25; Arce, "The Social Construction," p. 45; Arce, *Negotiating*, pp. 44, 85–93, 156; John J. Bailey, "Agrarian Reform in Mexico: The Quest for Self-Sufficiency," *Current History* 80, 469 (November 1981): 360; DeWalt, *Modernization*, pp. 2, 42.

40. "Interview with Julián Rodríguez Adame," p. 25; Gonzalo Arroyo and Mario Waissbluth, "Desarrollo biotecnológico en la producción agroalimentaria de México: Orientaciones de política" (unpublished report, United Nations, Comisión Económica para América Latina, March 10, 1988), pp. 35–48; Turrent Fernández, "Research," p. 309.

41. Pauline Madow, "Reminiscences of Elvin C. Stakman," June 6, 1970, p. 1068, RFA, RG 13, RAC; Gates, "Codifying Marginality: The Ev-

olution of Mexican Agricultural Policy and Its Impact on the Peasantry," *Journal of Latin American Studies* 20, 2 (November 1988): 303–4.

42. Eduardo Casas Díaz, "Presentación," in Jiménez Sánchez, *Las ciencias agrícolas*, p. x.; Carlos Manuel Castaños, *Testimonios de un agrónomo* (México, D.F.: Editorial Futura, S.A., 1981), pp. 205, 280–87.

43. Arce, *Negotiating*, pp. 30–31; Jaime Castrejón Diez, *La educación superior en México* (México, D.F.: Edicol, 1979), pp. 165–67; Hewitt de Alcantara, *Modernizing*, p. 20; Cleaves, *Professions and the State*, p. 27; William C. Cobb, "Interview with Norman Borlaug," June 1967, p. 157, RFA, RG 13, RAC; Castaños, *Testimonios*, p. 141; "Interview with de la Loma," in Jiménez Sánchez, *Las ciencias agrícolas*, p. 184; Felipe Santander, *El extensionista* (México, D.F.: CREA, 1978), pp. 18–19; Alvaro Echeverria Zuno, *Problema alimentario y cuestion rural* (México, D.F.: Editorial Nuevo Imagen, 1984), pp. 268–69; Alan Riding, *Distant Neighbors: A Portrait of the Mexicans*, 2nd ed. (New York: Vintage Books, 1989), pp. 275–76.

44. DeWalt, *Modernization*, p. 201; Ann L. Craig, *The First Agraristas: An Oral History of a Mexican Agrarian Reform Movement* (Berkeley: University of California Press, 1983), p. 221; Arce, *Negotiating*, p. 141; Gledhill, *Casi Nada*, p. 313.

45. Secretaría de Agricultura y Recursos Hidraulicos (SARH), *Desarrollo Rural Integral* (México, D.F.: Tiempo, S.A., 1988), p. 94; "Interview with de la Loma," in Jiménez Sánchez, *Las ciencias agrícolas*, p. 186; Arce, *Negotiating*, p. 77.

46. Arroyo and Waissbluth, "Desarrollo biotecnológico," p. 63; Castaños, *Testimonios*, pp. 77, 88, 129–50, 280–81; Bartra, "Colectivización o proletarización," p. 81; Martin H. Greenberg, *Bureaucracy and Development: A Mexican Case Study* (Lexington, Mass.: D.C. Heath, 1970), pp. 133; Grindle, *Bureaucrats, Politicians, and Peasants*, pp. 101, 167; Manuel Mesa A., *Relatos autobiográficos* (México, D.F.: Editorial Nuestro Tiempo, 1981), p. 98; Santander, *El extensionista*, p. 73; Long and Magdalena Villareal, "The Changing Life-Worlds of Women in a Mexican Ejido: The Case of the Bee-Keepers of Ayuquilla and the Issue of Intervention," in Long, ed., *Encounters at the Interface*, p. 130; Gates, *In Default*, p. 169; Arce, "The Social Construction," pp. 28, 45; Arce, *Negotiating*, pp. 31–38, 136–37.

47. Guillermo Knockenhauer, "La modernización del agro en Mexico," *Comercio Exterior* 40, 9 (September 1990): 832–38.

48. DeWalt, "Inequalities in Wealth, Adoption of Technology, and Production in a Mexican Ejido," *American Ethnologist* 2, 1 (February 1975): 149–64; Secretaría de Agricultura y Ganadería, "Dirección General de Divulgación," *Tierra* 26, 10 (October 1971): 724; "Drying and Dying: Deforestation, Overfarming, and Overgrazing Are Turning Once-Productive Mexican Farms into Wastelands," *Albuquerque Journal*, June 18, 2000; Bolvinik, "Estrategia de desarrollo rural," p. 822; George A.

Collier, "Rebellion against Economic Exile," *Los Angeles Times*, January 23, 1994; Carlos Videli, "Apuntes sobre la modernización del sector agropecuario," *El Día*, August 23, 1988; Mesa A., *Relatos*, p. 112; Castaños, *Testimonios*, p. 280; Tuckman, "The Green Revolution," pp. 17–24; Merilee Grindle, "Rhetoric, Reality, and Self-Sufficiency: Recent Initiatives in Mexico's Rural Development," *Journal of Developing Areas* 19 (January 1985): 178; Sergio de la Peña and Marcel Morales, "Las transformaciones capitalistas en la agricultura: El caso del maiz en Morelos," in Enrique C. Ochoa and David E. Lorey, eds., *Estado y agricultura en México: Antecedentes e implicaciones de las reformas salinistas* (Azcapotzalco, México, D.F.: Universidad Autónoma Metropolitana de Azcapotzalco, 1994), p. 60; Hewitt de Alcantara, *Modernizing*, pp. 178–80; Gates, *In Default*, p. 7; Arce, *Negotiating*, pp. 50–51.

49. Nugent, *Spent Cartridges of Revolution*, pp. 27, 121.

50. David L. Clawson and Don R. Hoy, "Nealtican, Mexico: A Peasant Community That Rejected the 'Green Revolution,'" *American Journal of Economics and Sociology* 38, 4 (October 1979): 374, 380–83; Alan R. Sandstrom, *Corn Is Our Blood: Culture and Ethnic Identity in a Contemporary Aztec Indian Village* (Norman: University of Oklahoma Press, 1991), pp. 102, 121, 148; Craig, *The First Agraristas*, pp. 195, 221; Gates, *In Default*, pp. 1, 118, 155; Arce, *Negotiating*, pp. 115, 132–33.

51. "Have-not Nations Now Have," *Boston Globe*, August 10, 1986; "Drying and Dying"; Jeffrey H. Cohen, *Cooperation and Community: Economy and Society in Oaxaca* (Austin: University of Texas Press, 1999), pp. 36–59; Gates, *In Default*, pp. 133–41, 225–26; Frans J. Schryer, *The Rancheros of Pisaflores: The History of a Peasant Bourgeoisie in Twentieth-Century Mexico* (Toronto: University of Toronto Press, 1980), p. 23.

52. "Mexico City Rat Plague Reflects Bigger Problems," *Los Angeles Times*, September 7, 1985; "Comments on the Emerging Mexico," *Los Angeles Times*, October 22, 1991.

53. "Border Towns Brace for Influx," *Los Angeles Times*, October 22, 1991.

54. "A Bad Year for Some Farms, a Great Year for Subsidies," *San Jose Mercury News*, September 22, 1986; "Toxins Found in Blood of Babies Stir Fears about Mexico's Future," *Manchester Herald*, November 5, 1992; Angus Wright, *The Death of Ramon Gonzalez: The Modern Agricultural Dilemma* (Austin: University of Texas Press, 1991); Feder, *Strawberry Imperialism*, p. 95; Wayne A. Cornelius, "The Scare Stories Don't Wash," *Los Angeles Times*, February 28, 1992; "From Lerma River Flows a Tale of Politics and Pollution," *Los Angeles Times*, October 22, 1991; Lane Simonian, "Pesticide Use in Mexico: Decades of Abuse," *The Ecologist* 18, 2–3 (1988): 82–87; "Mexican Agricultural Reforms Set Stage for a New

Revolution," *Christian Science Monitor*, November 21, 1991; Gates, *In Default*, p. 226.

55. Mercedes Jiménez Velázquez, "La Fundación Rockefeller y la investigación agrícola en América Latina," *Comercio Exterior* 40, 10 (October 1990): 974; Bruce F. Johnston, "Rural Development in the U.S., Mexico, Japan, and Taiwan," in Johnston et al., *U.S.–Mexico Relations*, p. 51; Cassio Lusselli, "The Way to Food Self-Sufficiency in Mexico and Its Implications for Agricultural Relations with the United States," in Johnston et al., *U.S.–Mexico Relations*, p. 338.

56. Gates, *In Default*, pp. 156–57; "Unión de Ejidos y Comunidades de Cafeticultores del Beneficio de Majomut," San Cristobal de los Casas, Chiapas, pamphlet, n.d.

57. Unión de Comunidades Indigenas de la Región del Istmo, "Unidos venceremos," Santa María Guienagati, Oaxaca, n.d.; Alejandro Angulo, "Productores, alimentos orgánicos y reglamentación," *La Trilla* 3, 24 (October–November 1992): 12; Gates, *In Default*, p. 227.

58. Gates, *In Default*, p. 1.

59. Rosario Robles and Julio Moguel, "Agricultura y proyecto neoliberal," *El Cotidiano* 34, 2 (March–April 1990): 3–12.

60. Mario García Sordo, "Se restructura la industria de fertilizantes," *La Trilla* 3, 24 (October–November 1992): 5–7; "Mexican Agricultural Reforms Set Stage for a New Revolution," *Christian Science Monitor*, December 21, 1991; Roberto Rodríguez and Patrisia Gonzales, "Chiapas Only a Continuation of the Revolution," *Santa Barbara News Press*, March 8, 1994; Videli, "Apuntes sobre la modernización" (part II); Rosario Varo Berra, *La reforma agraria en México desde 1853: Sus tres cíclos legales* (Guadalajara: Universidad de Guadalajara, 2002), pp. 63–65.

61. "Propuestas de política agropecuaria y reforma institucional," *La Trilla* 3, 24 (October–November 1992): 20–22; Michael W. Foley, "Agenda for Mobilization: The Agrarian Question and Popular Mobilization in Contemporary Mexico," *Latin American Research Review* 26, 2 (Spring 1991): 61.

62. Jaime de la Mora Gómez, "La banca del desarrollo en la modernización del campo," *Comercio Exterior* 40, 10 (October 1990): 948–50; Foley, "Agenda," pp. 61–62; Secretaría de Agricultura . . . , *Desarrollo*, pp. 3–14, 85; Secretaría de Agricultura . . . , "Programa Nacional," p. 1003; Gordillo de Anda, "La inserción," pp. 812–15; Arce, *Negotiating*, p. 3.

63. Angulo, "Productores," p. 12; Trujillo Arriaga, "Desarrollo," pp. 953–58; Turrent Fernández, "Research," p. 318; Winder and Eade, "Agricultural," p. 368; Barajas, "Biotecnología," pp. 75, 81.

64. Unión de Ejidos Professor Otilio Montaño, "Por la conservación de la tierra, la naturaleza, y la cultura," Motozintla, Chiapas, n.d.; Coor-

dinadora Nacional de Organizaciones Cafetaleras, "Mexico's Small Coffee Growers: Working Together for the Future," México, D.F., n.d.

65. SARH, "Programa Nacional de Modernización del Campo, 1990–1994," *Comercio Exterior* 40, 10 (October 1990): 994; Gustavo Gordillo de Anda, "La inserción de la comunidad rural en la sociedad global," *Comercio Exterior*, pp. 812–13; Daniel Goldrich and D.V. Caruthers, "Sustainable Development in Mexico," *Latin American Perspectives* 19, 1 (Winter 1992): 115–16; "Mexican Maize Farmers Loath to Abandon Fight for Land," *Financial Times of London*, January 24, 1992; "Mexican Reform Sparks Farm Investment Interest," *Journal of Commerce*, April 15, 1992; José Eduardo Zárate Hernández, "Agricultural Modernization and Political Recomposition in El Llano Grande," in Sergio Zendejas and Pieter De Vries, eds., *Rural Transformations Seen from Below: Regional and Local Perspectives from Western Mexico* (La Jolla: Center for U.S.–Mexican Studies, University of California, San Diego, 1995), p. 13.

66. García Sordo, "Desempleados o subempleados," p. 39; Gates, *In Default*, p. 239; "Drying and Dying," *Albuquerque Journal*, June 18, 2000.

67. Gates, *In Default*, p. 238; Fernando E. Vega and Javier Trujillo Arriaga, "Biotecnología agrícola, espejo de la revolución verde," *Comercio Exterior* 39, 11 (November 1989): 952; Angulo, "Productores," p. 13.

68. Irene Sánchez Gordillo, "Justicia y empleo al agro: Salinas de Gortari," *La Trilla* 3, 24 (October–November 1992): 2; "The Drive for Reform Leaves Some Behind," *Los Angeles Times*, October 22, 1991; "Mexico's Reforms Transform Agrarian Sector," *Christian Science Monitor*, November 12, 1992; Cony Sánchez Gordillo, "Indicadoras agropecuarios," *La Trilla* 3, 24 (October–November 1992): 29; Gates, *In Default*, p. 241.

69. Rodolfo García Zamora, "Crisis agrícola y nuevos subsidios al campo," in Ochoa and Lorey, eds., *Estado y agricultura en México*, p. 82; "Our World," *USA Today*, June 15, 1994, p. 12; Alain de Janvry, Gustavo Gordillo, and Elisabeth Sadoulet, *Mexico's Second Agrarian Reform: Household and Community Responses* (La Jolla: Center for U.S.–Mexican Studies, University of California, San Diego, 1997), pp. 75–86, 139; James W. Wilkie, "Prologo," in Varo Berra, *La reforma agraria en México desde 1853*, pp. 34–48; Margaret C. Boardman, "Sowing the Seeds of the Green Revolution: The Pivotal Role Mexico and Non-Profit Organizations Play in Making Biotechnology an Important Foreign Policy Issue for the 21st Century," *PROFMEX Journal* 4, 3 (Summer 1999): 3, 11, 14.

70. Arce, *Negotiating*, pp. 23, 75–76.

71. "Our World," *USA Today*, June 15, 1994; "A Farmer Learns about Mexico's Lack of the Rule of Law," *New York Times*, October 27, 2000.

Conclusions: Agronomy and Mexico's Revolutions

The *agrónomos* played important roles in the Mexican and Green Revolutions. Some of the first students of the Mexican Revolution described a heroic struggle of *campesinos* who won the right to "Peace and Bread." Revisionists rejected this view, claiming that the Revolution created a powerful State that controlled the masses by co-opting or destroying their leaders, but recent counter-revisionists empower the *campesinos* again by emphasizing their use of negotiation and resistance to attain their goals. Some scholars called the Green Revolution a devious assault on Third World peasant agriculture constructed by *Yanqui* capitalists and imperialists, but others found scientists who worked to defeat the Malthusian specter of world starvation, an example of humans' ability to solve insurmountable problems with ingenuity, innovation, and international cooperation. This author believes that Mexico's agricultural development during the twentieth century was a failure for most *campesinos* but also a triumph for *agrónomos*, farmers who got rich using scientific agriculture, the urban middle class that benefited from industrialization, and the Revolutionary Family that dominates politics and the economy. Like a classical tragedy, a revolution claimed to bring democracy and social justice to oppressed peasant farmers but produced neither and a second supposedly created to end poverty and hunger also failed. The Revolution and the agrarian reform shaped the early professionalization of Mexican agronomy, and Cárdenas' project to modernize *campesino* agriculture puts him on the "*política agraria*" side of Mexico's agricultural policy discourse.

Did *agrónomos'* commitment to social justice for the *campesinos* influence post-revolutionary regimes, as Rebeca de Gortari Rabiela suggests? Did "progressive agronomists" work in the Cárdenas administration, as Gledhill and others assert? Or, is Lorey more correct to claim that "a majority of agricultural engineers may not take up the career in order to work toward revolutionary aims of social transformation"?[1] Many scholars, pundits, and activists criticize the Rockefeller Foundation's contributions to Mexican agriculture and agronomy, and some of them are quite vehement. Was the MAP socially irresponsible science? Examining the foundation's records in the context of the *agrónomos'* professional history sheds new light on these issues.

As John H. Perkins notes, the foundation's directors knew that rural poverty was a central issue in the Mexican Revolution, that addressing it contributed to the Revolutionary Family's legitimacy, and that *agrónomos* played an important role in this process. The MAP's work on fish ponds and vegetables demonstrates a desire to improve the *campesino* diet, and it included a few *ejidatarios* when testing its new seeds and agrochemicals, but its commitment to hybrid corn shows a lack of interest in sustaining small-scale peasant cultivation. Critics claim that the foundation wanted to introduce U.S. farming in Mexico and blindly ignored the logic of peasant farming systems. Sánchez, Sauer, and Anderson warned the directors about the pitfalls of this strategy, and the MAP scientists understood the logic of the *campesinos'* practices, but their biases shaped their approach to modernizing Mexican agriculture. The foundation wanted to push Mexico through the transition from an agrarian to an industrial society like the transformation of the United States during the late nineteenth and early twentieth centuries and thus tried to create commercial farmers, not vibrant, autonomous communities of peasant corn growers. Evidence proves that the foundation had some inkling of the MAP's socioeconomic impact before the project began and gained a deeper understanding as it progressed during the 1940s, but using this evidence to conclude that the officers claimed that the project would meet the Mexican Revolution's goal of prosperity for the *campesinos* to divert attention from more sinister or self-serving motives and sell the program to the government and the *agrónomos* is too Machiavellian. The *agrónomos'* and the State's interest in the MAP was motivated partly by right-wing attacks on the agrarian reform and the political and economic problems caused by continuing to distribute high-quality farmland to *campesinos*. By 1950 the foundation decided that the issue of beneficiaries was not their concern, as Jennings asserts, but if the officers stressed matters like land reform or credit for peasants, this would have been out of touch with Mexico's political realities of the 1940s and 1950s.

Perkins' assertion that the foundation's desire to assist U.S. foreign policy by "keeping Mexico friendly to the United States" created a lack of concern about the equity of the MAP's research is valid in this sense, but it is important to remember that these realities developed in an undemocratic political system dominated by the Revolutionary Family, which designed agricultural policy to serve members' interests.[2] World War II policy makers like Wallace thought that the equity of agricultural assistance was crucial to U.S. interests because it would raise standards of living and win the sympathy of the poor, who would elect pro-U.S. leaders. The original plans for postwar agricultural assistance programs also gave some weight to the issue. Indeed, the Truman and Eisenhower administrations' shift toward supporting dictators or non-democratic regimes willing to contain Communism by suppressing the masses' demands, a by-product of which was to create a social and political climate beneficial to U.S. investors, exerted a crucial influence on the outcome of the Green Revolution. The foundation knew that science could promote inter-American collaboration. Alemán and Ruiz Cortines courted U.S. investment, Mexico's economy grew, and popular unrest was manageable thanks to Cárdenas' land reform and pro-labor policies that established State control over peasant organizations and trade unions. The Green Revolution assisted Mexico's moderately successful industrialization project by increasing the supply of inexpensive basic foods, which fostered political stability and sustained low wages, and by freeing farmland for export crops to trade for foreign machinery.

Responding to the new policy that emphasized economic assistance, structural reform, and democratization to avert another Cuban Revolution, Mexico complied with the U.S. plan to modernize Latin America to stop Communism. The PRI held elections that produced a new president every six years. López Mateos accelerated the pace of land reform and increased opposition party representation in Mexico's legislature. Indeed, some academics validated this by calling Mexico a "one-party democracy." Mexico avoided U.S. pressure (as did some even less democratic Latin American regimes) to implement political reform that might have led to a more equitable approach to agricultural research. The foundation's approach to rural development, which addressed the needs of commercial farmers, industrialists, and cities, reflected the containment agenda and by extension the *campesinos'* political disempowerment, as well as the apolitical orientation of the MAP's U.S. scientists. Science provided a way to deal with Latin America's food-population and industrial development problems without raising difficult issues like land reform, democratization, birth control, or the redistribution of income. The MAP's agenda addressed the objectives of Mexico's ruling PRI and those of U.S. foreign policy. By 1950 the foundation began to publicize the MAP, emphasizing research on corn and its role as the staple of most Mexicans, implying that their primary concern was peasant farming, but the Green Revolution did not target *campesinos* or make them more prosperous, rather, it contributed to their political and economic disempowerment. The foundation and its scientists' contributions to world food production were substantial and merit congratulations, but in this regard laudatory works like *Campaigns against Hunger* and histories that stress the MAP's humanitarian motivations are misleading when they mention poverty and peasant agriculture, at least in Mexico after 1950.[3]

The *agrónomos'* alliance with the winners of the Mexican Revolution by implementing agrarian reform, which peaked under Cárdenas, was their first step toward professionalization. Land distribution fostered rural peace, brought many *campesinos* under the State's wing, helped legitimate the Revolutionary Family's rule, and solidified the *agrónomos'* claim to membership. Some of the *cardenista agrónomos'* ideas about rural development, such as urging peasants to grow vegetables to improve their diet or their support for small-scale irrigation, are remarkably similar to the proposals of recent appropriate technology advocates, but *agrónomos* also rejected the *campesinos'* agricultural knowledge, even during the 1920s and 1930s, when the State celebrated other aspects of "traditional" culture. The MAP's scientists criticized peasant farmers, but *agrónomos* attacked corn-bean intercropping before the project began. Similar to the land reform, the BNCE, and rural education, the State rejected Zapata's goal of "*libertad*" by trying to deny the *campesinos* the right to keep their traditional farming practices, even though they might meet their needs more effectively than those advocated by French, U.S., or Mexican scientists. The *cardenistas'* effort to support the traditional sector of the economy was often informed by their rejection of its cultural foundations. If not in language or culture, in social class, professional, and intellectual outlook, the *agrónomos* had more in common with U.S. scientists than the *campesinos* whom they said they wanted to liberate. The *campesinos* refused to cooperate with the cultural campaign, rejecting the *agrónomos'* claim to professional expertise. During the 1920s and 1930s the SAF advanced land reform and less successfully the State's cultural project, but its research, extension, and pest and disease control branches fit Martin H. Greenberg's description of agencies that do "not fulfill any function in the sense of attaining prescribed goals," but rather "are designed with specific personnel in mind" who "require a job for political purposes . . . whether or not there is anything for them to do."[4]

Agrarismo established the conditions that led *agrónomos* to seek the foreign training and assistance that transformed them into *técnicos* after 1940 and gave them the means to obtain it. The agricultural crisis of the late 1930s led to attacks on the land reform and *agrónomos'* status as scientists, driving them to depoliticize their science, seek help to learn a body of specialized knowledge to validate their expertise, and find farmers who acknowledged their professional status. World War II ensured that this aid came from the United States. *Agrónomos* and other members of the Revolutionary Family recognized the obstacles to making *campesinos* more productive farmers. After 1940 *agrónomos* advocated urban industrial growth to support Avila Camacho and Alemán, who rewarded them by founding and funding the MAP. As Perkins notes, at this time almost all First and Third World politicians and development experts defined this transition as modernization. The approach to economic development proposed by *cardenistas* like Ramón Beteta, who said that Mexico could avoid the pitfalls of the advanced capitalist nations by industrializing with small-scale manufacturing in rural communities, seems an appealing anomaly when compared to today's *utopía industrial.* U.S. training and the internationalization of Mexican agronomy reinforced *agrónomos'* belief that scientific agriculture and urban industrialization equaled modernity, but contrary to the implications of Simonian's argument, *cardenista agrónomos* did not think that "small is beautiful"; their search for alternatives to *campesino* corn-bean-squash agriculture reflected the "tortilla discourse" and other goals of the cultural campaign, as well as their attempt to establish their hegemony as scientific professionals with the right to define the terms of peasant farming. Whatever changes the MAP produced, the *cardenista agrónomos* were not members of a "popular political movement with a different conception of the improvement of human life" that Wright and Hewitt de Alcantara claim shaped his rural development strategy; they were middle-class advocates of the hegemony of Western science seeking professional status and upward social, economic, and political mobility.[5] Given *agrónomos'* lack of jobs and status during the *Porfiriato,* their opinions about *campesinos* and their farming practices, and their waning enthusiasm for *agrarismo* and the *ejido* after 1940, when they obtained assistance for their professionalization agenda and these projects were less "politically correct," their alliance with the revolutionaries was motivated by professional interests at least as much as by concerns about social justice.

Deborah Fitzgerald claims that the scientific style that the MAP's scientists brought to Mexico was the principal reason that the Green Revolution reached more affluent and educated private farmers rather than *campesinos,* but the *agrónomos'* professional interests and class status also influenced its outcome. Thanks to their alliance with the State, the *agrónomos* worked in a bureaucracy "carefully attuned to the protection of elite interests" such as commercial farmers, industrialists, bankers, and politicians.[6] A few rose to the upper ranks of the Revolutionary Family, and many filled middle-level positions, joining other bureaucrats who "became the major instruments of cooptation and control," in the *agrónomos'* case by advancing the myth that the *ejido* brought *campesinos* social justice and democracy in a "uniquely Mexican" way.[7] Like historian Jürgen Kocka's account of the professionalization of the German "*Bürgertum*" in the nineteenth century, the *agrónomos* professionalized inside a "strictly hierarchical structure" that "required their obedience and limited their responsibility," producing a "bureaucratically crippled professionalization."[8]

Supporting Siegrist's analysis, *agrónomos* attained "higher cultural, social and economic status . . . [and] increased socio-political power of individual professionals and of the profession as a collective" and after 1940 reduced their revolutionary action as they joined "the dominant middle class."[9] Cleaves observed that "notwithstanding the *agrónomos'* traditional support for agrarian reform," their politics were "ambiguous, depending on the political winds of the time."[10] From 1950 to 1970 most *agrónomos* supported the Green Revolution, blamed the *campesinos* if they did not adopt it, and took pride in the political and economic benefits and international acclaim that it brought them. After 1970, reflecting the State's desire to increase production in the peasant sector, some *agrónomos* tried to find new ways of working with *campesinos*, but most still believed in Green Revolution agriculture.

Some charges against the Green Revolutionaries are not so solidly grounded. Contrary to the assertions of Wright and others, the MAP did not create Mexico's interest in agrochemical farming; *agrónomos* advocated it since the early twentieth century, Cárdenas wanted to manufacture them, and the presidents of the 1940s invested in fertilizer factories. The State and the *agrónomos* are at least equally culpable for the spread of agrochemical farming and for causing problems when it is practiced irresponsibly. The MAP's research and technologies facilitated the process, even though they also evaluated organic methods of enhancing soil fertility, and Borlaug's vehement criticism of environmentalists was rather mean-spirited. In 1941 Sauer warned the foundation about the "irreparable loss of native crop forms" with "rapid Americanization," but the MAP's scientists still urged farmers to use standardized varieties.[11] Like the case of agrochemicals, *agrónomos* advocated the same thing during the 1920s and 1930s. Although genetic erosion remains a serious concern, the foundation's scientists were aware of the problem by the 1950s and took steps to address it, and later they began to evaluate traditional farming systems like intercropping as advocates of appropriate technology urged.[12]

Charges that the Green Revolutionaries fostered complacency on the issue of birth control are also weak. In 1949 Anderson warned the foundation:

> My feeling has been that population pressure in Mexico is so strong and has always been so strong for so many centuries in spite of changes in

religion and ways of living, that we may expect the birth rate to keep right up with the increase in maize production, or, at most, to lag only a few years behind it. When we raise production 10%, population will come up another ten, and the same old scarcity of maize will still be facing us.[13]

Birth control and family planning were, however, very controversial issues in Mexico, a very Catholic country notwithstanding Anderson's opinion, and taking a stand would have alienated the vast majority of Mexicans, as the *agrónomos* understood. Although most of the Green Revolutionaries were swept up in the wave of optimism in the mid- to late 1960s, and Borlaug's 1965 prediction about technology and the world food supply was excessive, they also mentioned the need for birth control and thus did not dodge this controversial issue.[14] Further, without the Green Revolution's productivity gains, Mexico's "population bomb" would have made the food supply situation much worse.

This study demonstrates the dangers that scientific professionals face when they participate in controversial political programs. The *agrónomos* of the 1920s to 1940s had no clientele who accepted their credibility as scientists, but many people rightly saw *agrarismo* as politics, not science, which worked against their effort to professionalize by "reduc[ing] them to their common status of citizens" when they became a target for the enemies of land reform.[15] After 1940 and especially 1970 the *agrónomos'* contributions to both sides of the *política agraria–política agrícola* debate influenced public perceptions of their credibility as scientific professionals. In 1987 the urban people whom Cleaves interviewed considered agronomy the least prestigious of six Mexican professions, below medicine, engineering, law, architecture, and economics. Disillusioned with the State's efforts to modernize agriculture, Niblo claims that the money would have been better spent by sending it to each *ejido*, an attitude that some *agrónomos* expressed in other words. The foundation's training program helped them attain a high level of scientific proficiency, especially in comparison to scientists in other Third World nations, but unfortunately for the *campesinos*, the way that they established their professional credibility also exacerbated their differences with peasants and erected a wall between them. The gap between urban, middle-class and poor rural Mexicans was and still is quite wide.[16]

The massification of higher education may cause another employment crisis for the *agrónomos*. More people earn agronomy degrees, including the sons and even daughters of humble *campesinos*, but the reduction in government jobs after the 1982 economic collapse, the termination of land reform, and the privatization of parastatals suggest that the expansion of State employment for *agrónomos* is unlikely in the near future. Arce found that the SAF's district fieldworkers, the lowest rung of the agrobureaucracy, felt that their technical expertise had little or no impact on their job performance or career advancement. Further, inflation and devaluations eroded the purchasing power and perquisites of bureaucrats who kept their jobs, and anti-corruption campaigns made informal ways of supplementing income, such as using official vehicles for personal matters, more difficult. Neoliberal economics and genetic engineering also threaten the *agronomos'* future. The increased presence of agribusinesses that use farming technologies developed abroad may create proprietary concerns that stifle Mexican research or render it redundant or moot, as they did in the export strawberry industry during the 1960s and 1970s. Genetic engineering may make the plant-breeding methods that produced the Green Revolution obsolete. Further, *agrónomos* are losing the struggle to define the direction of biotechnology research to the much better funded and trained scientists in the developed nations, to scientists who work for transnational agribusiness, and to a lesser extent to Mexico's own biologists and biochemists. Ironically, if neoliberal economics brings a biotechnology-based second Green Revolution to Mexico, foreign scientists and technologies supplied by agribusinesses may reconvert the *agrónomos* into transmitters of foreign knowledge or, worse, into professionals without jobs. *Agrónomos* are caught between their *agrarista* heritage, the legacies of the cultural campaign, the foreign training that established their scientific credibility, and their inability to keep pace with the advance of international science. As Lorey suggests, there is a chance that a growing pool of unemployed professionals may threaten political stability, as they did when they joined the Mexican Revolution and overthrew the *Porfiriato*.[17]

What does the future hold for the *campesinos*? Although rural development remains contested turf in Mexico, the *campesinos* have defenders. Some Mexican and U.S. social scientists and environmentalists took their side by making frightening predictions about their future under NAFTA and neoliberalism, including the death of the *ejido* and the *milpa*, both fundamental pillars of Mexican culture, environmental disasters, and massive flight from the countryside, especially from land unsuitable for export crops that will not attract foreign or national capital, causing political unrest in urban areas. They achieved a marginal victory when corn received a 15-year reprieve from the free market, and when the new millennium began, these dire prognostications had not come to pass. Using the first Green Revolution as an example of the shortcomings of foreign-directed science, some of the *campesinos'* defenders urge the government to support more research on appropriate technologies for small-scale farming, especially to raise the productivity of peasant corn growers.[18] But some scholars argue that the Mexican State is inherently biased against peasants and always treats them in a domineering and patronizing manner. Others criticize the *agrónomos* of the 1990s for being too production-oriented and committed to the Green Revolution, unable to relate to *campesinos*, and unprepared to solve their agricultural problems. In addition to this author, a few other scholars have urged the government to prepare *agrónomos* specifically for this task, emphasizing sustainable farming systems in their training. As the new millennium began, some early advocates of appropriate technology fretted that the forces of neoliberalism and globalization, such as the World Bank, co-opted their agenda and distorted it to advance their own sinister objectives.[19]

Worried about the imminent demise of small-scale agriculture in the Third World as the biotechnology revolution proceeds, a group of Swedish scientists asks: "Could we not devise an agricultural system that allows more people to live on the land, providing a life and employment for themselves as well as food for others?"[20] In contrast, claiming that "science and technology per se have never solved social and political problems," Argentine biotechnology researcher and medical doctor Daniel J. Goldstein claims that Latin America's desire to use genetic engineering to "eradicate malnutrition, end endemic parasitic diseases, and improve the poor farmers' fare" rather than generate "profit through the invention of internationally competitive, high-value products" is a misguided policy that causes brain drain and inadequate funding for programs.[21] Mexican researchers Fernando E. Vega,

Javier Trujillo Arriaga, and Estela Cervantes Reyes worry that a biotechnology-based Second Green Revolution will bypass *campesinos*, stifle researchers' interest in traditional farming systems, and create greater wealth inequality in rural Mexico. Recognizing that "only Mexican laboratories can produce appropriate technologies that answer the country's ecological and economic demands, and such innovation can only be done on the basis of 'excellent science,' " political scientists N. Patrick Peritore and Ana Karina Galve-Peritore and Mexican sociologist Rosa Elvia Barejas urge the State to fund biotechnology research on new crop varieties for arid or semi-arid climates and argue that *campesinos* should be compensated for the genetic resources of their traditional crop varieties, the "product of art, not nature" thanks to "centuries of native cultivation, breeding and modification," rejecting the developed nations' claim that they are "the common heritage of mankind."[22]

Many Mexicans sympathize with the *campesinos*' plight, and biotechnology offers the chance to create new varieties that do not require agrochemicals or controlled irrigation; thus, the prospects for some peasants who remain in the countryside are not completely bleak. But how can these objectives be attained? Most important, the government must make sure that at least some biotechnology research addresses *campesino* agriculture and sustain policies to protect peasant corn growers until they become more competitive. The State should monitor the spread of new technologies closely to ensure that they do not displace environmentally sound peasant practices, at least in some regions, and should not react to productivity gains by abandoning technical support for peasant farmers again. Lastly, Mexico should join other Third World nations to construct international agreements that do not define genetic resources as the common heritage of humankind and give proprietary control over and monetary value to the crops developed by peasant plant breeders.

In her study of government agricultural programs in Campeche, when asked what could be done to improve them, the *campesinos* whom Gates interviewed said that it was "most important . . . [that] the agrarian bureaucracy should not only acknowledge but also learn from the local experience of farmers."[23] On the other side of the interface, *Procampo* agents who worked with *ejidatarios* in Oaxaca during the 1990s told Rosario Varo Berra that peasant distrust was the biggest obstacle in recruiting participants. The difficulties in establishing trust and cooperation between *agrónomos* and *campesinos* are deeply rooted in the history of Mexican agronomy. The professionalization of the discipline through its connections with U.S. agronomists exacerbated preexisting conditions. Like the researchers and extension agents whom Louk Box studied in the Dominican Republic, most *agrónomos* seek urban jobs in the agrobureaucracy, are tied to the technologies of the industrial nations like hybrids and agrochemicals that favor privileged farmers, and want to work on commercial crops rather than those of subsistence or small-scale cultivators because they offer the best chance of successful results that lead to upward career mobility. Further, although many senior *agrónomos* whom Gates interviewed believe that peasant farming cannot be modernized without abolishing the *ejido*, many *campesinos* want to preserve it. *Técnicos'* values must be transformed before they can work effectively with peasant farmers.[24]

I do not believe that Mexico's agricultural scientists should be completely transformed or that government programs should target only peasant farmers. Such recommendations are impractical and attack powerful interests, but this study suggests that a new type of *agrónomo* is needed. In keeping with Arce's findings, to make communication between peasants and *técnicos* more effective and involve *ejidatarios* in the process of agricultural modernization, the government should borrow an idea from the era of revolution and reform: training researchers and extension agents from peasant backgrounds who receive financial assistance in exchange for an obligation to work with *campesinos* in their home districts for several years. Ideally and in keeping with the commitment to provide more support for higher agricultural education and research, the State should establish new colleges in rural areas, unlike the ENA, which is so close to the bright lights of Mexico City, to train the new *agrónomos*. Meeting a professional objective that Gilberto Fábila proposed in 1954, to promote effective communication with female peasants, the government should help *campesinas* attend the schools. Recent government projects with women, such as beekeeping in Ayoquila, Jalisco, are a positive step forward. I believe that some people from non-peasant backgrounds can work closely with *campesinos* and appreciate the logic of their farming practices and that some males communicate effectively with females, but a new group of *agrónomos*, trained in agroecology and genetic engineering, who know the social, economic, and environmental conditions of small farmers of both sexes and are tied to life in the countryside by background and obligation, could develop appropriate technologies and become lobbyists with personal and professional stakes in advancing and defending the peasants' interests to ensure that they receive at least some support from the State.[25] Ideally, the new *agrónomos* would be an independent group of researchers with their own institutions so they do not have to compete with scientists working on commercial agriculture or transmit the latest developments in international agricultural science. If the government ignores the *campesinos*, or a second Green Revolution bypasses them, this will perpetuate their impoverishment and create more migrants who will be difficult to assimilate into the urban economy, even with the projected industrial growth that NAFTA might foster, and thus defeat the intentions of neoliberal reforms.

The struggle to define the agenda of scientific research on Mexican agriculture began before the Green Revolution, and it is still crucial issue as Mexico enters the twenty-first century with a much larger population. Today the prevailing trend in agricultural policy is to allow neoliberalism to bring scientific farming to the *campesinos*. This may lead to economic benefits for some peasants, and it might be done in environmentally responsible ways. The market for organically grown products in the developed nations may help peasants raise their income and preserve their rural environment. Neoliberalism will create employment opportunities for *campesinos*, such as jobs in new multinational-owned food-processing plants. President Vicente Fox's (2000–present) ministry of agriculture proposes transforming those peasant farmers who cannot modernize by forming ties with multinational agribusiness into manufacturers of tourist curios and nature tour guides. But the history of Mexican agriculture since 1940 suggests that the introduction of foreign scientific agriculture exacerbates rural income inequality and carries high environmental costs. In 1994 the outbreak of rebellion in Chiapas proves that some peasants have rejected the Mexican Revolution and lost their faith that the State it created would bring them prosperity and democracy, but Mexico is a different country, much more urban and "modern" than in 1910. Some urban Mexicans sympathize with the EZLN rebels, but after six years of struggle their chances of "defeating the Federal army" and establishing "democracy" as they "advance toward the Federal Capital" seem very remote, and the majority of the Mexican people probably would not support them.[26]

Constructing policies that meet the needs of both rural and urban Mexico has been a difficult task, but to this author the notion that science and/or free trade will create a modern, prosperous, and industrial nation with a rural sector like the United States (i.e., without peasant farmers) seems unrealistic. Miguel A. Velasco told a gathering of *agrónomos* in 1954: "Only a government in which workers and peasants play an important role will be able to bring the democratic Revolution to a conclusion and consummate and consolidate the agrarian reform . . . not only by democratizing land ownership, but also by the technification of agriculture, raising its productivity." Like Velasco, I hope that real political reform will give *campesinos* some control over their future by electing leaders who enact policies to preserve their culture and environment and raise their standard of living. Zedillo's 1996 electoral reforms offer a chance for Mexico to finally become a democracy, and the election of Fox, an opposition candidate, in 2000 suggests that the time may have arrived. I agree with Gates and sociologist Anthony Hall: *campesinos* "internalized the ideals of the Mexican Revolution," especially Zapata's and Cárdenas' vision of social justice, and "cannot be discounted as potential agents for radical change by virtue of their determination, the clarity of their vision, and their refusal to disappear."[27] With political empowerment the *campesinos'* wisdom may not be lost or stolen in a maelstrom of neoliberal-induced technological and socioeconomic change.

NOTES

1. Susan R. Walsh Sanderson, *Land Reform in Mexico: 1910–1980* (Orlando, Fla.: Academic Press, 1984), p. 57; John Gledhill, *Casi Nada: A Study of Agrarian Reform in the Homeland of Cardenismo* (Albany: State University of New York, Institute for Mesoamerican Studies, 1991), p. 115; Rebeca de Gortari Rabiela, "Educación y conciencia nacional: Los ingenieros despues de la revolución Mexicana," *Revista Mexicana de Sociología* 49, 3 (1989): 123–41; David E. Lorey, *The University System and Economic Development in Mexico since 1929* (Stanford, Calif.: Stanford University Press, 1993), p. 68.

2. Advisory Committee for Agricultural Activities, "Minutes of Annual Meeting," October 26, 1950, p. 1, RFA, RG 6.13, Series 1.1, Box 32, Folder 361, RAC; "Interview with Mangelsdorf," p. 67; Bruce H. Jennings, *Foundations of International Agricultural Research: Science and Politics in Mexican Agriculture* (Boulder, Colo.: Westview Press, 1988), p. 50; John H. Perkins, "The Rockefeller Foundation and the Green Revolution, 1941–1956," *Agriculture and Human Values* 26 (Summer–Fall 1990): 15.

3. "Rockefeller Fund to Help Mexico Greatly Increase Nation's Food," *New York Times*, March 16, 1950; C.M. Christensen, *E.C. Stakman, Statesman of Science* (St. Paul, Minn.: The American Phytopathological Society, 1984), pp. 104–106.

4. Jennings, *Foundations*, pp. 51, 68; Angus Wright, *The Death of Ramón González: The Modern Agricultural Dilemma* (Austin: University of Texas Press, 1991), p. 168; Martin H. Greenberg, *Bureaucracy and Development: A Mexican Case Study* (Lexington, Mass.: D.C. Heath, 1970), p. 54.

5. Antonio Turrent Fernández, "Research and Technology for Mexico's Small Farmers," in Bruce F. Johnston et al., *U.S.–Mexico Relations: Agriculture and Rural Development* (Stanford, Calif.: Stanford University Press, 1987), pp. 306–7; Cassio Luiselli, "The Way to Food Self-Sufficiency in Mexico and Its Implications for Agricultural Relations with the United States," in Johnston et al., *U.S.–Mexico Relations*, p. 339; Wright, *The Death*, p. 186; Cynthia Hewitt de Alcantara, *Modernizing Mexican Agriculture: Socioeconomic Implications of Technological Change, 1940–1970* (Geneva: United Nations Research Institute for Social Development, 1976), pp. 4, 19; Susantha Goonatilake, *Aborted Discovery: Science and Creativity in the Third World* (London: Zed Books, 1984), p. 88; Perkins, "The Rockefeller Foundation," p. 15; Emilio López Zamora, *El agua, la tierra, los hombres de México* (México, D.F.: Fondo de Cultura Económica, 1977), pp. 224–25; Edmundo Flores, *Tratado de economía agrícola* (México, D.F.: Fondo de Cultura Económica, 1961), pp. 188–90; Ramón Beteta, "Economic Aspects of the Six Year Plan," in Hubert Herring and Herbert Weinstock, eds, *Renascent Mexico* (New York: Couici Freide, 1935), pp. 88–109; Lane Simonian, *Defending the Land of the Jaguar: A History of Conservation in Mexico* (Austin: University of Texas Press, 1995), pp. 86–87.

6. Deborah Fitzgerald, "Exporting American Agriculture: The Rockefeller Foundation in Mexico, 1943–1953," in Marcos Cueto, ed., *Missionaries of Science: The Rockefeller Foundation & Latin America* (Indianapolis: Indiana University Press, 1994), pp. 72–96; Steven E. Sanderson, *The Transformation of Mexican Agriculture: International Structure and the Politics of Rural Change* (Princeton, N.J.: Princeton University Press, 1986), pp. 262; David Ronfeldt, *Atencingo: The Politics of Agrarian Struggle in a Mexican Ejido* (Stanford, Calif.: Stanford University Press, 1973), p. 236;

Manuel Moreno Sánchez, *México, 1968–1972: Crisis y perspectiva* (Austin: University of Texas Press, 1973), p. 16.

7. Ronfeldt, *Atencingo*, p. 236; Dana Markiewicz, *The Mexican Revolution and the Limits of Agrarian Reform 1915–1946* (Boulder, Colo.: Lynne Rienner, 1993), pp. 113–14.

8. Jürgen Kocka, " '*Bürgerstum*' and Professions in the Nineteenth Century: Two Alternative Approaches," in Michael Burrage and Rolf Torstendahl, eds., *Professions in Theory and History: Rethinking the Study of the Professions* (London: Sage Publications, 1990), p. 72.

9. Hannes Siegrist, "Professionalization as a Process: Patterns, Progression, and Discontinuity," in Burrage and Torstendahl, eds., *Professions in Theory and History*, pp. 177–78, 187.

10. Peter Cleaves, *Professions and the State: The Mexican Case* (Tucson: University of Arizona Press, 1987), p. 68.

11. Wright, *The Death*, p. 172; Sauer to Willits, March 12, 1941.

12. Rodolfo P. Peregrina, "La investigación agrícola en México," in Ateneo Nacional Agronómico, *Problemas agrícolas actuales* (México, D.F.: Ateneo Atenagro, 1955), p. 45.

13. Edgar Anderson to Weaver, September 21, 1949, RFA, RG 1.1, Series 323, Box 3, Folder 19, RAC.

14. Alan Anderson, "The Green Revolution Lives," *New York Times Magazine*, April 27, 1975, p. 80.

15. Magali Sarfatti Larson, "Expertise and Expert Power," in Thomas L. Haskell, ed., *The Authority of Experts: Studies in History and Theory* (Bloomington: University of Indiana Press, 1984), p. 65.

16. Cleaves, *The Professions*, p. 7; Stephen R. Niblo, *War, Diplomacy, and Development: The United States and Mexico, 1938–1954* (Wilmington, Del.: Scholarly Resources, 1995), p. 290.

17. Fernando del Río Físico, "Estado actual de la investigación científica," in Rosa María Farell, ed., *Los profesionales mexicanos y los desafíos de la modernidad* (México, D.F.: Editorial Diana, 1989), p. 207; Jeffrey H. Cohen, *Cooperation and Community: Economy and Society in Oaxaca* (Austin: University of Texas Press, 1999), p. 80; Alberto Arce, *Negotiating Agricultural Development: Entanglements of Bureaucrats and Rural Producers in Western Mexico* (Wageningen, the Netherlands: Agricultural University Wageningen, 1993), pp. 79, 95; Marilyn Gates, *In Default: Peasants, the Debt Crisis and the Agricultural Challenge in Mexico* (Boulder, Colo.: Westview Press, 1993), p. 99; Gerd Junne and Annemieke Roobeek, "Social and Economic Impact of Biotechnology," *Biofutur* (November 1989): 49; Ernest Feder, *Strawberry Imperialism: An Inquiry into the Mechanisms of Dependency in Mexican Agriculture* (The Hague, the Netherlands: Institute of Social Studies, 1977), p. 50; Daniel J. Goldstein, "Third World Biotechnology, Latin American Development, and the Foreign Debt Prob-

lem," in N. Patrick Peritore and Ana Karina Galve-Peritore, eds., *Biotechnology in Latin America: Politics, Impacts, and Risks* (Wilmington, Del.: Scholarly Resources, 1995), p. 37; Peritore and Galve-Peritore, "Mexican Biotechnology Policy and Decision Makers' Attitudes toward Technology Policies," in Peritore and Galve-Peritore, eds., *Biotechnology*, p. 73; Fawler C. Lachkovics et al., "Bio-farms: The End of the End. Third World Farmers and the New Plant Genetics," *Development Dialogue* 1–2 (1988): 61; Lorey, *The University System*, pp. 174–78; Greenberg, *Bureaucracy*, p. 53.

18. Luis Hernández, "Los convulsiones rurales," *El Cotidiano* 34 (March–April 1990): 15; Steven E. Sanderson, "Mexican Public Sector Food Policy under Agricultural Trade Liberalization," *Policy Studies Journal* 20, 2 (1992): 431–46; Kirsten Appendini, "Los campesinos maiceros frente a la política de abasto: Una contradicción pemanente," *Comercio Exterior* 41, 10 (October 1991): 976–84; "Trade Deal with the United States Puts Many Mexican Farmers at Risk," *Christian Science Monitor*, November 4, 1992; Rosario Robles and Julio Moguel, "Agricultura y proyecto neoliberal," *El Cotidiano* 34, 2 (March–April 1990): 12; Hewitt de Alcantara, "La economía política del maiz en México," *Comercio Exterior* 41, 10 (October 1991): 955–70; "For Mexicans Growing Corn Is a National Heritage," *Los Angeles Times*, February 18, 1992; David Barkin and Billie R. DeWalt, "Sorghum and the Mexican Food Crisis," *Latin American Research Review* 23, 3 (Summer 1988): 52; Ignacio Ovalle Fernández, "El ejido y sus perspectivas en enfoque jurídico," *Comercio Exterior* 40, 9 (September 1990): 845; Daniel Goldrich and D.V. Caruthers, "Sustainable Development in Mexico," *Latin American Perspectives* 19, 1 (Winter 1992): 97–122; Barkin and DeWalt, "Sorghum," pp. 47–49; Turrent Fernández, "Research and Technology," pp. 304–5; De Walt and Martha W. Rees, *The End of Agrarian Reform in Mexico: Past Lessons, Future Prospects* (La Jolla: Center for U.S.–Mexican Studies, University of California, San Diego, 1994), p. 58; Alain de Janvry, Gustavo Gordillo, and Elisabeth Sadoulet, *Mexico's Second Agrarian Reform: Household and Community Responses* (La Jolla: Center for U.S.–Mexican Studies, University of California, San Diego, 1997), pp. 213–14.

19. Barkin and DeWalt, "Sorghum," p. 45; Bruce F. Johnston, "Rural Development in the U.S., Mexico, Japan, and Taiwan," in Johnston et al., *U.S.-Mexico Relations*, p. 51; Turrent Fernández, "Research and Technology," pp. 301, 309; Micheal W. Foley, "Agricultural Policy and Politics: Theory and Practice," *Latin American Research Review* 24, 1 (Winter 1989): 236, 244; Guillermo Knochenhauer, "La modernización del agro en México," *Comercio Exterior* 40, 9 (September 1990): 832–35; Mercedes A. Jiménez Velázquez, "La Fundación Rockefeller y la investigación agrícola en América Latina," *Comerico Exterior* 40, 10 (October 1990): 975. For a

more optimistic assessment see Alain DeJanvry and Ann Vandeman, "The Macrocontext of Rural Development: A Second View of the U.S. Experience," in Johnston et al., *U.S.-Mexico Relations*, pp. 83–109. For the school see Javier Trujillo Arriaga, "Desarrollo de una agricultura sustentable en México," *Comercio Exterior* 40, 10 (October 1990): 933–58; Joseph Cotter, "Salinas de Gortari's Agricultural Policy and Scientific Exchange: Some Lessons from before and during the Green Revolution," in Enrique C. Ochoa and David E. Lorey, eds., *Estado y agricultura en México: Antecedentes e implicaciones de las reformas salinistas* (Azcapotzalco, México, D.F.: Universidad Autónoma Metropolitana de Azcapotzalco, 1994), p. 48.

20. Lachkovics et al., "Bio-farms," p. 60.

21. Goldstein, "Third World Biotechnology," p. 37.

22. David Carruthers, "From Opposition to Orthodoxy: The Remaking of Sustainable Development," *Journal of Third World Studies* 18, 2 (Fall 2001): 98–102; Fernando E. Vega and Javier Trujillo Arriaga, "Biotecnología agrícola: Espejo de la revolución verde," *Comercio Exterior* 39, 11 (November 1989): 947–52; Estela Cervantes Reyes, "Producción de semillas: Problema básico para la autosuficiencia," *El Cotidiano* 34 (March–April 1990): 27–33; Junne and Roobeek, "Social and Economic," p. 49; Rosa Elvia Barajas, "Biotecnología y Revolución Verde: Especificidades y divergencias," *Sociología* 6, 10 (May–August 1991): 73; N. Patrick Peritore, "Biotechnology: Political Economy and Environmental Impacts," in Peritore and Galve-Peritore, eds., *Biotechnology*, p. 17; Peritore and Galve-Peritore, "Mexican Biotechnology Policy and Decision Makers' Attitudes toward Technology Policy," in Peritore and Galve-Peritore, eds., *Biotechnology* p. 93.

23. Gates, *In Default*, p. 238.

24. Gates, *In Default*, p. 241; Louk Box, "Knowledge, Networks and Cultivators: Cassava in the Dominican Republic," in Norman Long, ed., *Encounters at the Interface: A Perspective on Social Discontinuities in Rural Development* (Wageningen, the Netherlands: Agricultural University Wageningen, 1989), pp. 172–78; Arce, *Negotiating Agricultural Development*, p. 4; Rosario Varo Berra, *La reforma agraria en México desde 1853: Sus tres cíclos legales* (Guadalajara: Universidad de Guadalajara, 2002), pp. 186–91.

25. Gustavo Esteva and James Austin claim that the *Sistema Alimentario Mexicano* was politically feasible because it did not "attack commercial agriculture, threaten the existing distribution of resources, or undermine the position of any well organized constituency"; see Foley, "Agricultural," p. 246; see also Barkin and DeWalt, "Sorghum," p. 48. Supported by Lombardo Toledano and the SEP, Cárdenas wanted to train *técnicos* committed to the goals of the Revolution, such as bringing

social justice and prosperity to peasants; see "Informe del Gobernardo ante la H XLIV Legislatura Local," September 16, 1932, in *Palabras y documentos públicos de Lázaro Cárdenas, informes del gobierno y mensajes presidenciales de año nuevo* (México, D.F.: Siglo Ventiuno Editores, 1978), p. 30; "Discurso en el Consejo Extraordinario de la C.T.M." (n.d.), Box 142, Document 550, ARFF; Jesús Diaz Barriga, "Inconvenientes de la supresión del Consejo Nacional de Educación Superior y de la Investigación Científica," 12 December 1938, Cárdenas papers, Expediente 534/100, AGN; Gilberto Fábila, "Primera intervención," in Ateneo Nacional Agronómico, *Problemas agrícolas actuales* (México, D.F.: Ateneo Atenagro, 1955), p. 162; Norman Long and Magdalena Villareal, "The Changing Life-Worlds of Women in a Mexican Ejido: The Case of the Bee-Keepers of Ayuquilla and the Issue of Intervention," in Long, ed., *Encounters at the Interface*, p. 142. Of agronomist Ezequiel M. Gracia, a Tlaxcalan of indigenous background, who "dressed, ate, and lived like the humble peasants" with whom he worked, something that he implied some of his colleagues had not done, Gómez wrote: "Never did I have as precise an image of the duty of the agronomist in Mexico." See *Biografías*, p. 265; Arce, *Negotiating Agricultural Development*, pp. 95–97.

26. James W. Wilkie, "Prologo," in Varo Berra, *La reforma agraria*, p. 25; Ejército Zapatista de Liberación Nacional, "Declaración de la Selva Lacondona; Hoy Decimos ¡Basta!," 1993.

27. Miguel A. Velasco, "Intervención," in Ateneo Nacional Agronómico, *Problemas agrícolas actuales* (México, D.F.: Ateneo Atenagro, 1955), p. 303; Gates, *In Default*, p. 244; Anthony Hall, "Community Participation and Development Policy: A Sociological Perspective," in Hall and James Midgley, eds., *Development Policies: Sociological Perspectives* (Manchester: Manchester University Press, 1988), p. 95.

Bibliography

ARCHIVES CONSULTED

Archivo Estatal de Michoacán, Morelia, Michoacán, Mexico.
Archivo General de la Nación (AGN), Mexico City.
  Fondo Presidentes.
    Miguel Alemán papers.
    Manuel Avila Camacho papers.
    Lázaro Cárdenas del Rio papers.
    Adolfo López Mateos papers.
    Abelardo Rodríguez papers.
    Adolfo Ruiz Cortines papers.
  Gonzalo Robles papers.
Archivo Histórico de Jalisco (AHJ), Guadalajara, Jalisco, Mexico.
Archivo Histórico de la Secretaría de Relaciones Exteriores (AHSRE), Mexico City.
Archivo de Ramón Fernández y Fernández (ARFF), Zamora, Michoacán, Mexico.
Archivo Municipal de Zamora, Zamora, Michoacán, Mexico.
National Archives (NA), Washington, D.C.
  Record Group (RG) 7, Bureau of Entomology and Plant Quarantine.
  Record Group 54, Bureau of Plant Industries.
  Record Group 59, State Department Decimal Files.
  Record Group 166, Office of Foreign Agricultural Relations.
  Record Group 229, Office of the Coordinator of Inter-American Affairs.
  Record Group 310, Agricultural Research Administration.

Record Group 353, Interdepartmental Committee on Technical Cooperation with Latin America.
Rockefeller Archive Center (RAC), North Tarrytown, N.Y.
Rockefeller Family Archives (RFA).
Record Group (RG) 4, Nelson Rockefeller papers.
Rockefeller Foundation Archives (RFA).
Record Group 1.1, Mexican Agricultural Program.
Record Group 1.2, Mexican Agricultural Program.
Record Group 6.13, Mexican Agricultural Program.
Record Group 13, Oral Histories.

UNPUBLISHED PRIMARY SOURCES

Aguirre, Arturo I. "El acido fosfórico en los suelos y los diferentes abonos fosfatos." Unpublished thesis, ENA, 1922.

Arroyo, Gonzalo, and Mario Waissbluth. "Desarrollo biotecnológico en la producción agroalimentaria de México: Orientaciones de política." Unpublished report, United Nations, Comisión Económica para América Latina, March 10, 1988.

Biblioteca del Colegio de Michoacán, Zamora, Michoacán, Mexico. Agrónomos. Bound Volume #168–40.

———. Azúcar. Bound Volume #10–13.

———. Enseñanza Agrícola Superior. Bound Volume #145–43.

———. Escuela Nacional de Agricultura. Bound Volume #168–2.

———. Hule. Bound Volume #43–3.

———. Política Agrícola. Bound Volume #179–53.

———. SAM Segundo Consejo Nacional Directivo. Bound Volume #186–5.

———. SAM V Consejo Nacional Directivo. Bound Volume #186–10.

Borlaug, Norman E. "Civilization Will Depend More upon Flourishing Crops than on Flowery Rhetoric." Alfred M. Landon Lecture Series on Public Issues, March 20, 1979, Kansas State University, Manhattan, Kansas.

Coordinadora Nacional de Organizaciones Cafetaleras. "Mexico's Small Coffee Growers: Working Together for the Future." México, D.F.: n.d.

Ejército Zapatista de Liberación Nacional. "Declaración de la Selva Lacondona; Hoy Decimos ¡Basta!" 1993.

Mora Olivera, Raziel. "Porvenir de la alfalfa en el Mexe, Edo. de Hgo." Unpublished thesis, ENA, 1930.

Osoyo Acala, Roberto. "Ensayos de maíz norteamericanos en León, Gto." Unpublished thesis, ENA, 1946.

Unión de Comunidades Indigenas de la Región del Istmo. "Unidos venceremos." Santa María Guienagati. Oaxaca: n.d.
Unión de Ejidos Professor Otilio Montaño. "Por la conservación de la tierra, la naturaleza, y la cultura." Motozintla, Chiapas: n.d.
Zálvidar Flores, José. "Un plan de explotación para la Cienega de Chapala." Unpublished thesis, ENA, 1942.

PERIODICAL PRIMARY SOURCES

Agricultura.
Agricultura Técnica en México.
Agronomía.
Agronómica.
Agros.
Albuquerque Journal.
American Journal of Industrial Medicine.
Anales de la Escuela Nacional de Ciencias Biológicas.
Anales de la Sociedad Mexicana de Historia de la Ciencia y de la Tecnología.
Así.
Bioscience.
Boletín de Informaciones (Liga de Agrónomos Socialistas).
Boletín de la Sociedad Científica Antonio Alzate.
Boletín de la Sociedad Mexicana de Geografía y Estadística.
Boletín Interior (Sociedad Agronómica Mexicana).
Boletín Platanero y Agrícola.
Bulletin of the Atomic Scientists.
Bulletin of the Pan American Union.
Campeche.
Carta Seminal.
Ceres.
Chapingo.
Chemistry.
Christian Science Monitor.
Columbia Journal of World Business.
Comercio Exterior.
The Conservationist.
Current History.
Development Digest (USAID).
Diario Oficial.
Discussión.
The Ecologist.
Economic and Political Weekly.
El Campesino (Durango).

El Cotidiano.
El Democrata.
El Día.
El Dictamen (Veracruz).
El Jaliscense (Guadalajara).
El Maestro Rural.
El Mundo.
El Nacional (Mexico City).
El Nacional Revolucionario.
El Popular.
El Porvenir (Monterrey).
El Siglo de Torreón.
El Sol (Monterrey).
El Tractor.
El Universal (Mexico City).
El Universal Gráfico.
Excelsior (Mexico City).
Extensión Agrícola.
Far Eastern Economic Review.
Farm Journal.
Financial Times of London.
Fitofilo.
Foreign Affairs.
Futuro.
Germinal (Sociedad Agronómica Mexicana).
Hortscience.
Hoy.
Intellect.
International Labor Review.
International Science and Technology.
Irrigación en México.
Issues in Science and Technology.
Jalisco Agrícola y Ganadero.
Jalisco Rural.
Journal of Commerce.
Journal of Economic Entomology.
La Hacienda.
Land Economics.
La Prensa.
La Tierra.
La Trilla.
Life.

Los Angeles Times.
Manchester Herald.
Medicina y Farmacía.
Memorias de la Sociedad Científica Antonio Alzate.
Memoria y Revista de la Academía Nacional de Ciencias Antonio Alzate.
Memoria y Revista de la Sociedad Científica Antonio Alzate.
Mercado Agrícola-Ganadero.
New Scientist.
New York Times.
Nosotros.
Novedades.
Organic Gardening and Farming.
Phytopathology.
Policy Studies Journal.
Problemas agrícolas e industriales de México.
Proceedings of the Soil Science Society of America.
Progressive Education.
Ramparts.
Revista Arrocera.
Revista de Economía.
Revista de Industria.
Revista del Banco Obrero.
Revista Mexicana de Economía.
San Jose Mercury News.
San Luis Potosí Agrícola-Ganadero.
Santa Barbara News Press.
Scientific American.
Siembra.
Sociología.
The Texas Quarterly.
Tierra.
Últimas Noticias de Excelsior.
UNESCO Courier.
U.S. News and World Report.
Washington Post.

PUBLISHED PRIMARY SOURCES

Aguirre Avellaneda, Jerjes. *La política ejidal en México.* México, D.F.: Instituto Mexicano de Sociología, 1976.

Aguirre, Norberto. *Necesidades y posibilidades agrarias de México.* México, D.F.: Talleres Gráficos de la Penitenciaría, 1938.

Alanis Patiño, Emilio. *Diversos aspectos de la situación agrícola de México.* México, D.F.: Instituto Mexicano de Estudios Agrícolas, 1934.

Ateneo Nacional Agronómico. *Problemas agrícolas actuales.* México, D.F.: Ateneo Atenagro, 1955.

Avila Camacho, Manuel. *Address to the Mexican Agronomists.* México, D.F.: Department of State for Foreign Affairs and the International Press Service Bureau, 1941.

Ávilez, Bernardo. *Cultivo y comercio de algodón en México.* México, D.F.: Secretaría de Agricultura y Fomento, Dirección de Economía Rural, 1939.

Banco Nacional de Crédito Agrícola. *Informe del Consejo de Administración a la Asamblea de Accionistas.* México, D.F.: 1939.

Banó, Eugenio. *Estudio sobre varias plantas tropicales y su cultivo en México.* México, D.F.: IFSF, 1911.

Barnard, Chester I. *The Rockefeller Foundation: A Review for 1950 and 1951.* New York: The Rockefeller Foundation, 1952.

Barreto, Antonio, ed. *A Study of the Social and Economic Implications of the Large Scale Introduction of High-Yielding Varieties of Food Grains: A Selection of Readings.* Geneva: United Nations Research Institute for Social Development, 1971.

Barriero, Adolfo D. *Reseña histórica de la enseñanza agrícola y veterinaria en México.* México, D.F.: Tipografía El Libro de Comercio, 1906.

Beltrán, Enrique. *Medio siglo de recuerdos de un biólogo mexicano.* México, D.F.: Sociedad Mexicana de Historia Natural, 1977.

Blumenkron, Julio. *Directorio de la industria azucarera en México.* México, D.F.: 1930.

Borlaug, Norman. "The Impact of Agricultural Research on Mexican Wheat Production." *Transactions of the New York Academy of Science* 20, 3 (January 1958): 279–80.

———, and J.A. Rupert. *Metodos que pueden aumentar la producción de trigo en México.* México, D.F.: Secretaría de Agricultura, 1949.

Brambila, Alejandro, Jr. *El analysis químico y la fertilidad de los suelos.* México, D.F.: Dirección General de Agricultura, 1928.

Brauer H., Oscar, ed. *Diagnóstico de las ciencias agrícolas en México.* México, D.F.: Ediciones Productividad, 1968.

Bukharin, Nicolai I. *Economics of the Transformation Period.* New York: Bergman, 1971, trans. from Russian original and authorized German edition, Moscow, 1920.

Bulnes, Francisco. *Los grandes problemas de México.* México, D.F.: Ediciones de "El Universal," 1926.

Cabañas, Lamberto. *Reflexiones sobre economía rural.* México, D.F.: Tipografía de la Secretaría de Fomento, 1885.

Calvino, Mario. *Cultivo de la morera y la higuera*. México, D.F.: IFSF, 1912.
———. *El jitomate*. México, D.F.: IFSF, 1910.
———. *Informe de trabajos, División de Horticultura, 1912*. México, D.F.: IFSF, 1913.
———. *Informe de la División de Horticultura, 1913*. México, D.F.: IFSF, 1914.
———. *La nutrición de las plantas*. San Jacinto, D.F.: 1909.
———. *Segundo Informe de la División de Horticultura*. México, D.F.: IFSF, 1914.
———. *Una revolución en arborcultura: La plantación de arboles con raices cortadas y en terrenos sin roturar, ni labrar, segun el Sr. Stringfellow*. México, D.F.: IFSF, 1910.
Campbell, Keith O. *Food for the Future: How Agriculture Can Meet the Challenge*. Lincoln: University of Nebraska Press, 1979.
Carpio, Manuel. *Una Escuela de Agricultura para agricultores y no para profesionales empleomanos*. México, D.F.: 1923.
Carson, Rachel. *Silent Spring*. Cambridge, Mass.: Riverside Press, 1962.
Castaños, Carlos Manuel. *Testimonios de un agrónomo*. México, D.F.: Editorial Futura, S.A., 1981.
Cavasos González, Manuel. *La ciencia y la técnica aplicada a la agricultura y sus relaciones con el derecho agrario*. México, D.F.: Tipografía Ortega, 1953.
Cedillo Vázquez, Vicente. *Principales factores del desarrollo agrícola en México 1925–1960*. México, D.F.: UNAM, 1964.
Centro de Estudios Sociales, ed. *Cuestiones Agrarias de México*. México, D.F.: El Colegio de México, 1945.
Centro de Investigaciones Agrícolas del Golfo Central, INIA, SARH. *Manual de producción de arroz de temporal en el Estado de Veracruz*. Veracruz: Campo Agrícola Experimental Cotaxtla, 1982.
Colegio de Ingenieros Agrónomos de México. *Directorio Agronómica Nacional*. México, D.F.: Talleres Gráficos de la Nación, 1954.
Colwell, W.E. "Studies on the Effect of Nitrogen, Phosphorus, and Potash on the Yield of Corn and Wheat in Mexico." *Proceedings of the Soil Science Society of America* 11 (1946): 332–40.
Compañía Constructadora Richardson. *Indicador de productors y manual de información agrícola para El Valle del Yaqui, Sonora, Mexico*. Esperanza, Sonora: Compañía Constructadora Richardson, 1911.
Comsión Agraria del Estado de Chihuahua. *Memorandum para los topógrafos y agrónomos*. Chihuahua: Imprenta del Gobierno, 1915.
Comisión Nacional de Irrigación. *Estudio Agrológico de la Cienega de Chapala, Proyecto de Los Corrales, Michoacán-Jalisco*. México, D.F.: Editorial Cultura, 1932.

Confederación Nacional Campesina. *Bases generales que en materia agraria fueron aprobados para el Plan Sexenal.* México, D.F.: 1939.

Consejo Nacional de Agricultura (CNAG). *Memoria de la Primera Convención Nacional de Cafeticultores.* México, D.F.: CNAG, 1937.

———. *Memoria de la Primera Convención Nacional de Productores de Naranja.* México, D.F.: CNAG, 1936.

———. *Memoria de la Primera Convención Nacional de Productores de Oleaginosas.* México, D.F.: CNAG, 1939.

———. *Memoria de la Primera Convención Nacional de Productores de Papa.* México, D.F.: CNAG, 1939.

Crowther, J.G. *Soviet Science.* London: Kegan Paul, Trench, Trubrer & Company, 1936.

Decreto No. 33, Que reglamenta el Articulo Constitucional sobre el ejercicío de los profesiones técnicos-científicos. San Luis Potosí: 1939.

De Dios Bojórquez, Juan. *Los agrónomos de México.* México, D.F.: Ediciones del Boi, 1959.

De la Loma, José Luís. *Experimentación agrícola.* México, D.F.: Unión Tipográfica Editorial Hispano America, 1955.

De la Peña, Moises T. *La producción y beneficio del cacao.* México, D.F.: 1953.

Dirección General de Agricultura, ed. *Concurso del mejor mazorca del maiz.* Tacubaya, D.F.: Imprenta de la Dirección de Estudios Geográficos y Climatológicos, 1926.

Durán, Marco Antonio. *Apuntes de agrobiología.* México, D.F.: Instituto Mexicano de Estudios Agrícolas, 1940.

———. *Del agrarismo a la revolución agrícola.* México, D.F.: Talleres Gráficos de la Nación, 1947.

———. *El agrarismo mexicano.* México, D.F.: Siglo Ventiuno, 1967.

———. *Los sofismos de la reforma agraria.* México, D.F.: Liga de Agrónomos Socialistas, 1939.

Ecos del aniversario de plata de la Escuela Particular de Agricultura de Ciudad Juárez. Ciudad Juárez, México: Escuela Particular de Agricultura, 1931.

Escobar, Romulo. *The Agricultural Resources of the United States of Mexico.* Buffalo, N.Y.: The White-Ebans-Penfield Company, 1904.

———. *El esparago.* México, D.F.: Imprenta de la Dirección General de Agricultura, 1922.

Exposición del criterio ideológico del Gremio Agronómico Mexicano. Puebla: Confederación Campesina Emiliano Zapata, 1932.

Fernández del Campo, Luis. *Cultivo de la caña de azúcar.* México, D.F.: IFSF, 1913.

Fernández y Fernández, Ramón. *Chapingo hace 50 años.* Chapingo, México: ENA, Colegio de Postgraduados, 1976.

———. *El autoabastecimiento nacional de copra*. México, D.F.: BNCA, 1941.
———. *Los agrónomos*. México, D.F.: Impresa Secretaría de Maquinas, 1933.
———, and Ricardo Acosta. *Política agrícola, ensayo sobre normas para México*. México, D.F.: Fondo de Cultura Económica, 1961.
Flores, Edmundo. *Historias de Edmundo Flores, autobiografía 1919–1950*. México, D.F.: Martín Casillas Editores, 1983.
———, et al. *La ciencia y la tecnología en México*. México, D.F.: CONACYT, 1982.
———. *Tratado de economía agrícola*. México, D.F.: Fondo de Cultura Económica, 1961.
———. *Vieja revolución, nuevas problemas*. México, D.F.: Cuaderno de Joaquín Mortiz, 1970.
Fosdick, Raymond B. *The Rockefeller Foundation: A Review for 1944*. New York: The Rockefeller Foundation, 1944.
———. *The Rockefeller Foundation: A Review for 1945*. New York: The Rockefeller Foundation, 1945.
———. *The Rockefeller Foundation: A Review for 1946*. New York: The Rockefeller Foundation, 1946.
———. *The Rockefeller Foundation: A Review for 1948*. New York: The Rockefeller Foundation, 1948.
———. *The Rockefeller Foundation: A Review for 1949*. New York: The Rockefeller Foundation, 1949.
Fourtón, León. *¿Que abonos necesitan las tierras? Medios de que dispone un agricultor para saberlo, campos de experimentación*. México, D.F.: Secretaría de Agricultura y Fomento, 1930.
———. *Utilidad de los analysis de tierras, reglamento del laboratorio de química, nomenclatura de tierras*. Tacubaya, D.F.: Talleres Gráficos de la Secretaría de Agricultura y Fomento, 1929.
Freeman, Orville. *World Without Hunger*. New York: Praeger, 1968.
Frente Revolucionario de Agrónomos Mexicanos. *Programa de acción para la Sociedad Agronómica Mexicana, 1939–1940 (una contribución inicial de los agrónomos para la elaboración del Segundo Plan Sexenal)*. México, D.F.: Gráficos, S.A., 1938.
Gama, Valentín. *Consideraciones sobre la produccion agrícola e industrial de México*. México, D.F.: Empresa Editorial de Ingenería y Architectura, 1936.
Gándara, Guillermo. *Uso y aplicación de la Preparación Bordelesa*. México, D.F.: IFSF, 1907.
García Oses, Ramón. *Cultivo de la alfalfa*. México, D.F.: IFSF, 1912.
Gardiner, C. Harvey, ed. *Mexico 1825–1828: The Journal and Correspondence of Edward Thornton Tayloe*. Chapel Hill: University of North Carolina Press, 1959.

Gobierno del Estado de México. *Campo Experimental Santa Elena, Informe Num. 1, 1952–1954*. Toluca: Dirección de Agricultura y Ganadería, 1954.

Gómez, Gabriel. *Empleo agrícola de algunos residuos industriales*. México, D.F.: IFSF, 1911.

———. *La higuerilla y el aciete de ricino*, 2nd ed. México, D.F.: Imprenta Bucareli, 1921.

Gómez, Marte R. *Biografías de agrónomos*. Chapingo, México: ENA, Colegio de Postgraduados, 1976.

———. *Tendencias, medios, y fines de la política agrícola de México*. México, D.F.: 1945.

González H., Gonzalo. *La producción del ajonjolí en México*. México, D.F.: BNCA, 1941.

———. *La producción de cacao en México*. México, D.F.: BNCA, 1939.

Guardiola, Jorge. *El cultivo del frijol*. México, D.F.: SAF, 1921.

Harrar, J. George. *Mexican Agricultural Program: A Review of the First Six Years of Activity under the Joint Auspices of the Mexican Government and the Rockefeller Foundation*. New York: The Rockefeller Foundation, 1950.

———. *Strategy toward the Conquest of Hunger*. Selected papers of J. George Harrar. New York: The Rockefeller Foundation, 1967.

Hernández Xolocotzi, Efrain, ed. *Agroecosistemas de México: Contribuciones a la enseñanza, investigación, y divulgación agrícola*. Chapingo, México: ENA, Colegio de Postgraduados, 1977.

Herring, Hubert, and Herbert Weinstock, eds. *Renascent Mexico*. New York: Couici Friede, 1935.

Ibarra, David S. *Enfermedades que se transmiten a las plantas con su semilla y metodos para prevenirlas*. Guanajuato, México: Gobierno del Estado, 1930.

Instituto Biotécnico. *Campaña Pro-Maiz, el primer paso*. México, D.F.: D.A.P.P., 1939.

Instituto Mexicano de Recursos Naturales Renovables. *Trigesimo quinto y trigesimo sexto informe anual*. México, D.F.: 1988.

Jiménez Sánchez, Leobardo. *Las ciencias agrícolas y sus protagonistas*. Chapingo, México: Colegio de Postgraduados, 1984.

Johnston, Bruce F., Cassio Luiselli, Celso Cartas Contreras, and Roger D. Norton, eds. *U.S.–Mexico Relations: Agriculture and Rural Development*. Stanford, Calif.: Stanford University Press, 1987.

Khankhoje, Pandurang. *Maiz Granada "Zea Mays Digitata": Su origen, evolución, y cultivo*. México, D.F.: Talleres Gráficos de la Oficina de Publicaciones y Propaganda, 1936.

———. *Nuevas variedades de maiz*. Tacubaya, D.F.: Talleres Gráficos de la Secretaría de Agricultura y Fomento, 1930.

Landaverde, Arnulfo. *El maiz y su cultivo en México.* México, D.F.: Editores José C. Torres y Cia., 1949.

Lázcano Romero, José. *Analysis de la situación algodonera de México.* México, D.F.: Instituto Mexicano de Estudios Agrícolas, 1938.

Ley que creo el Consejo Nacional de la Educación Superior y la Investigacion Científica. México, D.F: 1936.

Lira López, Salvador, Ramón Fernández y Fernández, and Quintín Ochoa. *La pobreza rural en México.* México, D.F.: mimeograph, 1945.

López Zamora, Emilio. *El agua, la tierra, los hombres de México.* México, D.F.: Fondo de Cultura Económica, 1977.

———. *La situación del Distrito de Riego de El Mante.* México, D.F.: Editorial Ramírez Alonso, 1939.

Luiselli, Cassio. "The Way to Food Self-Sufficiency in Mexico and Its Implications for Agricultural Relations with the United States." In Bruce F. Johnston, Cassio Luiselli, Celso Cartas Contreras, and Roger D. Norton, eds., *U.S.–Mexico Relations: Agriculture and Rural Development.* Stanford, Calif.: Stanford University Press, 1987, pp. 319–39.

MacDonald, William. *El cultivo de secano.* Ignacio Flores Sánchez, trans. México, D.F.: IFSF, 1911.

Mallen, Bernardo. *¿En donde esta nuestra riqueza?* México, D.F.: 1908.

Malo Álvarez, Ignacio. *La burgesia y la Reforma Agraria.* México, D.F.: Unión Cooperativa de Artes Gráficos del D.F, 1940.

Marín, Luis. *El cultivo del algodón.* México, D.F.: Talleres Gráficos de la SAF, 1931.

Martínez, Leandro. *Cultivo y beneficio del cacaotero,* 2nd ed. México, D.F.: IFSF, 1912.

McKelvey, J.J., and Dorothy Parker. *Nuevas insecticidas.* México, D.F.: Oficina de Estudios Especiales, Secretaría de Agricultura y Ganadería, 1948.

Mesa A., Manuel. *Agricultural Resources of Mexico.* Washington, D.C.: Interamerican Development Commission, 1946.

———. *Relatos autobiográficos.* México, D.F.: Editorial Nuestro Tiempo, 1981.

Mosher, Arthur T. *Technical Cooperation in Latin American Agriculture.* Chicago: University of Chicago Press, 1957.

Motz, Fred A., and Lester D. Mallory. *The Fruit Industry of Mexico.* Washington, D.C.: Office of Foreign Agricultural Relations, OFAR Foreign Agriculture Report No. 9, 1944.

Muñoz, Roberto. *El arroz: Monografía.* México, D.F.: D.A.P.P., 1939.

Ojeda Paullada, Pedro. *Política Pesquera 1983–1985.* México, D.F.: Secretaría de Pesca, 1987.

Olson-Seffer, Pehr. *La agricultura en varios paises tropicales y subtropicales: Organización de la agricultura en diversos paises extranjeros*. Mexico, D.F.: IFSF, 1910.

Orive Alba, Adolfo. *La irrigación en México*. México, D.F.: Editorial Grijalva, 1970.

Paddock, William, and Paul Paddock. *Famine, 1975! America's Decision: Who Shall Survive*? New York: Little, Brown & Company, 1967.

———. *Time of Famines*. New York: Little, Brown & Company, 1976.

Palabras y documentos públicos de Lázaro Cárdenas, Informes Presidenciales y Mensajes de Año Nuevo, 1938–1940, vol. 2. México, D.F.: Siglo Ventiuno Editoriales, 1978,

Pardo y Urbina, Manuel. *Cultivo del garbanzo*. México, D.F.: IFSF, 1911.

Partido Nacional Revolucionario (PNR), ed. *Los problemas agrícolas de México*. México, D.F.: PNR, 1934.

Pettus, Daisy Caden, ed. *The Rosalie Evans Letters from Mexico*. Indianapolis: Bobbs-Merrill Company, 1926.

Pittner, John B. *Trebol Huban*. México, D.F.: Oficina de Estudios Especiales, Secretaría de Agricultura y Ganadería, 1948.

Portes Gil, Emilio. *The Mexican Schools and the Peasantry*. México, D.F.: Press of the Ministry of Foreign Relations, 1936.

Powell, Jane Swift. *Agriculture in Mexico*. Washington, D.C.: Pan American Union, 1942.

Proposiciones del Partido Comunista Mexicano para el Plan Sexenal de 1940–1946. México, D.F.: 1939.

Prospects of the World Food Supply. Proceedings of a Symposium during the 103rd Annual Meeting of the National Academy of Sciences. Washington, D.C.: National Academy of Sciences, 1966.

Reeves, Frank. *Hacienda de Atotonilco*. Yerbanis, Durango: Cia. Ganadero de Atotonilco, 1936.

Rivas Tagle, Antonio. *El cultivo racional del maiz*. Tacubaya, D.F.: Talleres Gráficos de la Secretaría de Agricultura y Fomento, 1929.

———. *La selección de la semilla de maiz*. Guanajuato, México: Talleres Linotipográficos del Estado, 1929.

Robles, Leonel. *Agricultural Education, Investigation, and Extension in Mexico: A Preliminary Study*. México, D.F.: The National Association of Directors of Agricultural Schools, 1964.

Rockefeller Foundation. *President's Review, 1966*. New York: The Rockefeller Foundation, 1966.

———. *The President's Review and Annual Report, 1970*. New York: The Rockefeller Foundation, 1970.

Rodríguez Adame, Julián. *La Secretaría de Agricultura y Ganadería y su función en el desarrollo agropecuario nacional*. México, D.F.: Guanos y Fertilizantes de México, 1964.

Rodríguez, Eduardo. *El cultivo del lupulo*, 2nd ed. México, D.F.: Talleres Gráficos de la Nación, 1920.

Rosso, José R. *El alfafero mexicano: El manual de los colonos y campesinos*. México, D.F.: D.A.P.P., 1936.

Rubenstien, M. "Relations of Science, Technology, and Economics under Capitalism and in the Soviet Union." In *Science at the Crossroads*. London: Kniga, 1931, pp. 46–59.

Ruiz y Sandoval, Alberto. *El algodón en México*. México, D.F.: Oficina Tipografía de la Secretaría de Fomento, 1884.

Russell, Sir John. *World Population and World Food Supplies*. London: George Allen & Unwin, 1954.

Sahagún, Bernardino de. *General History of the Things of New Spain*. Volume 12, Book 11, "The Earthly Things." Charles E. Dibble and Arthur J.O. Anderson, trans. Provo: University of Utah Press, 1963.

Secretaría de Agricultura y Fomento. *Características de los trigos cultivados en la República Mexicana*. México, D.F.: D.A.P.P., 1937

———. *Estudio agro-económico del maiz*. México, D.F.: Oficina de Información y Propaganda, 1940.

———. *Informe de labores 1940–1941*. México, D.F.: Editorial Cultura, 1941.

———. *Informe de labores 1941–1942*. México, D.F.: Editorial Cultura, 1942.

———. *Informe de labores 1942–1943*. México, D.F.: Editorial Cultura, 1943.

———. *Informe de labores 1943–1944*. México, D.F.: Editorial Cultura, 1944.

———. *Informe de labores 1944–1945*. México, D.F.: Editorial Cultura, 1945.

———. *Informe de labores 1945–1946*. México, D.F.: Editorial Cultura, 1946.

———. *La organización agraria en 1927*. Tacubaya, D.F.: Talleres Gráficos de la Secretaría de Agricultura y Fomento, 1928.

———. *Memoria de la 1o Convención Triguera Nacional*. Tacubaya, D.F.: Talleres Gráficos de la Oficina de Publicaciones, 1934.

———. *Memoria de la Segunda Convención Resolutiva del Problema Platanero*. México, D.F.: 1939.

———. *Memoria 1924–1925*. Tacubaya, D.F.: Imprenta de la Dirección de Estudios Geográficos y Climatológicos, 1927.

———. *Memoria 1936–1937*. México, D.F.: D.A.P.P., 1937.

———. *Memoria 1937–1938*. México, D.F.: D.A.P.P., 1938.

———. *Memoria 1938–1939*. México, D.F.: D.A.P.P., 1939.

———. *Memoria 1939–1940*. México, D.F.: D.A.P.P., 1940.

———. *Memoria de trabajos de las Direccciones de Agricultura y Ganadería dependiente de la Secretaría de Agricultura y Fomento durante el periodo presidencial de Lázaro Cárdenas*. México, D.F.: D.A.P.P., 1940.

———. *Monografía del jitomate*. San Jacinto, D.F.: 1932.

———. *Monografía de la naranja*. México, D.F.: 1936.

———. *Plan de Movilización Agrícola*. México, D.F.: 1942.

———. *Programa de labores, 1939.* México, D.F.: mimeograph, 1939.
———. *Programa de labores, 1940.* México, D.F.: mimeograph, 1940.
———. *Resultados obtenidos con la applicación de abonos en el cultivo de chile.* México, D.F.: 1944.
———. *Sugestiones para la elaboración del Plan Sexenal 1934–1940.* México, D.F.: mimeograph, 1933.
Secretaría de Agricultura y Recursos Hidraulicos. *Desarrollo rural integral.* México, D.F.: Tiempo, S.A., 1988.
Secretaría de Economía Nacional. *El café: Aspectos económicos de su producción y distribución en México y en el extranjero.* México, D.F.: Editorial Cultura, 1933.
Secretaría de Educación Pública. *Instituto Politécnico Nacional, Anuario 1939.* México, D.F.: 1939.
———. *La resurrección de la tierra.* Cuaderno de Cultura Popular #8. México, D.F.: n.d.
———. *Plan de Acción de la Escuela Elemental Agrícola.* México, D.F.: 1938.
Secretaría de Fomento. *Algunos documentos para la historia de la enseñanza agrícola en México.* México, D.F.: IFSF, 1912.
Secretaría de Hacienda y Crédito Público. *Presupuesto general de egresos de la Federación 1938.* México, D.F.: 1938.
Sociedad Agrónomica Mexicana. *Primer ciclo de conferencias.* México, D.F.: D.A.P.P., 1938.
———. *Resoluciones y memoria de la Segunda Reunión del Consejo Nacional Directivo.* Mexico, D.F.: Sociedad Agrónomica Mexicana, 1941.
Stakman, Elvin C., Richard Bradfield, and Paul C. Mangelsdorf. *Campaigns against Hunger.* Cambridge, Mass.: Harvard University Press, 1967.
Symposium on Science and Foreign Policy—The Green Revolution. Proceedings before the Subcommittee on National Security Policy and Scientific Developments, House of Representatives, 91st Congress, December 5, 1969. Washington, D.C.: U.S. Government Printing Office, 1970.
Symposium on the Strategy for the Conquest of Hunger. Proceedings of a Symposium Convened by the Rockefeller Foundation, April 1 and 2, 1968, at the Rockefeller University. New York: The Rockefeller Foundation, 1968.
Taboada R., Edmundo. *Algunos datos relativos al metodo de Hoffer para investigar deficiencias de elementos fertilizantes en los suelos.* México, D.F.: 1941.
Trevino Martínez, Roberto. *El problema agrario de México y la solución revolucionaria (sugestión de una economía agrícola colectivista).* México, D.F.: Editorial Nuevo Era, 1936.
Turrent Fernández, Antonio. "Research and Technology for Mexico's

Small Farmers." In Bruce F. Johnston, Cassio Luiselli, Celso Cartas Contreras, and Roger D. Norton, eds., *U.S.–Mexico Relations: Agriculture and Rural Development*. Stanford, Calif.: Stanford University Press, 1987, pp. 301–18.

Unión Agronómica Mexicana. *Agrónomos y pasantes de agrónomo de la Escuela P. de Agricultura de Ciudad Juárez, Chih.* México, D.F.: 1934.

Uribe Ruiz, Jesús. *Problemas y soluciones en el desarrollo agrícola de México.* México, D.F.: Academía Mexicana de Ciencias Agrícolas, 1964.

Vargas, Rafael. *Procedimiento que debe seguirse para hacer menos inseguras las cosechas del maiz.* México, D.F.: IFSF, 1912.

U.S. Department of Commerce. "Insecticide Market in Mexico." *Industrial Reference Service, Part II, Chemicals, Drugs, and Pharmaceuticals* 4, 11 (January 1946): 1–4.

World War on Hunger. Hearings before the Committee on Agriculture, House of Representatives, 89th Congress, 2nd Session. Washington, D.C.: U.S. Government Printing Office, 1966.

PUBLISHED SECONDARY SOURCES

Aboites Aguilar, Luís. "Política de irrigación y cambio tecnológico en México, 1920–1940." *Papeles de la Casa Chata* 2, 3 (1987): 3–12.

Acosta Proudinat, Rodrigo. *La productividad agrícola en México.* México, D.F.: 1970.

Alba, Victor. *The Mexicans: The Making of a Nation*. New York: Praeger, 1967.

Alonzo, Armando C. *Tejano Legacy: Rancheros and Settlers in South Texas, 1734–1900*. Albuquerque: University of New Mexico Press, 1998.

Ankerson, Dudley. *Agrarian Warlord: Saturnino Cedillo and the Mexican Revolution in San Luis Potosí*. De Kalb: Northern Illinois University Press, 1984.

Arce, Alberto. *Negotiating Agricultural Development: Entanglements of Bureaucrats and Rural Producers in Western Mexico*. Wageningen, the Netherlands: Agricultural University Wageningen, 1993.

———. "The Social Construction of Agrarian Development: A Case Study of Producer–Bureaucrat Relations in an Irrigation Unit in Western Mexico." In Norman Long, ed., *Encounters at the Interface: A Perspective on Social Discontinuities in Rural Development*. Wageningen, the Netherlands: Agricultural University Wageningen, pp. 11–51.

Ashby, Joe C. *Organized Labor and the Mexican Revolution under Lázaro Cárdenas*. Chapel Hill: University of North Carolina Press, 1967.

Bailey, John J. "Agrarian Reform in Mexico: The Quest for Self-

Sufficiency." *Current History* 80, 469 (November 1981): 357–60, 391–92.

Bantjes, Adrian A. *As If Jesus Walked on Earth: Cardenismo, Sonora, and the Mexican Revolution*. Wilmington, Del.: Scholarly Resources, 1998.

———. "Burning Saints, Molding Minds: Iconoclasm, Civic Ritual, and the Failed Cultural Revolution." In William H. Beezley, Cheryl English Martin, and William E. French, eds., *Rituals of Rule, Rituals of Resistance: Public Celebrations and Popular Culture in Mexico*. Wilmington, Del.: Scholarly Resources, 1994, p. 69.

Barkin, David. "Agricultural Development in Mexico: A Case Study of Income Concentration." *Social Research* 37, 2 (Summer 1970): 306–20.

———. "Food Production, Consumption, and Policy." In Michael S. Werner, ed., *Encyclopedia of Mexico: History, Society and Culture*. Chicago: Fitzroy Dearborn, 1997, pp. 493–96.

———, and Billie R. DeWalt. "Sorghum and the Mexican Food Crisis." *Latin American Research Review* 23, 3 (1988): 30–59.

Bartra, Armando. "Colectivización o proletarización: el caso del Plan Chontalpa." *Cuadernos Agrarios* 1, 4 (October–December 1976): 56–101.

Becker, Marjorie. *Setting the Virgin on Fire: Lázaro Cárdenas, Michoacán Peasants, and the Redemption of the Mexican Revolution*. Berkeley: University of California Press, 1995.

Beckman, Svante. "Professionalization: Borderline Authority and Autonomy in Work." In Michael Burrage and Rolf Torstendahl, eds., *Professions in Theory and History: Rethinking the Study of the Professions*. London: Sage Publications, 1990, pp. 115–38.

Beezley, William H. *Judas at the Jockey Club and Other Episodes of Porfirian Mexico*. Lincoln: University of Nebraska Press, 1987.

———, Cheryl English Martin, and William E. French, eds. *Rituals of Rule, Rituals of Resistance: Public Celebrations and Popular Culture in Mexico*. Wilmington, Del.: Scholarly Resources, 1994.

Beltrán, Enrique. "Fuentes mexicanas en la historia de la ciencia." *Anales de la Sociedad Mexicana de Historia de la Ciencia y de la Tecnología* 2 (1970): 57–118.

———. "La Dirección de Estudios Biológicos de la Secretaría de Fomento y el Instituto de Biología de la Universidad Nacional Autónoma." *Anales de la Sociedad Mexicana de Historia de la Ciencia y de la Tecnología* 1 (1969): 105–141.

Bennet, Douglas, and Keith Sharpe. "The State as Banker and Entrepreneur: The Last Resort Character of the Mexican State's Economic Intervention 1917–1970." In Sylvia Ann Hewlitt and Richard S. Weinert, eds., *Brazil and Mexico: Patterns of Late Development*. Philadelphia: Institute for the Study of Human Issues, 1982, pp. 176–195.

Boardman, Margaret C. "Sowing the Seeds of the Green Revolution: The Pivotal Role Mexico and Non-Profit Organizations Play in Making Biotechnology an Important Foreign Policy Issue for the 21st Century." *PROFMEX Journal* 4, 3 (Summer 1999): 1–24.

Boltvinik, Julio. "Estrategia de desarrollo rural, economía campesina e innovación tecnológica en México." *Comercio Exterior* 26, 7 (July 1976): 813–26.

Booth, George C. *Mexico's School Made Society*. Stanford, Calif.: Stanford University Press, 1941.

Box, Louk. "Knowledge, Networks and Cultivators: Cassava in the Dominican Republic." In Norman Long, ed., *Encounters at the Interface: A Perspective on Social Discontinuities in Rural Development*. Wageningen, the Netherlands: Agricultural University Wageningen, 1989, pp. 165–182.

Boyer, Christopher R. "Old Loves, New Loyalties: Agrarismo in Michoacán, 1920–1928." *Hispanic American Historical Review* 78, 3 (August 1998): 419–55.

Brown, Lester R. *By Bread Alone*. New York: Praeger Publishers for the Overseas Development Council, 1974.

———. *The Seeds of Change: The Green Revolution and Development in the 1970s*. New York: Praeger, 1970.

———. *The Twenty Ninth Day*. New York: W.W. Norton & Company, 1978.

Brown, Richard Harvey. "Modern Science: Institutionalization of Knowledge and Rationalization of Power." *The Sociological Quarterly* 34, 1 (1993): 153–68.

Burrage, Michael, and Rolf Torstendahl, eds. *Professions in Theory and History: Rethinking the Study of the Professions*. London: Sage Publications, 1990.

———. "The Professions in Sociology and History." In Michael Burrage and Rolf Torstendahl, eds., *Professions in Theory and History: Rethinking the Study of the Professions*. London: Sage Publications, 1990, pp. 1–27.

Callcott, Wilfred Hardy. *Liberalism in Mexico, 1857–1929*. Hamden, Conn.: Archon Books, 1965 (orig. 1931).

Calva Téllez, José Luis, and Carlos Monzón. "La economía agrícola mexicana y el consumo nacional de fertilizantes." In Jesús Gracia Fadrique, ed., *Estado y fertilizantes*. México, D.F.: Fondo de Cultura Económica, 1988, pp. 16–27.

Camp, Roderic A. "A Middle Level Technocrat in Mexico." *Journal of Developing Areas* 6, 4 (July 1972): 571–81.

Carlos, Manuel L. *Politics and Development in Rural Mexico: A Study of Socioeconomic Modernization*. New York: Praeger, 1974.

Carroll, Patrick. "Science, Power, Bodies: The Mobilization of Nature as State Foundation." *Journal of Historical Sociology* 9, 2 (1996): 139–67.

Castrejón Diez, Jaime. *La educación superior en México*. México, D.F.: Edicol, 1979.

Chambers, David Wade. "Period and Process in Colonial and National Science." In Nathan Reingold and Marc Rothenberg, eds., *Scientific Colonialism: A Cross Cultural Comparison*. Washington, D.C.: Smithsonian Institution, 1987, pp. 297–321.

Channing, Margaret. "The Contours of the Post-1810 Depression in Mexico: A Reappraisal from a Regional Perspective." *Latin American Research Review* 27, 2 (1992): 119–150.

Christensen, C.M. *E.C. Stakman, Statesman of Science*. St. Paul, Minn.: The American Phytopathological Society, 1984.

Clawson, David L., and Don R. Hoy. "Nealtican, Mexico: A Peasant Community That Rejected the 'Green Revolution.' " *American Journal of Economics and Sociology* 38, 4 (October 1979): 371–87.

Cleaves, Peter S. *Professions and the State: The Mexican Case*. Tucson: University of Arizona Press, 1987.

Cline, Howard F. *The United States and Mexico*. Cambridge, Mass.: Harvard University Press, 1953.

Cohen, Jeffrey H. *Cooperation and Community: Economy and Society in Oaxaca*. Austin: University of Texas Press, 1999.

Collins, Randall. "Market Closure and the Conflict Theory of the Professions." In Michael Burrage and Rolf Torstendahl, eds., *Professions in Theory and History: Rethinking the Study of the Professions*. London: Sage Publications, 1990, pp. 24–43.

Córdoba, Arnaldo. *La política de mases del cardenismo*. México, D.F.: Ediciones Era, 1974.

Cotter, Joseph. "Salinas de Gortari's Agricultural Policy and Scientific Exchange: Some Lessons from before and during the Green Revolution." In Enrique Ochoa and David E. Lorey, eds., *Estado y agricultura en Mexico: Antecedentes e implicaciones de las reformas salinistas*. México, D.F.: Universidad Autonoma Metropolitana de Azcapotzalco, 1994, pp. 39–55.

Craig, Ann L. *The First Agraristas: An Oral History of a Mexican Agrarian Reform Movement*. Berkeley: University of California Press, 1983.

Crosland, Maurice. *Science under Control: The French Academy of Sciences 1795–1914*. Cambridge: Cambridge University Press, 1992.

Cueto, Marcos, ed. *Missionaries of Science: The Rockefeller Foundation & Latin America*. Indianapolis: Indiana University Press, 1994.

DeWalt, Billie R. "Inequalities of Wealth, Adoption of Technology and

Production in a Mexican Ejido." *American Ethnologist* 2, 1 (February 1975): 149–68.

———. *Modernization in a Mexican Ejido: A Study in Economic Adaptation.* Cambridge, Mass.: Cambridge University Press, 1979.

———, and Martha W. Rees. *The End of Agrarian Reform in Mexico: Past Lessons, Future Prospects.* La Jolla: Center for U.S.–Mexican Studies, University of California, San Diego, 1994.

Douwe van de Ploeg, Jan. "Knowledge Systems, Metaphor and Interface: The Case of Potatoes in the Peruvian Highlands." In Norman Long, ed., *Encounters at the Interface: A Perspective on Social Discontinuities in Rural Development.* Wageningen, the Netherlands: Agricultural University Wageningen, 1989, pp. 145–63.

Dunbar, Gary S. " 'The Compass Follows the Flag': The French Scientific Mission to Mexico, 1864–1867." *Annals of the Association of American Geographers* 78, 2 (1988): 229–40.

Esteva, Gustavo. *The Struggle for Rural Mexico.* South Hadley, Mass.: Bergin & Garvey, 1983.

Farell, Rosa María, ed. *Los profesionales mexicanos y los desafíos de la modernidad.* México, D.F.: Editorial Diana, 1989.

Feder, Ernest. *Strawberry Imperialism: An Inquiry into the Mechanisms of Dependency in Mexican Agriculture.* The Hague, the Netherlands: Institute of Social Studies, 1977.

Fedoseyev, P.N. "The Social and Ideological Foundations for Drawing Together of Nations and Nationalities." In G.D. Konkov, ed., *Science in the U.S.S.R.* Moscow: Progress Publishers, 1973, pp. 69–81.

Fitzgerald, Deborah. "Exporting American Agriculture: The Rockefeller Foundation in Mexico, 1943–1953." In Marcos Cueto, ed., *Missionaries of Science: The Rockefeller Foundation & Latin America.* Indianapolis: Indiana University Press, 1994, pp. 72–96.

Foley, Michael W. "Agricultural Policy and Politics: Theory and Practice." *Latin American Research Review* 24, 1 (Winter 1989): 233–49.

Freidson, Eliot. "Are Professions Necessary?" In Thomas L. Haskell, ed., *The Authority of Experts: Studies in History and Theory.* Bloomington: University of Indiana Press, 1984, pp. 3–27.

French, William E. "Prostitutes and Guardian Angels: Women, Work, and the Family in Porfirian Mexico." *Hispanic American Historical Review* 72, 4 (1992): 529–53.

Friedlander, Judith. *Being Indian in Hueyapan: A Study of Forced Identity in Contemporary Mexico.* New York: St. Martin's Press, 1975.

Friedrich, Paul. *Agrarian Revolt in a Mexican Village.* Englewood Cliffs, N.J.: Prentice-Hall, 1970.

García de León, Antonio. "Las grandes tendencias de la producción agraria." In Paul Friedrich, Enrique Semo, and Ricardo Gamboa Ramírez, eds., *Historia de la cuestión agraria mexicana, vol. 1: El siglo de la hacienda*. México, D.F.: Siglo Ventinuo Editores, 1988, pp. 13–85.

———, Enrique Semo, and Ricardo Gamboa Ramírez, eds. *Historia de la cuestión agraria mexicana, vol. 1: El siglo de la hacienda*. México, D.F.: Siglo Ventinuo Editores, 1988.

García Sordo, Mario. "Desempleados o subempleados, más de 90 mil agrónomos." *El Financiero*, October 5, 1988, p. 39.

García Zamora, Rodolfo. "Crisis agrícola y nuevos subsidios al campo." In Enrique C. Ochoa and David E. Lorey, eds., *Estado y agricultura en México: Antecedentes e implicaciones de las reformas salinistas*. México, D.F.: Universidad Autonoma Metropolitana de Azcapotzalco, 1994, pp. 73–100.

Gates, Marilyn. *In Default: Peasants, the Debt Crisis and the Agricultural Challenge in Mexico*. Boulder, Colo.: Westview Press, 1993.

Gladwin, Christina H. "A View of the Plan Puebla: An Application of Hierarchical Decision Models." *American Journal of Agricultural Economics* 58 (December 1976): 881–87.

Gledhill, John. *Casi Nada: A Study of Agrarian Reform in the Homeland of Cardenismo*. Albany: State University of New York, Institute for Mesoamerican Studies, 1991.

Goldrich, Daniel, and D.V. Caruthers. "Sustainable Development in Mexico." *Latin American Perspectives* 19, 1 (Winter 1992): 97–122.

González López, Guillermo. "Lecciones del Plan Puebla para el cambio tecnológico en el campo Mexicano." *Comercio Exterior* 40, 10 (October 1990): 962–67.

González y González, Luis. *Los dias del Presidente Cárdenas*. México, D.F.: El Colegio de México, 1981.

———. *San José de Gracia: Mexican Village in Transition*. John Upton, trans. Austin: University of Texas Press, 1974.

Goonatilake, Susantha. *Aborted Discovery: Science and Creativity in the Third World*. London: Zed Books, 1984.

Gortari, Eli de. *La ciencia en la historia de México*. México, D.F.: Fondo de Cultura Económica, 1963.

———. *La ciencia en la historia de México*, 2nd ed. México, D.F.: Tratados y Manuales Grijalbo, 1980.

Gortari Rabiela, Rebeca de. "Educación y conciencia nacional: Los ingenieros despues de la revolución Mexicana." *Revista Mexicana de Sociología* 49, 3 (1989): 123–41.

Gracia Fadrique, Jesús, ed. *Estado y Fertilizantes en Mexico, 1760–1985*. México, D.F.: Fondo de Cultura Económica, 1988.

Graham, Richard, ed. *The Idea of Race in Latin America, 1870–1940*. Austin: University of Texas Press, 1990.

Greenberg, Martin H. *Bureaucracy and Development: A Mexican Case Study*. Lexington, Mass.: D.C. Heath, 1970.

Griffin, Keith. *The Political Economy of Agrarian Change: An Essay on the Green Revolution*. Cambridge, Mass.: Harvard University Press, 1974.

Grindle, Merilee. *Bureaucrats, Politicians, and Peasants in Mexico: A Case Study in Public Policy*. Berkeley: University of California Press, 1977.

———. "Power, Expertise, and the Tecnico: Suggestions from a Mexican Case Study." *Journal of Politics* 39, 2 (May 1977): 399–426.

———. "Rhetoric, Reality, and Self-Sufficiency: Recent Initiatives in Mexico's Rural Development." *Journal of Developing Areas* 19 (January 1985): 171–84.

Gruening, Ernest. *Mexico and Its Heritage*. New York: The Century Company, 1928.

Hall, Anthony. "Community Participation and Development Policy: A Sociological Perspective." In Ernest Gruening and James Midgley, eds., *Development Policies: Sociological Perspectives*. Manchester: Manchester University Press, 1988, pp. 91–107.

———, and James Midgley, eds. *Development Policies: Sociological Perspectives*. Manchester: Manchester University Press, 1988.

Hall, Linda B. *Alvaro Obregón: Power and Revolution in Mexico, 1911–1920*. College Station: Texas A&M University Press, 1981.

Hamilton, Nora. *The Limits of State Autonomy: Post Revolutionary Mexico*. Princeton, N.J.: Princeton University Press, 1982.

Hanson, Mark. "Organizational Bureaucracy in Latin America and the Legacy of Spanish Colonialism." *Journal of Inter-American Studies and World Affairs* 16, 2 (May 1974): 199–219.

Harris, Charles H., III. *A Mexican Family Empire: The "Latifundio" of the Sánchez Navarros, 1765–1867*. Austin: University of Texas Press, 1975.

Haskell, Thomas L., ed. *The Authority of Experts: Studies in History and Theory*. Bloomington: University of Indiana Press, 1984.

Hertford, Reed. *Sources of Change in Mexican Agricultural Production, 1940–1965*. Washington, D.C.: Economic Research Service, U.S. Department of Agriculture, 1971.

Hewitt de Alcantara, Cynthia. *Modernizing Mexican Agriculture: Socioeconomic Implications of Technological Change, 1940–1970*. Geneva: United Nations Research Institute for Social Development, 1976.

Hewlitt, Sylvia Ann, and Richard S. Weinert, eds. *Brazil and Mexico: Patterns of Late Development*. Philadelphia: Institute for the Study of Human Issues, 1982.

Hudson, Charles, ed. *Black Drink*. Athens: University of Georgia Press, 1979.

Iszacvich, Abraham. *Modernización campesina*. México, D.F.: Editorial Edicol, 1980.

Janvry, Alain de. *The Agrarian Question and Reformism in Latin America.* Baltimore: Johns Hopkins University Press, 1981.

———, Gustavo Gordillo, and Elisabeth Sadoulet. *Mexico's Second Agrarian Reform: Household and Community Responses.* La Jolla: Center for U.S.–Mexican Studies, University of California, San Diego, 1997.

Jennings, Bruce H. *Foundations of International Agricultural Research: Science and Politics in Mexican Agriculture.* Boulder, Colo.: Westview Press, 1988.

Jennings, Francis. *The Invasion of America: Indians, Colonialism, and the Cant of Conquest.* Chapel Hill: University of North Carolina Press, 1975.

Johnson, Stanley. *The Green Revolution.* London: Hamish Hamilton, 1972.

Junne, Gerd, and Annemieke Roobeek. "Social and Economic Impact of Biotechnology." *Biofutur* (November 1989): 48–50.

Katz, Friedrich. *The Secret War in Mexico: Europe, the United States and the Mexican Revolution.* Chicago: University of Chicago Press, 1981.

Keldysh, M.V. "The Multinational Union of Soviet Socialist Republics and the Development of Science." In G.D. Konkov, ed., *Science in the U.S.S.R.* Moscow: Progress Publishers, 1973.

Kneller, George A. *The Education of the Mexican Nation.* New York: Columbia University Press, 1951.

Knight, Alan. "Cardenismo: Juggernaut or Jalopy." *Journal of Latin American Studies* 26, 1 (February 1994): 73–107.

———. "Racism, Revolution, and *Indigenismo*: Mexico, 1910–1940." In Richard Graham, ed., *The Idea of Race in Latin America, 1870–1940.* Austin: University of Texas Press 1990, pp. 71–113.

———. "Revolutionary Project, Recalcitrant People: Mexico, 1910–1940." In Jamie Rodríguez O., ed., *The Revolutionary Process in Mexico: Essays on Political and Social Change, 1880–1940.* Los Angeles: UCLA Latin American Center Publications, 1990, p. 245.

———. "The United States and the Mexican Peasantry, c. 1880–1940." In Daniel Nugent, ed., *Rural Revolt and U.S. Intervention.* La Jolla: U.C. San Diego, Center for U.S.–Mexican Studies, 1988, pp. 73–97.

Kocka, Jürgen. " '*Bürgertum*' and Professions in the Nineteenth Century: Two Alternative Approaches." In Michael Burrage and Rolf Torstendahl, eds., *Professions in Theory and History: Rethinking the Study of the Professions*, London: Sage Publications, 1990, pp. 62–74.

Konkov, G.D., ed. *Science in the U.S.S.R.* Moscow: Progress Publishers, 1973.

Konrad, Herman W. *A Jesuit Hacienda in Colonial Mexico: Santa Lucia, 1576–1767.* Stanford, Calif.: Stanford University Press, 1980.

Kroeber, Clifton B. *Man, Land, and Water: Mexico's Farmlands Irrigation Policies 1885–1911.* Berkeley: University of California Press, 1983.

Kutzner, Patricia. *World Hunger: A Reference Handbook.* Santa Barbara, Calif.: ABC-CLIO, 1994.

Lachkovics, Fawler C., et al. "Bio-farms: The End of the End. Third World Farmers and the New Plant Genetics." *Development Dialogue* 1–2 (1988): 60–71.

Lappé, Francis Moore, and Joseph Collins. *Food First: Beyond the Myth of Scarcity.* New York: Ballantine Books, 1977.

Larson, Magali Sarfatti. "Expertise and Expert Power." In Thomas L. Haskell, ed., *The Authority of Experts: Studies in History and Theory.* Bloomington: University of Indiana Press, 1984, pp. 28–80.

Leonard, Irving A. *Hispanic America and Science.* Miami: University of Miami Press, 1949.

Lerner, Victoria. *La educación socialista.* México, D.F.: El Colegio de México, 1979.

Long, Norman, ed. *Encounters at the Interface: A Perspective on Social Discontinuities in Rural Development.* Wageningen, the Netherlands: Agricultural University Wageningen, 1989.

———. "Introduction: The Raison D'Être for Studying Rural Development Interface." In Norman Long, ed., *Encounters at the Interface: A Perspective on Social Discontinuities in Rural Development.* Wageningen, the Netherlands: Agricultural University Wageningen, 1989, pp. 1–10.

———. "Sociological Perspectives on Agrarian Development and State Intervention." In Anthony Hall and James Midgley, eds., *Development Policies: Sociological Perspectives.* Manchester: Manchester University Press, 1988, pp. 108–33.

———, and Magdalena Villareal. "The Changing Life-Worlds of Women in a Mexican Ejido: The Case of the Bee-Keepers of Ayoquilla and the Issue of Intervention." In Norman Long, ed., *Encounters at the Interface: A Perspective on Social Discontinuities in Rural Development.* Wageningen, the Netherlands: Agricultural University Wageningen, 1989, pp. 103–44.

Lorey, David E. *The University System and Economic Development in Mexico since 1929.* Stanford, Calif.: Stanford University Press, 1993.

Lynch, William T. "Ideology and the Sociology of Scientific Knowledge." *Social Studies of Science* 24, 2 (May 1994): 197–227.

Mabry, Donald J. *The Mexican University and the State: Student Conflicts, 1920–1971.* College Station: Texas A&M University Press, 1982.

Macías, Anna. *Against All Odds: The Feminist Movement in Mexico to 1940.* Westport, Conn.: Greenwood Press, 1982.

Maddox, James G. "Point IV in Mexico." *American Universities Field Staff Reports, Mexico and Caribbean Area* 4, 3 (1957): 8.

Markiewicz, Dana. *The Mexican Revolution and the Limits of Agrarian Reform 1915–1946*. Boulder, Colo.: Lynne Rienner, 1993.

May, Stacey, and Galo Plaza. *The United Fruit Company in Latin America*. Washington, D.C.: National Planning Association, 1958.

McCook, Stuart. *States of Nature: Science, Agriculture and Environment in the Spanish Caribbean, 1760–1930*. Austin: University of Texas Press, 2002.

Medina, Luís. *Del cardenismo al avilacamachismo*. México, D.F.: El Colegio de México, 1978.

Merrill, William L. "The Beloved Tree: *Ilex vomitoria* among the Indians of the Southeast and Adjacent Regions." In Charles M. Hudson, ed., *Black Drink*. Athens: University of Georgia Press, 1979, pp. 40–82.

Moreno Sánchez, Manuel. *México, 1968–1972: Crisis y perspectiva*. Austin: University of Texas Press, 1973.

Myers, Charles N. *Education and National Development in Mexico*. Princeton, N.J.: Princeton University Press, 1965.

Nandy, Ashis. "Science as a Reason of State." In Ashis Nandy, ed., *Science, Hegemony, and Violence: A Requiem for Modernity*. Delhi: Oxford University Press, 1988.

Niblo, Stephen R. *War, Diplomacy, and Development: The United States and Mexico, 1938–1954*. Wilmington, Del.: Scholarly Resources, 1995.

Nicolson, Harold. *Dwight Morrow*. New York: Harcourt, Brace and Company, 1935.

Nugent, Daniel, ed. *Rural Revolt and U.S. Intervention*. La Jolla: U.C. San Diego, Center for U.S.–Mexican Studies, 1988.

———. *Spent Cartridges of Revolution: An Anthropological History of Namiquipa, Chihuahua*. Chicago: University of Chicago Press, 1993.

Ochoa, Enrique C., and David E. Lorey, eds. *Estado y agricultura en México: Antecedentes e implicaciones de las reformas salinistas*. Azcapotzalco, México: Universidad Autonoma Metropolitana de Azcapotzalco, 1994.

Pearse, Andrew. *Seeds of Plenty, Seeds of Want*. New York: Oxford University Press, 1980.

Peña, Sergio de la, and Marcel Morales. "Las transformaciones capitalistas en la agricultura: El caso del maiz en Morelos." In Enrique C. Ochoa and David E. Lorey, eds., *Estado y agricultura en México: Antecedentes e implicaciones de las reformas salinistas*. México, D.F.: Universidad Autonoma Metropolitara de Azcapotzalco, 1994, pp. 57–72.

Peritore, N. Patrick, and Ana Karina Galve-Peritore, eds. *Biotechnology in Latin America: Politics, Impacts, and Risks*. Wilmington, Del.: Scholarly Resources, 1995.

Perkins, John H. "The Rockefeller Foundation and the Green Revolution, 1941–1956." *Agriculture and Human Values* 26 (Summer–Fall 1990): 6–18.

Pilcher, Jeffrey M. *¡Que vivan los tamales! Food and the Making of Mexican Identity*. Albuquerque: University of New Mexico Press, 1998.

Plenn, John H. *Mexico Marches*. Indianapolis: Bobbs-Merrill; 1939.

Poleman, Thomas T. *The Papaloapan Project: Agricultural Development in the Mexican Tropics*. Stanford, Calif.: Stanford University Press, 1964.

Rabe, Stephen G. *Eisenhower and Latin America: The Foreign Policy of Anticommunism*. Chapel Hill: University of North Carolina Press, 1988.

Reingold, Nathan, and Marc Rothenberg, eds. *Scientific Colonialism: A Cross Cultural Comparison*. Washington, D.C.: Smithsonian Institution, 1987.

Reyes Osorio, Sergio, et al. *Estructura agraria y desarrollo agrícola en México: Estudio sobre las relaciones entre la tenencia y el uso de la tierra y el desarrollo agrícola en México*. México, D.F.: Fondo de Cultura Económica, 1974.

———, and Salomon Eckstein. "El desarrollo polarizado de la agricultura mexicana." In Miguel Wionczek et al., *¿Crecimiento o desarrollo económico?* México, D.F.: SepSetentas, 1971, pp. 21–43.

Río Físico, Fernando del. "Estado actual de la investigación científica." In Rosa María Farell, ed., *Los profesionales mexicanos y los desafíos de la modernidad*. México, D.F.: Editorial Diana, 1989, pp. 205–18.

Rodríguez O., Jamie, ed. *The Revolutionary Process in Mexico: Essays on Political and Social Change, 1880–1940*. Los Angeles: UCLA Latin American Center Publications, 1990.

Ronfeldt, David. *Atencingo: The Politics of Agrarian Struggle in a Mexican Ejido*. Stanford, Calif.: Stanford University Press, 1973.

Rounds, Christopher R. *From Hacienda to Ejido: Land Reform and Economic Development in Yautepec, Morelos, 1920–1970*. New York: Garland, 1987.

Ruiz, Ramón Eduardo. *Mexico: The Challenge of Poverty and Illiteracy*. San Marino, Calif.: The Huntington Library, 1963.

———. *Triumphs and Tragedy: A History of the Mexican People*. New York: W.W. Norton & Company, 1992.

Safford, Frank. *The Ideal of the Practical: Columbia's Struggle to Form a Technical Elite*. Austin: University of Texas Press, 1976.

Sagasti, Francisco R. *Ciencia, tecnología, y desarrollo latinoamericano*. México, D.F.: Fondo de Cultura Económica, 1981.

Sánchez, George I. *The Development of Higher Education in Mexico*. Westport, Conn.: Greenwood Press, 1970.

———. *Mexico: A Revolution by Education*. New York: Viking Press, 1936.

Sánchez Durón, Nicolas. "La investigación agrícola y el desarrollo agropecuario de México." *Anales de la Sociedad Mexicana de Historia de la Ciencia y de la Tecnología* 2 (1970): 175–86.

Sanderson, Steven E. *The Transformation of Mexican Agriculture: International Structure and the Politics of Rural Change*. Princeton, N.J.: Princeton University Press, 1986.

Sanderson, Susan R. Walsh. *Land Reform in Mexico: 1910–1980*. Orlando, Fla.: Academic Press, 1984.

Sandstrom, Alan R. *Corn Is Our Blood: Culture and Ethnic Identity in a Contemporary Aztec Indian Village*. Norman: University of Oklahoma Press, 1991.

Santander, Felipe. *El extensionista*. México, D.F.: CREA: 1978.

Sarukhán, José. "The Status of Mexican Science and Technology Research: Potential Avenues for Collaborative Programs with the U.S." *Mexican Studies* 5, 2 (Summer 1989): 266.

Scarry, Margaret C., and Elizabeth J. Reitz, "Herbs, Fish, Scum, and Vermin: Subsistence Strategies in Sixteenth-Century Spanish Florida." In David Hurst Thomas, ed., *Columbian Consequences, vol. 2: Archaeological and Historical Perspectives on the Spanish Borderlands Southeast*. Washington, D.C.: Smithsonian Institution Press, 1990, pp. 343–54.

Schryer, Frans J. *The Rancheros of Pisaflores: The History of a Peasant Bourgeoisie in Twentieth-Century Mexico*. Toronto: University of Toronto Press, 1980.

Schuler, Friedrich E. *Mexico between Hitler and Roosevelt: Mexican Foreign Relations in the Age of Lázaro Cárdenas*. Albuquerque: University of New Mexico Press, 1998.

Scott, James C. *Weapons of the Weak: Everyday Forms of Peasant Resistance*. New Haven, Conn.: Yale University Press, 1985.

Scott, Roy V. *The Reluctant Farmer: The Rise of Agricultural Extension to 1914*. Urbana: University of Illinois Press, 1971.

Selander, Staffan. "Associative Strategies in the Process of Professionalization: Professional Strategies and the Scientification of Occupations." In Michael Burrage and Rolf Torstendahl, eds., *Professions in Theory and History: Rethinking the Study of the Professions*. London: Sage Publications, 1990, pp. 139–50.

Semo, Enrique. "Haciendas, campesinos, y rancheros." In Antonio García de León, Enrique Semo, and Ricardo Gamboa Ramírez, eds., *Historia de la cuestión agraria mexicana, vol. 1: El siglo de la hacienda*. México, D.F.: Siglo Ventinuo Editores, 1988, pp. 86–164.

Shadle, Stanley F. *Andrés Molina Enríquez, Mexican Land Reformer of the Revolutionary Era*. Tucson: University of Arizona Press, 1994.

Sherman, John W. *The Mexican Right: The End of Revolutionary Reform, 1929–1940*. Westport, Conn.: Praeger, 1997.

Shumacher, E.F. *Small Is Beautiful: A Study of Economics as If People Mattered*. London: Blond & Briggs, 1973.

Siegrist, Hannes. "Professionalization as a Process: Patterns, Progression, and Discontinuity." In Michael Burrage and Rolf Torstendahl, eds., *Professions in Theory and History: Rethinking the Study of the Professions*. London: Sage Publications, 1990, pp. 177–202.

Silva Herzog, Jesús. *El agrarismo mexicano y la reforma agraria: Exposición y critica*. México, D.F.: Fondo de Cultura Económica, 1959.

Simonian, Lane. *Defending the Land of the Jaguar: A History of Conservation in Mexico*. Austin: University of Texas Press, 1995.

Simpson, Eyler N. *The Ejido: Mexico's Way Out*. Chapel Hill: University of North Carolina Press, 1937.

Smith, Peter H. *Labyrinths of Power: Political Recruitment in Twentieth Century Mexico*. Princeton, N.J.: Princeton University Press, 1979.

———. *Talons of the Eagle: Dynamics of U.S.–Latin American Relations*. New York: Oxford University Press, 1996.

Stepan, Nancy Leys, *"The Hour of Eugenics": Race, Gender, and Nation in Latin America*, Ithaca, N.Y.: Cornell University Press, 1991.

Tannenbaum, Frank, *Mexico: The Struggle for Peace and Bread*. New York: Alfred A. Knopf, 1950.

Torres, Blanca. *Hacia la utopía industrial*. México, D.F.: El Colegio de México, 1984.

Townsend, William Cameron. *Lázaro Cárdenas: Mexican Democrat*. Ann Arbor, Mich.: George Wahr Publishing Company, 1952.

Trabulse, Elias. *El circulo roto: Estudios históricos sobre la ciencia en México*. México, D.F.: Fondo de Cultura Económica, 1982.

Tuckman, Barbara H. "The Green Revolution and the Distribution of Agricultural Income in Mexico." *World Development* 4 (1976): 17–24.

Tutino, John. *From Insurrection to Revolution in Mexico: Social Bases of Agrarian Violence*. Princeton, N.J.: Princeton University Press, 1986.

Varo Berra, Rosario. *La reforma agraria en México desde 1853: Sus tres cíclos legales*. Guadalajara: Universidad de Guadalajara, 2002.

Vaughan, Mary Kay. *Cultural Politics in Revolution: Teachers, Peasants, and Schools in Mexico, 1930–1940*. Tucson: University of Arizona Press, 1997.

———. *The State, Education, and Social Class in Mexico, 1880–1928*. De Kalb: Northern Illinois University Press, 1982.

Vucinich, Alexander. *Empire of Knowledge: The Academy of Sciences in the U.S.S.R. 1917–1970*. Berkeley: University of California Press, 1984.

Walker, Samuel J. *Henry A. Wallace and American Foreign Policy*. Westport, Conn.: Greenwood Press, 1976.

Weismann, Steve. "Why the Population Bomb Is a Rockefeller Baby." *Ramparts* (May 1970): 43.

Werner, Michael S., ed. *Encyclopedia of Mexico: History, Society and Culture*. Chicago: Fitzroy Dearborn, 1997.

Weyl, Nathaniel, and Sylvia Weyl. *The Reconquest of Mexico: The Years of Lázaro Cárdenas*. New York: Oxford University Press, 1939.

Whetten, Nathan. *Rural Mexico*. Chicago: University of Chicago Press, 1948.

Wilken, Gene C. *Good Farmers: Traditional Resource Management in Mexico and Central America*. Berkeley: University of California Press, 1987.

Wilkie, James W. "Prologo." In Rosario Varo Berra, *La reforma agraria en México desde 1853: Sus tres cíclos legales*. Guadalajara: Universidad de Guadalajara, 2002, pp. 15–53.

Wilkie, Raymond. *San Miguel: A Mexican Collective Ejido*. Stanford, Calif.: Stanford University Press, 1971.

Wionczek, Miguel, et al. *¿Crecimiento o desarrollo económico?* México, D.F.: SepSetentas, 1971.

Wolf, Eric. *Sons of the Shaking Earth: The People of Mexico and Guatemala—Their Land, History, and Culture*. Chicago: University of Chicago Press, 1959.

Wright, Angus. *The Death of Ramón González: The Modern Agricultural Dilemma*. Austin: University of Texas Press, 1991.

Zárate Hernández, José Eduardo. "Agricultural Modernization and Political Recomposition in El Llano Grando." In Sergio Zendejas and Pieter De Vries, eds., *Rural Transformations Seen from Below: Regional and Local Perspectives from Western Mexico*. La Jolla: Center for U.S.–Mexican Studies, University of California, San Diego, 1995, pp. 9–21.

Zendejas, Sergio. "Appropriating Government Reforms: The Ejido as an Arena of Confrontation and Negotiation." In Sergio Zendejas and Pieter De Vries, eds., *Rural Transformations Seen from Below: Regional and Local Perspectives from Western Mexico*. La Jolla: Center for U.S.–Mexican Studies, University of California, San Diego, 1995, pp. 23–48.

———, and Pieter De Vries, eds. *Rural Transformations Seen from Below: Regional and Local Perspectives from Western Mexico*. La Jolla: Center for U.S.–Mexican Studies, University of California, San Diego, 1995.

UNPUBLISHED SECONDARY SOURCES

Brown, Lyle C. "General Lázaro Cárdenas and Mexican Presidential Politics: A Study in the Acquisition and Manipulation of Political Power." Unpublished Ph.D. dissertation, University of Texas, Austin, 1964.

Cleaver, Harry M. "Origins of the Green Revolution," Unpublished Ph.D. dissertation, Stanford University, 1974.

Dovring, Folke. "Land Reform and Productivity: The Mexican Case, a Preliminary Analysis." Unpublished paper, Department of Agricultural Economics, Agricultural Experimentation Station, University of Illinois, November 1966.

Erb, Claude Curtis. "Nelson Rockefeller and U.S.–Latin American Relations, 1940–1945." Unpublished Ph.D. dissertation, Clark University, 1982.

Index

Acevedo, Juan Ballesteros, 37, 52, 242
Acosta, Andres, 137, 248–49
Acosta, Rodolfo, 239
Agentes de Información y Propaganda Agrícola, 61
Agrarian Codes, 133, 242
Agraristas, 5, 49, 50, 53, 54, 126, 242
Agribusiness, 302, 305–6, 329
Agribusiness Council, 250–51
Agricultural Adjustment Act, 140
Agricultural Machinery Trains. See *Trenes de Maquinaria Agrícola*
Agriculture, crisis in, 10, 125, 129, 130, 132, 140, 325. *See also* Food
Agrochemicals, 183; and agricultural crisis, 129–30; and *agrónomos*, 294; and Alemán, 199; and Avila Camacho, 150–51, 153; and Calvino, 51; and *campesinos*, 299, 300; cost of, 282, 283; criticism of, 258, 283; debate over, 63; and *Escuela Nacional de Agricultura*, 67; and Green Revolution, 13, 233, 235, 236, 237, 260–61; and *hacendados*, 26; and health, 294, 302; and Mexican Agricultural Program, 198, 199–200, 327; and organic farming, 284; and profitable farming, 98, 99; replacement of, 99; and *Secretaria de Agricultura y Fomento*, 182, 199–200, 203; support for, 286; and traditional agriculture, 23; and U.S. aid, 58, 206
Agrónomos: and *agraristas*, 49, 50, 54; and agricultural crisis, 133, 134; and agrochemicals, 294; and Alemán, 195, 325; and American education, 189–90; and appropriate technology, 333; and authority, 9; and Avila Camacho, 325; and bias, 243; and biotechnology, 329; and *campesinos*, 8–9, 11, 12, 13, 32, 50, 54, 56–57, 63–64, 70–71, 82–83, 84, 126, 134, 135, 136, 151, 155–57, 183, 191–92, 204, 210,

245–46, 256, 257, 291, 292, 294, 295–96, 297, 303, 306, 324–26, 327, 332; and class, 30–32, 70, 191, 243; and communication, 7–8; and credibility, 3–4, 8, 10, 33, 39, 68, 82, 134, 151, 155, 156, 181, 192, 193, 203, 239, 296; and crop profit, 98–99; and cultural campaign, 61, 62, 126, 136, 326; and culture, 202–3; and Díaz, 20, 31; and Economic Miracle, 234; and education, 156–57, 198, 203, 210, 329; and expertise, 100; and exports, 307; and family haciendas, 33; and foreign education, 85; and foreign scientists, 83–84; and France, 29, 30, 57; and genetics, 129; and government, 20, 36, 39, 55, 82, 202, 291; and Green Revolution, 12, 238, 253–54, 260, 263–64, 327; and *hacendados*, 20, 34, 50, 55, 56–57, 70, 83, 84; and jobs, 6, 52–53, 65, 68, 70, 83, 106, 132, 191, 241, 329; and land distribution, 54; and land reform, 50, 52–53, 81, 83; and liberal professions, 36; and Mexican Agricultural Program, 11, 239; and Mexican Revolution, 5, 49, 50–55, 324; and modernization, 256; and nationalism, 10, 291–92; and Native Americans, 82; origins of, 31–32; and politics, 83, 106, 254–55, 291; and prestige, 244; as professionals, 2, 131–37, 326, 328, 332; and race, 39; and research, 106; and Revolutionary Family, 69, 70, 106, 137, 290, 308, 326; and Rockefeller Foundation, 157, 296; and Roman Catholic Church, 62; and science, 94, 181, 241, 326, 328; and *Secretaria de Agricultura y Fomento*, 83; and seed selction, 37; and *Sistema Alimentario Mexicano*, 295; and *Sociedad Agronómica Nacional*, 6; and Soviet Union, 81; success of, 321; as *técnicos*, 308, 325; and traditional farming, 39, 56–57, 63–64, 91, 324–25, 326; and United States, 57–58, 81, 139, 145, 179, 180, 210, 239–40; and urban population, 31, 126, 132, 298
Agrónomos Regionales, 62, 63, 68
Aguilar, Armando Martínez, 94
Aguilar, Carlos J., 69, 204
Aguilar, Javier Ramírez de, 51
Aguilar, Rafael, 204
Aguillón, Jacobo Aragón, 91
Aguirre, Manuel Bernardo, 298
Aguirre, Norberto, 203, 294
Alacantara, Hewitt de, 210, 284
Alam, Carlos Orozco, 307
Alarcón, Jesus, 190
Alba, Enrique Martínez de, 31
Alba, Ernesto Martínez de, 57
Alba, Victor, 86
Alcala, Roberto Osoyo, 183
Alcazar, Manuel, 57
Alcocer, Daniel Rios, 241
Alemán, Miguel, 210, 323; and agrochemicals, 199, 200; and *agrónomos*, 325; and *ejidos*, 179–80; and *Instituto de Investigaciones Agrícolas*, 239; and Mexican Agricultural Program, 196, 197; and modernization, 193–95
Allaby, Michael, 259
Allied Chemical and Dye Company, 182
Almazán, Juan Andreu, 64, 68, 134
Alsberg, Carl, 89

Alzate y Ramírez, José Antonio, 27
Amaro, Santos, 150
Ambrosio, Antonio Marino, 203
Amin, Samir, 258
Anderson, Benedict, 4
Anderson, Clinton H., 205
Anderson, Edgar, 196, 327–28
Angulo, Alejandro, 305
Ankerson, Dudley, 104
Antonio Narro Agricultural School, 83, 84–85, 132, 183, 191, 202
Anzures, Carlos, 152, 156
Aragón León, Agustin, 28, 29, 34, 50
Arce, Alberto, 9, 295, 296, 308, 329, 332
Ariziaga, Pedro, 291
Armour Research Foundation, 208
Arriaga, Javier Trujillo, 306, 331
Arrieta M., Bernardo L., 84, 192
Arrieta M., Dario L., 190
Arrom, Daniel L., 284
Arzave, Genero, 59, 68, 242
Athwal, D.S., 287
Atwood, R.S., 184
Avellaneda, Jerjes Aguirre, 291–92
Avila Camacho, Manuel, 10, 11, 126, 133; and *agrónomos*, 325; and Alemán, 194; and fertilizers, 153; and food imports, 181; and pesticides, 182; and plows, 153; and politics, 151; and privileged farmers, 189; and production, 192; and research, 191; and *Sociedad Agronómica Mexicana*, 134, 150; and soil conservation, 199; support for, 154; and U.S. wartime assistance, 143, 145–46, 186, 190; and World War II years, 182–83
Avila Camacho family, 136
Baca, Gustavo Martínez, 51
Baenz, Adrián Lozano, 307
Baker, A.C., 141, 184
Banco de México, 183, 187, 208
Banco Nacional de Crédito Agrícola (National Agricultural Credit Bank), 6, 87–88, 101, 130, 136, 180
Banco Nacional de Crédito Ejidal (National Ejidal Credit Bank), 7, 139, 141, 183, 185, 238
Bangladesh, 288
Banó, Eugenio, 33
Bantjes, Adrian, 4
BAR. See *Bloque de Agrónomos* (Revolutionary Agronomists Block)
Barba, Rafael, 28–29
Barejas, Rosa Elvia, 331
Barker, Randolph, 287
Barra, Rosario Varo, 331
Barreda, Gabino, 28, 33
Barreda, Leopoldo de la, 134
Barrus, M.F., 197, 198
Barta, Armando, 238
Barusta, Gonzalo Torres, 95, 129
Baum, Warren G., 286
Bautista, Gonzalo, 130
Beans, 98, 188, 195. *See also* Crops
Becerra, Rene A., 88
Becerril, Carlos, 96
Beckman, Svante, 8
Becquerel, Paul, 206
Beezley, William, 23
Bell, B.R., 184
Bell, David E., 252
Beltrán, Enrique, 68, 84, 90, 91, 93, 94, 138, 293
Beltrán, Pedro G., 286
Bennett, H.L., 250
Berlandier, Jean Louis, 27
Bermúdez, José López, 242
Berry, Wendell, 284, 285

Beteta, Ramón, 86, 325
Biotechnology, 14, 296, 305, 329, 330, 331. *See also* Genetics
Birth control, 234, 247, 249, 259, 262, 284, 286, 288, 327–28. *See also* Population
Black, J.N., 283
Black, Roe, 286
Bloque de Agrónomos (Revolutionary Agronomists Block), 145, 150, 151
Blumenkron, Julio, 93
BNCA. See *Banco Nacional de Crédito Agrícola* (National Agricultural Credit Bank)
BNCE. See *Banco Nacional de Crédito Ejidal* (National Ejidal Credit Bank)
Bonacee, Silvio, 37
Booth, George, 91
Borlaug, Norman, 1, 12, 209, 234, 235, 238, 252–53, 262–63, 281, 286, 287, 288
Bowman, Thomas D., 105
Box, Louk, 332
Brackel-Welda, Othon E., 32
Bradfield, Richard, 21, 149, 198, 199, 208, 251–52
Brambila, Alejandro, Jr., 63–64, 67
Brambila, Alejandro, Sr., 28, 34, 37, 51, 55
Brandenberg, Frank, 5
Brannan, Charles F., 207
Brauer H., Oscar, 254, 255, 291
Bravo, Guillermo Quesada, 186
Bressman, Earl N., 148
Brigadas de Promocion Agropecuarias, 245
Briggs, Ellis O., 141
Brigham, Ruben, 143
Brister, W.C., 184
Brondo Whitt, Guillermo, 51, 153, 192
Brown, Lester, 252, 261, 288
Brown, Richard Harvey, 6
Bukasov, S., 60
Bulnes, Francisco, 33, 55
Burbach, Hans, 86
Burbank, Luther, 36
Busch, August, 59

Cabrera, Luis, 82, 86
Cadeña, Manuel, 53
Cadet, José Arturo Jiménez, 106
C. Aguilar, Agustín, 34
Calles, Plutarco Ellias: and *agrónomos*, 83; and corn imports, 63; and de Dios Bojórquez, 106; and education, 65; and *Escuelas Centrales Agrícolas*, 57; and irrigation, 291; and land reform, 82, 133; and León, 53, 70; and Mexican Revolution, 5; and modernization, 6, 49, 60, 62; and *Oficina Federal de Defensa Agrícola*, 64; and reform, 81; and research, 68; and Roman Catholic Church, 64; and United States, 50, 60, 82
Calles, Rodolfo Ellias, 86, 201
Calverton, Victor F., 86
Calvino, Mario, 37, 38, 51
Camacho, Roberto Gómez, 63
Campa, Arturo García, 244
Campaña en Pro del Maiz, 57, 96, 97, 127–28, 152, 191
Campaña Nacional de Fertilizantes, 199
Campaña pro Nogal, 97
Campbell, Keith O., 286
Campesinos (peasants): and agribusiness, 302, 305–6; and agrochemicals, 299, 300; and *agrónomos*, 8–9, 11, 12, 13, 32, 50, 54, 56–57, 63–64, 70–71, 82–83, 84, 126, 134, 135, 136, 151,

155–57, 183, 191–92, 204, 210, 245–46, 256, 257, 291, 292, 294, 295–96, 297, 303, 306, 324–26, 327, 332; and American crops, 95, 96; as backward, 3, 60; bias against, 105, 204, 243; and Cárdenas, 81, 82; Cervantes on, 56; continuing problems of, 321; control of, 69; and cooperation, 9, 101–3, 106–7, 135, 155–56, 183, 325; and cultural campaign, 6, 127; and desertification, 301; and Díaz, 20; and economy, 282, 324; and education, 65, 329, 332; and *Escuela Nacional de Agricultura*, 65; future of, 329–34; and Green Revolution, 235–36, 237, 238, 252, 253–54, 255, 257, 258, 260, 262, 263, 288, 299–303, 324, 330; and *Huerto Familiar*, 239; improvement of, 84, 192, 195; and jobs, 65; and land, 2, 3; and Mexican Revolution, 4, 14, 321; and modernization, 90–106, 256; organization of, 97–98; and poverty, 13; and *ranchos*, 26; and religion, 107, 295; and research, 106; and Revolutionary Family, 324; and Salinas, 13, 304; and science, 94; and *Secretaria de Agricultura y Fomento*, 84, 192; and seed selection, 191; and social justice, 54; and *Sociedad Agronómica Mexicana*, 192; and Soviet Union, 85–86; and stereotypes, 56; and tradition, 19, 22–23, 56–57, 196; and traditional farming, 63–64, 135, 183, 188, 289, 294, 295, 303, 324–25, 326; and U.S. wartime assistance, 143, 144–45. *See also* Diet/nutrition; Farming

Campo, Luís Fernández del, 29
Canada, 96
Capitalism, 20, 86
Cárdenas, Bernardo González, 200
Cárdenas, Lázaro: and agricultural crisis, 132; and agrobureaucracy, 7; and *agrónomos*, 84; and *campesinos*, 81, 82; and cultural campaign, 101, 126–27, 131, 136; and de Dios Bojórquez, 106; and disease, 130; and Echavarrí, 87; and *Escuela Nacional de Agricultura*, 136; and foreign business, 9; and foreign study, 85; and foreign training, 6; and government control, 323; and Harrison, 88; and irrigation, 99; and land reform, 125; and nationalism, 10; and peasant organization, 97; and plows, 153; and populism, 81, 82, 95; and profit, 98; and research, 137; and Rockefeller, 90; and *Secretaria de Agricultura y Fomento* purges, 104; and Soviet Union, 138; and sunflower campaign, 92; and technological colonialism, 291; and United States, 139, 141, 190
Cardenismo, 7, 86, 281, 308, 325
Carpio, Manuel, 54
Carranza, Venustiano, 49, 51–52, 53, 61
Carson, Rachel, *Silent Spring*, 258, 286
Castaños, Carlos Manuel, 257, 292, 293, 295, 297, 298–99
Castaños, Manuel, 153
Castellanos, Guillermo Maqueo, 105
Castillo, Alfonso, 234
Castillo, Salvador del, 155

Castor beans, 98, 147. *See also* Crops
Castro, Fidel, 249
Cedillo, Saturnino, 81, 91, 101, 103, 104
Centrales, 62, 65
Centro de Investigación y de Estudios Avanzados (Center of Investigation and Advanced Studies), 296
Centro Internacional del Mejoramiento de Maiz y Trigo (International Center for Corn and Wheat Improvement), 287, 290
Cervantes, Enrique A., 56
Chamberlain, Charles J., 38
Chandler, Robert, 199, 252, 261, 287
Channing, Margaret, 25
Chávez, Eduardo, 36
Chávez, José O., 131
China, 96, 288, 292
Ch. Ramírez, José, 155, 204
CIANO. *See* Construction Industry Association of New Orleans
Cienfuegos, Hilario Mesa, 192
CIMMYT. See *Centro Internacional del Mejoramiento de Maiz y Trigo* (International Center for Corn and Wheat Improvement)
CINVESTAV. *See Centro de Investigación y de Estudios Avanzados* (Center of Investigation and Advanced Studies)
Class, 29, 30–33, 70, 105, 191, 241
Cleaves, Peter, 3, 327, 328
Clubes de Fomento Agrícola, 97
CNA. See *Comisión Nacional Agraria* (National Agrarian Commission)
CNAG. See *Consejo Nacional de Agricultura* (National Council on Agriculture)
CNC. See *Confederación Nacional Campesino* (National Peasant Confederation)
CNI. See *Comisión Nacional de Irrigación* (National Irrigation Commission)
Cold War, 11, 208–9
Colegio de Ingenieros Agrónomos, 240
Colín, Salvador Sánchez, 193
Colín, Sánchez, 183, 254
Collectivization, 85, 138, 153, 192, 281
Collins, Randall, 12
Colonialism, 19, 24, 26–27, 32, 38, 291–92, 292
Colwell, William E., 182
Comisiónes Locales Agrarias, 53
Comisión Internacional de Plagas, 142
Comisión Nacional Agraria (National Agrarian Commission), 6, 53, 54, 195, 242, 298–99
Comisión Nacional de Irrigación (National Irrigation Commission), 6, 94, 141, 153, 201, 298
Comisión Nacional del Maiz, 194
Communism, 107, 154, 234, 247, 263, 286, 297, 308, 323
Compania Nacional de Subsistencias Populares (National Company of Popular Subsistences), 294, 295
Complementary Crops Program, 141
Comte, Auguste, 20, 28
CONACYT. *See Consejo Nacional de Ciencia y Technología* (National Council of Science and Technology)
CONASUPO. See *Compania Nacional de Subsistencias Populares* (National Company of Popular Subsistences)

Confederación de Trabajadores Mexicanos (Confederation of Mexican Workers), 136
Confederación Nacional Campesino (National Peasant Confederation), 134, 137, 138, 154, 186, 195, 200, 238, 245
Consejo Nacional de Agricultura (National Council on Agriculture), 97–98, 103
Consejo Nacional de Ciencia y Technología (National Council of Science and Technology), 290
Consejos Locales de Fomento Agrícola, 98
Constitutionalists, 49, 52
Constitution of 1917, 5, 52
Construction Industry Association of New Orleans, 292
Contreras, Eustado L., 55, 66
Contreras, Juan E., 30, 34, 37
Córdova, Arnaldo, 97
Cordóva, Pablo, 36, 37, 86, 105, 136
Corn, 98, 181, 188; and crisis, 127–28; and diet, 92, 129, 195; and Green Revolution, 194, 233, 235, 246, 256–57, 262; hybrid, 186, 198, 322; and imports, 62, 63; and *Instituto de Investigaciones Agrícolas*, 210; and intercropping, 57, 188; production of, 194. *See also* Crops
Corn (maize), 128, 234; hybrid, 199; and Khankhoje, 67; and nutrition, 56; and traditional agriculture, 23–24
Corn Commission, 197, 198, 201, 235, 238
Correu, Damién, 150, 239
Corruption, 38, 104, 183
Cortés, Nardo Mansilla, 31
Cota, Felix Agramont, 203
Cowan, Mark, 284
Crawford, R.F., 139
Creel, Enrique C., 36
Crespo, Romulo Delgado, 94
Crops: and American influence, 35, 86; and biotechnology, 296; and Calvino, 37, 51; colonial, 24; and cultural campaign, 61, 62, 128, 129; and diet, 92, 192; and education, 65; and experimentation, 67–68; export, 125; and food shortages, 281; foreign, 63; and Gómez, 152; and Green Revolution, 233, 234, 235, 237, 246, 256–57, 262, 281; and *hacendados*, 26; hybrid, 67–68, 294; and profit, 98–99, 128; and science, 94; and *Secretaria de Agricultura y Fomento*, 152, 181, 298–99, 306; success with, 132; traditional, 23, 24, 57; and *Union de Ejidos "Professor Otilio Montaño,"* 305; and United States, 51, 146–48, 184; and U.S. aid, 58, 59, 95, 96; and World War II, 182
Cruz, Alfonso, 53
Cruz, Miguel García, 136
CTM. *See Confederación de Trabajadores Mexicanos* (Confederation of Mexican Workers)
Cuba, 185
Cuellar, Arturo, 96
Cuevas R., Atanasio, 246, 253
Cultural campaign, 6, 60–69, 81–107, 136, 151, 326
Cultural Missions, 83
Cultural Relations Program for Latin America, 140, 142, 148, 187

Dahne, Rodolfo Moreno, 190
Dalrymple, Lena, 258
Dampf, Alfonso, 64, 85, 87, 88

Daniels, Josephus, 142
Darwin, Charles, 247
Davis, Chester C., 205
DCA. See *Dirección General de Agricultura* (General Office of Agriculture)
De Dios Bojórquez, Juan, 51, 52, 53, 62, 106, 241
De la Garza, Jesús M., 53
De la Loma, José Luís, 203, 243, 293–94, 297
De la Madrid, Roberto, 282, 291, 308
De la Peña, Moises T., 243
De Lasse, Rolando, 306
De la Torre, Enrique, 101
Denmark, 96
DeWalt, Bille R., 7, 285
DGA. See *Dirección General de Agricultura* (General Office of Agriculture)
Díaz, Porfirio: and activism, 39; and *agrónomos*, 31, 36; contradictions of, 2; and diet, 33; and Durán, 51; and exports, 26; and France, 29; and jobs, 50; and liberalism, 20; and Mexican Revolution, 4, 49; and Olson-Seffer, 38
Díaz de León, Jésus, 28, 29, 32
Díaz del Pino, Alfonso, 132, 240, 243, 246, 260
Díaz Ordaz, Gustavo, 1, 233, 246, 254, 257, 259, 260
Dickey, John S., 204–5
Diet/nutrition: and beans, 195; and *Campaña en Pro del Maiz*, 152; and corn, 92, 195; and crops, 192; and cultural campaign, 61, 62, 128; and maize, 56; and Mexican Agricultural Program, 189; and National Institute of Nutrition, 181; and race, 56; and racial improvement, 92; and stereotypes, 32–33; and traditional agriculture, 23; and U.S. wartime assistance, 143, 187. See also *Campesinos* (peasants); Food
Dirección de Estudios Biológicos (Office of Biological Studies), 52, 67
Dirección General de Agricultura (General Office of Agriculture), 52, 95
Disease, 33, 204, 207; and education, 97; and farming crisis, 130–31; and Gándara, 38; and government, 62; and Green Revolution, 238; and profitable farming, 100; and traditional farming, 135; and U.S. aid, 58; and U.S. Department of Agriculture, 36; and World War II, 180
Division of Agricultural Cooperation, 58
Domínguez, Francisco Reyes, 204
Domínguez, Nícolas, 297
Draper, William H., 249
D'Shaw, R., 258
Dueñas, Julian, 26
Dugan, Laurence, 141
Dumont, Rene, 284
DuPont, John J., 284
Du Pont company, 207
Durán, Gustavo, 32, 50–51, 149, 150, 151, 192, 245, 256
Durán, Marco Antonio, 91, 137

East Indies, 96, 140
Echavarrí, Jesús, 87, 294
Echeverría, Luís, 13, 281, 290–91, 295, 297, 308
Economic Miracle, 10, 11, 233, 234, 242, 257

Economy, 2, 89, 93, 257, 258, 259, 282, 303–4, 306–7, 324
Education: adequacy of, 84–85; and *agrónomos*, 156–57, 189–90, 198, 203, 210, 329; and Calles, 65; and *campesinos*, 329, 332; and class, 30; credibility of, 33; and cultural campaign, 61, 127; and disease, 97; and Echeverría, 290–91; foreign, 6, 85, 138, 149; and France, 30; and Gómez, 183; and Green Revolution, 235; and Mexican Agricultural Program, 191, 195; and Mexican Revolution, 4; and *Porfiriato*, 28; and positivism, 28; and Salinas, 304; and science, 90–91, 297; and *Secretaria de Educación Pública*, 105; and socialism, 104; and *Sociedad Agronómica Mexicana*, 190–91; and United States, 35, 58, 60, 140, 141, 149, 189–90, 191, 207. See also *Escuela Nacional de Agricultura* (National School of Agriculture); *Secretaria de Educación Pública* (Ministry of Public Education)
Edwards, Harry T., 143
Ehrlich, Paul, 259
Eisenhower, Dwight D., 247, 249, 323
Ejército Zapatista de Liberación Nacional (Zapatista Army for National Liberation), 14
Ejidal parcelas, 8
Ejido/ejidatorios, 27, 49, 194; and Alemán, 179–80; and Avila Camacho, 126; bias against, 244; and collectivization, 138, 281; and economy, 306–7; and fertilizers, 9; and government, 12, 183; and Green Revolution, 234, 237, 299, 301; and privatization, 304, 306; and U.S. assistance, 143, 145
Elias, Fancisco S., 81, 96
Elias, José Rodríguez, 255
Elorduy, Samuel Torres, 52, 88
ENA. See *Escuela Nacional de Agricultura* (National School of Agriculture)
Enlightenment, 27
Enríquez, Andres Molina, 50
Enríquez, Ignacio L., 54
Environment, 21, 306
Environmentalism, 283, 285
EPA. See *Escuela Particular de Agricultura* (Private Agricultural School)
Escobar, Romulo, 33, 56, 136
Escobar brothers, 87
Escuela Nacional de Agricultura (National School of Agriculture): and agrochemicals, 67; and agroecology, 305; and American education, 88; and American influence, 36; and appropriate technology, 293, 305; and Barreda, 28; and biotechnology, 296; and *campesinos*, 65; and class, 30; and cultural campaign, 62, 128; and curriculum, 28, 29, 53; early graduates of, 33; and experimentation, 69; and Fernández y Fernández, 190; and fertilizer, 67, 182; and foreign educators, 292; and foreign scientists, 138; and France, 55; graduation from, 83, 191; and Green Revolution, 245, 256; and hybrids, 67; improvement of, 136; and industry, 307; and local conditions, 66; and Mexican Agricultural Program, 11, 195; and Obregón, 61; and politics, 50–55; and religion versus sci-

ence, 91; and research, 100, 151–52; and Rockefeller Foundation, 187; and transmission, 3; and Uribe Ruiz, 255; and U.S. aid, 190, 240. *See also* Education
Escuela Nacional Preparatoria, 28
Escuela Particular de Agricultura (Private Agricultural School), 33
Escuelas Centrales Agrícolas, 57
Esquivel, Armando, 100
Europe, 37, 139
Evans, L.E., 186–87
Evans, Rosalie, 55
EXIM Bank, 208
Experimentation: and Barreda, 33; and Durán, 151; and education, 33; and *Escuela Nacional de Agricultura*, 69; and Extension Service, 66. *See also* Research
Experimentation stations: and Calles, 62, 67; and Calvino, 51; and Creel, 36; deficiencies of, 100, 151; and *Dirección General de Agricultura*, 95; and Herrera, 51; and Molina, 37; and sunflowers, 92
Exports, 125, 130–31, 154, 181, 254, 302, 307
Extension Service, 57, 66, 84, 96–97
EZLN. See *Ejército Zapatista de Liberación Nacional* (Zapatista Army for National Liberation)

Fábila, Gilberto, 31, 53, 59, 192, 203–4, 332
Fagoaga, Lorenzo Flores, 104
Famine Emergency Committee, 205
FAO. *See* Food and Agriculture Organization
Farming: dry, 38; Mesoamerican, 22; organic, 283, 284–85; small-scale, 53–54, 85; traditional, 56–57, 63–64, 91, 93–94, 96, 135, 196, 287–88, 292, 294, 295, 303, 305, 306, 322, 326. See also *Campesinos* (peasants)
Ferguson Tractors, 207
Fernández, Aaron Merino, 136, 254–55
Fernández, Antonio Turrent, 296
Fernández, Carlos, Jr., 30
Fernández, Jesús Merino, 194
Fernández y Fernández, Ramón: and birth control, 248; and corruption, 104; and cultural campaign, 136; and English, 187; and *Escuela Nacional de Agricultura*, 65, 190; and field work, 105; and Flores, 149; and foreign study, 138; and Green Revolution, 248–49; and image, 192; and Japan, 138–39; and population control, 244; and research, 100; and U.S. aid, 186
Ferrell, John A., 89–90
Fertilizer, 25–26; and Avila Camacho, 153; chemical, 26, 99, 129–30, 150–51, 153, 182, 183, 233; and expertise, 100; and González, 63; green, 99, 198–99; and Green Revolution, 233, 236, 237, 281; and health, 302; and profitable farming, 98, 99; shortage of, 283; and traditional agriculture, 23. *See also* Agrochemicals
Figueroa, Ignacio L., 60, 106, 180–81
Figueroa, José, 92, 130, 137
Figueroa, Virgilio, 34, 37, 50
Fishing, 141, 195, 293
Fitzgerald, Deborah, 326
Flores, Edmundo, 6, 86, 100, 149, 242, 260, 295
Flores, José, 102

Food: distribution of, 125; importation of, 181; price of, 20; production of, 197; shortage of, 180, 249, 250, 281, 282, 283–84, 286, 288, 289. *See also* Agriculture, crisis in
Food and Agriculture Organization, 187, 205, 206, 246
Forbes, Don Mc H., 58, 181
Fosdick, Raymond B., 144, 188, 190, 197
Fourtón, León, 37, 63–64, 67
Fox, Charles M., 156
Fox, Vicente, 333, 334
FRAM. See *Frente Revolucionario de Agrónomos Mexicanos* (Revolutionary Front of Mexican Agronomists)
France: and *agrónomos*, 57; and colonialism, 19, 27; contribution of, 27–33; and henequen industry, 37; influence of, 50; and *Secretaria de Agricultura y Fomento*, 55–56; and seeds, 96; and taxonomy, 85; worship of, 20
Freebairn, Donald K., 258
Freeman, Orville, 249, 252
Frente Revolucionario de Agrónomos Mexicanos (Revolutionary Front of Mexican Agronomists), 135, 138, 145, 194
Friedlander, Judith, 292
Friedrich, Paul, 22
Frogs, 129
Fruit fly, 59, 88, 139, 145. *See also* Pests
Fruit trees, 59, 68, 128–29, 192–93, 235. *See also* Crops
Frutchey, Frederick P., 187

Galán, Rafael Garcia, 67
Gallardo, González, 206
Gamio, Manuel, 87, 91
Gándara, Guillermo, 38, 60, 85, 93, 94, 129, 130
Gándara, Manuel J., 210
García, León, 83
García de la Cadena, Gumaro, 93, 104
García Teruel, Alberto, 26, 30, 34
García y García, José, 245
Garrido Canabal, Tomás, 81, 87, 91, 95
Gates, Marilyn, 295, 296, 307, 331, 332, 334
Gaud, William, 252
General Agreement on Tariffs and Trade, 282
General Products Corporation, 186
Genetics, 1, 33, 55, 191, 258, 288–89, 290, 293, 296, 327, 329, 330. *See also* Biotechnology; Crops
Germany, 30, 86, 96, 138, 141
Ghandi, Indira, 259
Gibboney, C.N., 184
Girón, Alfredo Téllez, 88
Gledhill, John, 85, 105, 293, 297–98, 322
Goldstein, Daniel J., 330
Gómez, Gabriel, 30, 35
Gómez, Jaime de la Mora, 304
Gómez, José, 200
Gómez, Marte R., 30, 60, 64, 70, 106, 237; and Alemán, 194; appointment of, 10, 11; and *Campaña en Pro del Maiz*, 191; and *campesinos*, 192; and collectivization, 153; and *Confederación Nacional Campesino*, 154; and corn, 181; and Economic Miracle, 242; and education, 183; and experimentation stations, 151; and genetics, 191; and haciendas, 25; opposition to, 155; and reforms, 150–56; and *Secretaria de Agri-*

cultura y Fomento, 151, 193; and small farmer, 53–54; and United States, 143, 146, 147, 152, 190
González, Carlos "Hank," 306
González, Juan A., 35, 59, 61, 63, 87
González, Manrique, 88
González, Manuel, 34, 35
González, Pablo Tamez, 100
González Gallardo, Alfonso, 190, 201
González H., Gónzalo, 52, 54, 55, 60, 241
Government: and agricultural programs, 35; and *agrónomos*, 20, 36, 39, 55, 82, 202, 291; and biotechnology, 331; and Mexican Revolution, 321; and productivity programs, 36
Gracia, Ezequiel M., 64–65
Grady, M.N., 206
Gramsci, Antonio, 4
Graton, L.C., 187
Great Depression, 82, 87, 89
Greenberg, Martin H., 298, 325
Greenland, D.J., 287
Green Revolution, 1, 233–64, 250; aftermath of, 299–303; and *agrónomos*, 12, 238, 253–54, 260, 263–64, 327; and biotechnology, 331; and *campesinos*, 235–36, 237, 238, 252, 253–54, 255, 257, 258, 260, 262, 263, 288, 299–303, 324, 330; cause of, 126; criticism of, 12, 283, 292, 293–94, 321, 327–28; and crop research, 194; development of, 179; and Economic Miracle, 11; and economy, 258, 259; and food shortages, 281; and industry, 323; lessons of, 289; and politics, 238, 300, 323; second, 14, 288–89, 329, 333; and *Secretaria de Agricultura y Fomento*, 298; success of, 259, 295; support for, 286–87
Griffin, Keith, 1
Guanos y Fertilizantes de México, 99, 182, 183, 184, 191, 199, 236
Guevara, Cecilio Cuanalo, 203
Gurria, Francisco Trujillo, 141
Gutiérrez, Alberto A., 93
Gutiérrez, Efrain A., 53
Gutiérrez, Leonel Robles, 100

Hacendados: and *agrónomos*, 3–4, 20, 34, 50, 55, 56–57, 70, 83, 84; as backward, 60; and Calvino, 38; and Durán, 51; and land reform, 200; methods of, 24–26; and modernization, 53; and paternalism, 188–89; and *Plan de Ayala*, 49; and *ranchos*, 26; and science, 93, 94; and *Secretaria de Agricultura y Fomento*, 105; and *Sociedad Agronómica Nacional*, 54; and tradition, 56–57; and U.S. aid, 58
Haciendas, 3, 20, 39
Hall, Anthony, 334
Hamilton, Nora, 137
Hardin, Clifford, 253
Hardin, Garret, 250
Haro, Alfonso Magellón, 69
Harrar, J. George, 186, 187, 188, 190, 195, 196, 198, 202, 248, 251, 261–62
Harris, Marvin, 1, 283
Harris, Robert S., 189
Harrison, J.H., 88
Hatch, Spencer, 188
Hawaiian Sugar Growers Association, 208
Henequeñeros de Yucatán, 186
Herbicides, 182, 199–200. *See also* Agrochemicals

Hermán, Luis, 94
Herrera, Alonso, 29, 35, 51–52, 62
Herzog, Jesús Silva, 149
Hicks, W. Whitney, 258
Hinojosa, Augusto, 83
Hope y Hope, Pablo, 91
Hopper, W. David, 251
Howard, Leland Ossian, 36
Huerta, Adolfo de la, 70
Huerta, Victoriano, 49
Huerto Familiar, 239, 300
Hull, Cordell, 141

Ibarra, David S., 63, 95
Ibarra, Hilario López, 50
IHD. *See* International Health Division
IIA. See *Instituto de Investigaciones Agrícolas* (Agricultural Research Institute)
Ildefonso, Rogelio López, 292
Imports, 62, 63, 181, 281
Inda, Julio Riquelme, 28, 36, 85, 243
India, 284, 288
Industry, 323–24, 325–26
INIA. See *Instituto Nacional de Investigaciones Agrícolas* (National Agricultural Research Institute)
Instituto Biotécnico, 82, 92, 93, 94, 128, 130, 191
Instituto de Investigaciones Agrícolas (Agricultural Research Institute), 194, 201–2, 210, 239, 245
Instituto Minería, 27
Instituto Nacional de Investigaciones Agrícolas (National Agricultural Research Institute), 245, 246, 257, 259
Instituto Tecnico Henequeñero, 148
Integrated Rural Development, 82
Inter-American Development Bank, 140
Intercropping, 57, 188, 287
Interdepartmental Advisory Council on Technical Cooperation, 142
International Food Policy Institute, 288
International Health Division, 142, 144, 189
International Plagues Commission, 89
International Rice Research Institute, 248, 251, 252, 287
Investment, foreign, 3, 4–5, 20
IRRI. *See* International Rice Research Institute
Italy, 37, 86, 96
Ítie, Gabriel, 37, 93–94, 95, 100

Jackson, Dorothy, 249
Jackson, Herbert S., 195
Jaimes brothers, 30, 70
Janet, Pierre, 56
Japan, 86, 138–39, 140, 141
Jasso, Salvador Uribe, 98
Jenkins, William, 58
Jennings, Bruce H., 90
Jennings, Peter R., 287
J.G. White company, 87
Jiménez, Bruno, 133
Jiménez, Felipe Chacón, 52
Jiménez S., Leobardo, 245–46
Johnson, Lyndon B., 249, 250
Johnson, Stanley, *The Green Revolution*, 283
Jones-Hettelsater Construction Company, 208
Júarez, Benito, 28
Juzepezuck, S., 60

Kempton, James H., 143
Kennedy, John F., 249
Kesselbruner, Elias, 85
Khankhoje, Pandurang, 67, 95

Klatt, Art, 261
Knapp, Seaman A., 70
Knight, Alan, 4, 7
Knockenhauer, Guillermo, 298–99
Knowles, John H., 287
Kocka, Jürgen, 326
Kroeber, Clifton, 36
Kuenzer, Luis Tellez, 307
Kyle, E.J., 148

Landaverde, Arnulfo, 204
Land/land reform, 2, 82; and agricultural crisis, 10, 133, 180; and *agrónomos*, 50, 54, 83; and Calles, 82, 133; and Calles era, 81; and *campesinos*, 84; and Constitution of 1917, 5; and crisis, 6; and Flores, 242; and Green Revolution, 234, 238, 284; and *hacendados*, 200; and jobs, 52; and prosperity, 192; and Revolutionary Family, 11; and Salinas, 13, 304; and Third World, 285; and U.S. wartime assistance, 141; and Wallace, 89; and Zapata, 49
Lappe, Francis Moore, 284; *Diet for a Small Planet*, 283
Larrainzar, Andres Basurto, 28
Larrainzar, Basurto, 30, 34, 35, 37, 55
Larson, Magali Sarfatti, 7, 9
LAS. See *Liga de Agrónomos Socialistas* (Socialist Agronomists League)
Latin America, 208, 234, 241, 247, 263, 323, 330
Lavoisier, Antoine, 27
Lázcano Romero, José, 87, 139
Lazos, Hector, 148, 193
Left (political), 81
Lemus, Leonel, 91
Leñero, Diego Moreno, 26
Lenin, V.I., 86
León, J.J. Islas, 102, 133
León, Luís L.: and *agrónomos*, 83; and Calles appointments, 53, 70; and Green Revolution, 293; and *hacendados*, 34, 54–55; and land reform, 82, 192; and private farmers, 136; and rural reform, 242–43; and U.S. wartime assistance, 146
Liberalism, 2, 3, 19, 20, 38. *See also* Politics
Liera B., Guillermo, 106
Liga de Agrónomos Socialistas (Socialist Agronomists League), 86, 130, 137, 138
Limón G., Eduardo, 65, 88, 148, 152, 156, 190, 193
Linder, Milton J., 141
Llave, Ricardo G. de la, 155
Lloyd, W.A., 87
Loma, José Luís de la, 138, 240
Lombardo Toledano, Vicente, 56, 83, 85–86, 138
Long, Norman, 7
López, Adam Ramírez, 194
López, Ambrosio, 31
López, Elpidio, 58
López, Francisco Cabanillas, 192
López, Rubén, 192
López, Salvador Lira, 186
López Mateos, Adolfo, 233, 235, 245, 255, 324
López Portillo, José, 282, 292, 298
López Zamora, Emilio, 240, 242, 243, 254, 260, 291
Lorey, David E., 6, 30, 65, 322, 329
Loyo, Gilberto, 86
Lozoya L., Justo, 62
Lucio, Gabriel, 90
Lynch, William T., 5
Lysenko, T.D., 192

Machinery, 89
Macias, Andres, 88
Macias, Blanco, 244, 248, 260–61
Macias, José N., 51
Macias V., Mario, 93
Mackie, W.W., 35
Madero, Francisco, 49, 51, 257
Madriaga, Alfonso, 52, 66
Madsen, Louis L., 147
Maize. *See* Corn (maize)
Mallen, Bernardo, 32
Mallory, Lester D., 145, 155–56
Mangelsdorf, Paul C., 21, 197, 202, 251–52, 262
Mann, A.R., 144, 189, 196
MAP. *See* Mexican Agricultural Program
Marín, Luis, 96
Marino A., Antonio, 190
Markiewicz, Dana, 65
Marshall, George C., 206
Martínez, Hildebrando Montemayor, 52
Martínez, Justino T., 51, 69
Martínez, Nabor Cuervo, 68, 104–5, 183
Martino, Cesar, 150, 151
Marxism, 56, 86, 138, 285
Mattieu, Juan M., 292
McDonald, R.E., 139
McGovern, George, 249
McNamara, Robert, 259
Mendoza, Andres, 104
Mendoza Vargas, Gilberto, 65, 135, 240, 244
Mendoza Vargas, Leandro, 51
Mesa A., Manuel: and *campesinos*, 204; and *Comisión Nacional Agraria*, 298–99; and cultural campaign, 136; and experimentation stations, 151; and France, 56; and job, 52, 54; and Mexican Agricultural Program, 202; and politics, 31, 83; and poverty, 244–45, 248; and research, 100, 137; and Soviet Union, 138, 192; and United States, 59, 141–42, 239
Messersmith, George, 142, 148, 180, 184, 187
Mexican Agricultural Program: and agrochemicals, 198, 327; and *agrónomos*, 239; and Alemán, 180, 196, 197; and colonialism, 292; and credibility, 179; critique of, 251–52; and crops, 182, 194; and education, 191, 195; and *Escuela Nacional de Agricultura*, 195; failure of, 322–23; and farming organizations, 200–201; and fertilizer, 199; and financing, 201; and Green Revolution, 12, 233, 234, 245, 260, 261, 262, 263, 264, 284, 294; and hybrids, 189, 322; impact of, 204–5; influence of, 209–10, 296; and *Instituto de Investigaciones Agrícolas*, 201–2; and Latin America, 208; and malnutrition, 189; and Mesa A., 202; and Mexican Revolution, 197; and pesticides/herbicides, 199–200; and politics, 157, 323; and population, 247; and productivity, 251; and research, 194; and Rockefeller Foundation, 11, 234; and science, 209, 324; and *Secretaria de Agricultura y Fomento*, 153, 154, 186, 188, 189, 190; and seeds, 197, 202; and traditional farming, 196; and UNAM, 187; and U.S. wartime assistance, 144–45, 147–48; and Uribe Ruiz, 255. *See also* Rockefeller Foundation
Mexican-American Agricultural Commission, 184

Mexican Revolution, 1, 4, 14, 38, 49, 50–55, 69, 197, 242, 321, 324
Mexico City, 302
Meyer, Bob, 288
Military, 31, 33
Miller, Henry M., 143, 144, 188
Ministry of National Economy, 126
Mireles, Alfonso Támez, 255
Modernization, 2, 25, 58
Moguel, Jorge Rodríguez, 54
Molina, Olegario, 36–38
Monroy, Rufino, 56, 57–58, 66
Monzón, Carlos, 69
Moore, R.E., 148, 184
Morales, Guillermo Aguirre, 238
Morales, Jesús Garza, 152, 255
Morales, Juan Garza, 59
Morales, Juan Manuel Corona, 87
Morán, Enrique, 31
Moreno, Jesús Alarcón, 294
Moreno, Manuel Maldonado, 135
Moreno, Nicolas, 182
Moro, Carlos Terrazas, 56, 106
Morocco, 96
Morrow, Dwight, 82
Muñoz, Ricardo Flores, 239
Murphy, William B., 250
Myers, William I., 189
Myrdal, Gunnar, 258
Myron, Delbert, 261

NAFTA. *See* North American Free Trade Agreement
National Institute of Nutrition, 181, 195
Nationalism, 131, 282, 291–92
Native Americans: and *agrónomos*, 82; Cervantes on, 56; and colonialism, 24; and cultural campaign, 6, 129; and diet, 92; and education, 31; and tradition, 23; and U.S. wartime assistance, 144
Navarette, Federico Sánchez, 203
Neo-Lamarkianism, 55, 61
Neoliberalism, 303–7, 308, 329, 333
New Alchemy Institute, 285
Niblo, Stephen R., 210, 328
Nixon, Richard, 1, 253, 261
Noble, Gontran, 191
Noriega, Carlos, 31, 92
Noriega, Juan de, 58
North American Free Trade Agreement, 13, 14, 282
Noye, Alejandro, 180
Noyola, Juan Francisco, 31
Nugent, Daniel, 8, 300

Obregón, Alvaro, 5, 6, 49, 50, 53, 54, 57, 60, 61, 68, 70
Obregón, Enrique Baron, 70
Ochoa, Quintín, 186
OFAR. *See* Office of Foreign Agricultural Relations
Office of Biological Studies. See *Dirección de Estudios Biológicos* (Office of Biological Studies)
Office of Foreign Agricultural Relations, 208
Oficina Federal de Defensa Agrícola, 63
Oil, 13, 137, 141, 282
Oilseed, 125, 129
Olivera, Raziel Mora, 66
Olives, 98, 152
Olmedo, Ignacio Hernández, 59
Olson-Seffer, Pehr, 37, 38
Olvera, Rodrigo, 151
Organic Law for Education, 138
Origel, Carlos Ayala, 145
Orive Alba, Adolfo, 257, 260
Orozco, Wistano Luís, 50
Orr, John Boyd, 206, 253

Ortega brothers, 26
Ortiz, Eulogio, 187
Osborne, Fairfield, 206
Oses, Ramón García, 51
Osler, Robert, 261
Osorio, Sergio Reyes, 294
Osoyo, Roberto, 259

Paczka, Rafael Ortega, 292
Paddock, Paul, 250
Paddock, William, 250, 259
Palácios, Enrique Navarro, 238
Palacios, Leopoldo, 29
Palacios, Roberto E., 105, 203
Pan American Union, 58, 139
Paredes, Rafael Batiz, 52, 62
Paris Institute of Agronomy, 38
Parres, José G., 81, 127, 128, 130, 132, 136
Partido Institucional Revolucionario (Institutional Revolutionary Party), 2, 323–24
Patiño, Alanis, 86, 138, 201, 293
Paulín, Alfonso Ramírez, 92
Payne, George C., 142
Peregrina, Rodolfo, 246
Pérez, Felipe Salcedo, 245
Pérez, Luís Vera, 204
Pérez, Manuel, 58
Peritore, Ana Karina Galve, 331
Peritore, N. Patrick, 331
Perkins, John H., 322, 323, 325
Pesticides: and Avila Camacho, 153; criticism of, 258; and Green Revolution, 233, 236, 260–61; and health, 302; and Mexican Agricultural Program, 199–200; and organic farming, 284; and profitable farming, 98, 99; and traditional agriculture, 23. *See also* Agrochemicals
Pests: and agrochemicals, 199–200; and farm crisis, 125; and Gándara, 38; and government, 62; and health improvement, 93; and local conditions, 66; natural control of, 63, 130; and profitable farming, 100; and *Secretaria de Agricultura y Fomento*, 100; and traditional farming, 23, 287; and United States, 36, 58–59, 85, 88–89, 139, 141, 145, 147, 206–7
PIDER. See *Proyecto Internacional de Desarrollo Rural* (International Rural Development Project)
Pilcher, Jeffery M., 22, 24, 32
Pimentel, David, 283
Pink bollworm, 58–59, 88, 100, 102, 139, 145. *See also* Pests
Pittner, John B., 199
Plan de Ayala, 49, 50
Plan de la Barranca, 25
Plan Sexenal, 84
Point IV Program, 247, 248
Poleman, Thomas, 243
Politics, 19, 82; and *agrónomos*, 54, 83, 106, 254–55, 291; and Avila Camacho, 151; and Green Revolution, 238, 300, 323; and industry, 323–24; and Mexican Agricultural Program, 157, 323; and pest control, 63; reform in, 324; and *Secretaria de Agricultura y Fomento*, 136; and *Sociedad Agronómica Mexicana*, 244; and Stakman, 239
Population, 251, 253. *See also* Birth control
Porfiriato, 3; and education, 28; and France, 30; and modernization, 26; and United States, 35–36. *See also* Díaz, Porfirio
Portes Gil, Émilio, 70, 85, 91
Posada, Angel, 31, 106
Positivism, 28, 29
Poverty, 302, 303, 308, 323

Preciado, Juan Gil, 254
PRI. See *Partido Institucional Revolucionario* (Institutional Revolutionary Party)
Primera Convención Nacional Hulera, 37–38
Prister, Augusto, 57
Procampo program, 306, 307, 331
Productora Nacional de Semillas, 293, 296, 300
Promotora Comercial de Cafes Suaves Mexicanos, S.A., 305
PRONASE. *See Productora Nacional de Semillas*
Proyectismo, 7
Proyecto Internacional de Desarrollo Rural (International Rural Development Project), 281–82, 308

Quevedo, Miguel Angel de, 30, 52
Quintanar, Francisco, 92

Rabiela, Rebeca de Gortari, 322
Race, 24, 32–33, 39, 56, 61, 92
Railroads, 20
Ramírez, Rafael, 90
Ramírez A., 134, 135
Ramírez M., Salvador, 127, 131
Ramos, Ramón, 106
Ramos Millán, Gabriel, 194, 197
Ranchos/rancheros, 3, 26–27, 31, 33, 39, 87, 94, 234, 237
Rebonato, Egidio Gaetano, 85
Regolos, Manuel de, 131
Religion, 25, 27, 90, 91, 107, 246, 295. *See also* Roman Catholic Church
Renaudet, M. Georges, 30
Rendón, Heliodoro, 53
Research: and Avila Camacho, 191; and Cárdenas, 137; and cultural campaign, 100; and *Escuela Nacional de Agricultura*, 151–52; and farming systems, 289–90; and Green Revolution, 194; importance of, 106, 181; and Salinas, 304–5; and *Secretaria de Agricultura y Fomento*, 93, 94–95; and *Sociedad Agronómica Mexicana*, 151; and U.S. wartime assistance, 143, 144. *See also* Experimentation
Retana, Ismael Reyes, 54
Reves, Frank, 87
Revolutionary Family, 6–7, 192, 193, 263; and *agraristas*, 54; and *agrónomos*, 69, 70, 106, 290, 308, 326; and *campesinos*, 324; and cultural campaign, 131; and Economic Miracle, 10; and Green Revolution, 263, 264; and land reform, 11, 82, 242; and *Oficina Federal de Defensa Agrícola*, 64; and *Plan Sexenal*, 84; and production, 194; and *Sociedad Agronómica Mexicana*, 134, 150
Reyes, Estela Cervantes, 331
Rice, 251, 252
Richardson, Ralph W., 239
Richardson Construction Company, 35, 58
Riding, Alan, 297
Rivas, Enrique Coppel, 60
Rivas Tagle, Antonio, 56, 63, 66, 68, 96, 97
Rivera, Alfredo Valle, 92
Rivera, Diego, 90
Rivera, Jesús Valenzuela, 156
Roberts, Lewis, 286–87
Robledo, Francisco García, 100, 242
Robles, Aurelio Castillo, 189
Robles, Colina, 102, 105
Robles, Gonzalo, 57, 61, 62, 66, 138
Robles, Rodolfo Peregrina, 199

Robles, Sixto Colina, 94
Rochow, Eugene G., 206
Rockefeller, John D., III, 148, 246
Rockefeller, Nelson, 148, 149, 184, 208
Rockefeller family, 9
Rockefeller Foundation, 2, 184; and *agrónomos*, 157, 296; and Cultural Relations Program for Latin America, 187; and *Escuela Nacional de Agricultura*, 136, 187; failure of, 322–23; and Ferrell, 89–90; and Green Revolution, 12, 247–48, 250, 263, 281, 284, 286, 287, 289, 295; and Mexican Agricultural Program, 11, 234; and postwar programs, 207, 208; and science, 324; and technological colonialism, 291–92; and U.S. wartime assistance, 142. *See also* Mexican Agricultural Program
Rodale, J.I., 199
Rodale, Robert, 285
Rodríguez, Abelardo, 83, 106
Rodríguez, Eduardo, 29
Rodríguez, José, 149, 189
Rodríguez Adame, Julián, 97, 240, 242, 243, 245, 246, 293, 296
Rodríguez G., Guillermo, 133
Rodríguez León, Panfilo, 248
Rodríguez y Rodríguez, Conrado, 129
Rojo, Baltazar Isaguirre, 93
Roland, Modesto C., 99
Roldán, Emilio Gutiérrez, 150, 253
Roldán, Gutiérrez, 97
Roldán, Pascual Gutiérrez, 87, 147, 239
Roman Catholic Church, 4, 6, 8, 19, 22, 25, 39, 62, 63, 91, 234, 328. *See also* Religion
Romero, Alfonso, 191
Romero, Francisco MacGregor, 52, 53
Romo, Basilio, 34, 37, 50
Romo, Luis Arturo, 237
Ronfeldt, David, 8
Roosevelt, Franklin D., 10, 89, 140, 142
Rosas, Benito Juárez, 301
Rosso, José R., 94
Rowe, L.S., 89
Rubber, 37–38, 89, 140, 141, 142, 145–46, 185, 186
Rubio, Pascual Ortiz, 70, 82
Ruiz, Ramón Eduardo, 60
Ruiz Cortines, Adolfo, 233, 234, 239, 241, 245, 323
Ruiz y Sandoval, Alberto, 35
Rulfo, Juan, 95
Russell, Sir John, 206, 247

SAF. See *Secretaria de Agricultura y Fomento* (Ministry of Agriculture and Development)
Safa, Morilla, 151
Safford, Frank, 32
Safford, W.E., 38
Sahagún, Bernardino de, 22, 24
Salazar, Alejandro Santiago, 299
Salinas, Faustino, 306–7
Salinas de Gortari, Carlos, 2, 13, 282, 303–4, 305, 308
SAM. See *Sociedad Agronómica Mexicana* (Mexican Agronomic Society)
SAN. See *Sociedad Agronómica Nacional* (National Agronomic Society)
Sanabria, Ramiro, 183
Sánchez, Arcadio, 55–56
Sánchez, Carlos Gajón, 99
Sánchez, George I., 143
Sánchez Durón, Nicolas, 67, 260
Sanderson, Steven E., 300

Sandoval, Antonio, 100
Sandstrom, Alan R., 22
Santa Anna, Antonio López de, 27
Santillán, Miguel, 88
Santuario, Manuel, 135
Sauer, Carl O., 81, 143, 189, 322, 327
Schuler, Friedrich, 82
Schultz, Miguel E., 151, 182
Schultz, Theodore, 286
Science: agenda of, 1, 2, 333; and *agrónomos*, 94, 181, 241, 326, 328; autonomy in, 37; and Barreda, 33; and credibility, 192; and cultural campaign, 60, 61; and Díaz, 20; and economy, 93; and education, 33, 90–91, 297; foreign, 10, 137, 145; and Mexican Agricultural Program, 209; and modernization, 39; and politics, 324; and religion, 90, 91; and *Secretaria de Agricultura y Fomento*, 193; and *Secretaria de Educación Pública*, 137–38; and seeds, 57; and seed selction, 37; and society, 4; and Stakman, 201; and U.S. Department of Agriculture, 36
Scott, James C., 288
Secretaria de Agricultura y Fomento (Ministry of Agriculture and Development): and agrochemicals, 198–99, 203; and *agrónomos*, 6, 83, 132; and *campesinos*, 84, 98, 102, 103, 192; and *Comisión Nacional de Irrigación*, 201, 298; and Corn Commission, 201; and corruption, 104; and crops, 181, 298–99, 306; and cultural campaign, 63, 127, 128, 129–30, 131–32; and diet/nutrition, 92, 181; and disease, 130; and Echeverría, 290–91; and experimentation, 68; and extension service, 96–97; and fertilizer, 130, 153, 199; and financial problems, 201; and food imports, 181; and foreign scientists, 138; and France, 55–56; and Gómez, 10, 151; and Green Revolution, 234, 245, 254, 255, 259, 260, 298; and *hacendados*, 104; and Hawaiian Sugar Growers Association, 208; and Herrera, 51–52; and hybrids, 294; and incompetence, 297–98; and jobs, 242; and local conditions, 66; and Mexican Agricultural Program, 153, 154, 186, 188, 189, 190; and Obregón, 61; and pesticides/herbicides, 199; and pests, 100; and politics, 136; problems of, 153; and reform, 54; and religion versus science, 91; and research, 69, 93, 94–95; and Rodríguez, 84; and science, 193; successes of, 132; and traditional farming, 96, 135; and United States, 58, 59, 88, 139, 141, 142, 143, 144, 146, 147, 148, 184–85, 187; and Uribe Ruiz, 255, 256
Secretaria de Educación Pública (Ministry of Public Education): and *agrónomos*, 6; and Alemán, 195; and class, 105; and cultural campaign, 50, 127, 129; and education, 105; and fertilizer, 199; and Green Revolution, 294; and jobs, 65; problems with, 82–83; and refugee scientists, 137–38; and religion versus science, 91; and Soviet Union, 138; and United States, 138, 140
Secretaria de Fomento, 38
Seed, 200, 202, 294, 300; selection of, 37, 56, 57, 96, 129, 132, 191

Segura, José del Carmen, 30, 33, 34, 35
Seitz, Frederick, 250
Sen, Binay Rangan, 249
SEP. See *Secretaria de Educación Pública* (Ministry of Public Education)
Serrano, Gustavo P., 190
Shultz, Miguel E., 95
Shumacher, E.F., 285
Siegrist, Hannes, 5, 327
Simonian, Lane, 153
Simpson, Eyler N., 7, 106
Sinclair Oil, 141
Sistema Alimentario Mexicano, 282, 295, 296, 298, 299, 301, 308
Smith, Erwin F., 38
Smith, Peter, 150
Soberón, Waldo, 53
Socialism, 86, 90, 91, 104
Sociedad Agronómica Mexicana (Mexican Agronomic Society): and Avila Camacho, 150; and *campesinos*, 192; and *Confederación Nacional Campesino*, 134; and Echeverría, 291; formation of, 6; and politics, 244; and religion versus science, 92; and research, 151; and Revolutionary Family, 134; and U.S. education, 190–91
Sociedad Agronómica Nacional (National Agronomic Society), 6, 54
Sociedad Científica Antonio Alzate, 29, 30
Sociedad Mexicana de Historia Natural, 28
Soil, 153, 182, 199, 207, 287
Sonora, 236–37, 254, 260, 296, 299
Soto, Blas E. Romo, 70
Soto, Ramon Corral, 69
Sousa, Mario, 85
Soviet Union, 50, 66–67, 81, 85–86, 96, 138, 179, 192, 206
Spain/Spaniards, 19, 96
Spanish Civil War, 138
Stakman, Elvin C., 21; and Cold War, 208–9; and culture, 202–3; and *hacendados*, 188–89; and Mexican Agricultural Program, 187, 198, 201, 251–52, 296; and politics, 239; and population, 253; and *Sociedad Agronómica Mexicana*, 190
Stalin, Joseph, 138, 192
Standard Oil, 9
Stern, Jeannot, 130
Stevens, Donald H., 142
Strong, Lee A., 139
Suárez, Eduardo, 149
Sunflowers, 92

Taboada, Rodolfo Sánchez, 210, 293
Taboada R., Edmundo, 88, 137, 148, 150, 152
Tamayo, Jorge, 133
Taylor, R.M., 189
Técnicos, 126, 242, 308
Téllez, José Luis Calva, 69
Tellez, Olivero, 67
Tello, Rafael M., 32
Terrazas, Luis, 31
Third World, 246, 263, 283–84, 285, 288, 328, 331
Tizzrard, Sir Henry, 206
Tlatelolco Massacre, 13, 290
Toro, Augusto Pérez, 148, 183
Tortilla discourse, 24, 33, 92, 128, 129, 152, 326. *See also* Diet/nutrition
Trenes de Maquinaria Agrícola, 52, 55
Trevino, ManueI Pérez, 81
Trotman, O.C., 59

Trujillo, Victor M., 304
Truman, Harry S., 207, 247, 323
Turcott B., Eduardo, 139

UFCO. *See* United Fruit Company
UNAM. *See Universidad Nacional Autónoma de México* (National Autonomous University of Mexico)
UNFASA. *See Union de Fabricantes de Conservos S.A.* (Union of Canned Goods Manufacturers)
Union de Ejidos "Professor Otilio Montaño," 305
Union de Fabricantes de Conservas S.A. (Union of Canned Goods Manufacturers), 201
Union Nacional de Vetranos de la Revolución, 126
Unión Nacional Sinaquarista, 133
United Fruit Company, 125–26, 153
United States, 157; and agribusiness, 302; and *agrónomos*, 57–58, 81, 179, 180, 210, 239–40; and Calles, 50; discontent with, 185; and education, 35, 190, 191, 207; and *Escuela Nacional de Agricultura*, 240; and food shortages, 180; and genetics, 85; and Green Revolution, 247–48, 261, 263; and *hacendados*, 26; influence of, 67, 86–90; interest in, 3; and market, 89; and Obregón, 50; and pests, 85; and political reform, 324; and *Porfirato*, 35–36; and postwar programs, 187, 205–9, 210–11; and poverty, 246–47; reliance on, 10; and Soviet Union, 179; and wartime programs, 11, 139–49, 179, 183–91, 323; and World War II, 140
Universidad Nacional Autónoma de México (National Autonomous University of Mexico), 139, 148, 187, 305
Uranga M., Javier, 82
Urban population, 26, 31, 132, 298, 302
Uribe Ruiz, Jesús, 255–56
Ursua, José Jorge de C., 51
U.S. Bureau of Fisheries, 141
U.S. Department of Agriculture: and *campesino* cooperation, 102; and corruption, 38; and crops, 184; and dry farming, 38; and Gómez, 152; and Green Revolution, 252; and Herrera, 51; and hybrid corn, 63; influence of, 35–36; and organic farming, 285; and pests, 89, 206–7; and postwar aid, 205; and *Secretaria de Agricultura y Fomento*, 58, 59, 187; and *Secretaria de Educación Pública*, 138; seeds from, 87; and wartime programs, 139–40, 142, 143, 146, 147, 149, 184–85
USDA. *See* U.S. Department of Agriculture

Van der Ploeg, Jan Douwe, 8
Vanegas, Jesús, 103
Vargas, Marcué, 244
Varsavsky, Oscar, 292
Vasconcelos, José, 60
Vaughan, Mary Kay, 4, 9, 22–23, 57
Vega, Fernando E., 306, 330–31
Velasco, Miguel A., 334
Vera, Manuel, 35, 55
Viadas, Lauro, 29, 34, 35, 36
Victoria, Guadalupe, 27
Vidali, Carlos, 293
Villa, Pancho, 49, 53, 54
Villada, Manuel M., 28
Villaseñor, Eduardo, 82, 180

Vogt, William, 182, 258
Vogtmann, Hardy, 284
Voronoff, G.N., 60

Wald, George, 259
Wallace, Henry A., 59, 89, 139–40, 141, 142, 143, 144, 145, 190, 197, 323
War of La Reforma, 19
Waters, Herbert J., 250
Weaver, Warren, 182, 189, 196, 209
Wellhausen, Edwin J., 23, 195, 196, 197, 198, 200, 202, 209, 259, 262, 287–88, 289
Went, Fritz, 206
Wernimont, Kenneth, 238, 255
Weyl, Nathaniel, 91
Weyl, Sylvia, 91
Wharton, Clifton R., Jr., 284
Wheat, 181, 198, 199, 233, 235, 236, 237, 246. *See also* Crops
Whetten, Nathan, 143, 188, 196
Willits, Joseph, 143, 189
Wilson, M.L., 139
Winkelman, Don, 287
Wittner, S.H., 284
Women, 32, 156–57, 202, 246, 332–33
Woods Hole Oceanographic Institute, 285
World Bank, 290
World Neighbors Aid, 290
World War II, 11, 126, 140, 180, 182
Worldwatch Institute, 288, 290
Wortman, Sterling, 250–51, 287, 288
Wright, Norman, 249, 327

Xolocotzi, Hernández, 292–93

Zálvidar Flores, José, 155, 156
Zapata, Emiliano, 4, 5, 49, 53, 150, 157, 325
Zayas, Joaquín de, 134
Zedillo, Ernesto, 307, 334
Zendejas, Sergio, 8
Zepeda, Ignacio E., 238
Zuno, Álvaro Echeverría, 297

About the Author

JOSEPH COTTER was Associate Professor of History at Augusta State University prior to his death in September 2002. He completed his Ph.D. in Latin American History at the University of California, Santa Barbara, in 1994. He also taught at the University of West Florida and the University of Arizona.

CPSIA information can be obtained at www.ICGtesting.com
Printed in the USA
LVOW101316291012

304902LV00004B/1/P

9 780313 325151